Periodic table of the elements and element atomic weights (adapted from IUPAC 1991 values)

[illegible]	2 IIA IIA	3 IIIA IIIB	4 IVA IVB	5 VA VB	6 VIA VIB	7 VIIA VIIB	8 VIIIA VIIIB	9 VIIIA VIIIB	10 VIIIA VIIIB	11 IB IB	12 IIB IIB	13 IIIB IIIA	14 IVB IVA	15 VB VA	16 VIB VIA	17 VIIB VIIA	18 VIIIB VIIIA
[illegible]																1 **H** 1.008	2 **He** 4.003
[illegible] **Li** 6.941	4 **Be** 9.012											5 **B** 10.811	6 **C** 12.011	7 **N** 14.007	8 **O** 15.999	9 **F** 18.998	10 **Ne** 20.180
11 **Na** 22.990	12 **Mg** 24.305											13 **Al** 26.982	14 **Si** 28.086	15 **P** 30.974	16 **S** 32.066	17 **Cl** 35.453	18 **Ar** 39.948
19 **K** 39.098	20 **Ca** 40.078	21 **Sc** 44.956	22 **Ti** 47.88	23 **V** 50.942	24 **Cr** 51.996	25 **Mn** 54.938	26 **Fe** 55.847	27 **Co** 58.933	28 **Ni** 58.693	29 **Cu** 63.546	30 **Zn** 65.39	31 **Ga** 69.723	32 **Ge** 72.61	33 **As** 74.922	34 **Se** 78.96	35 **Br** 79.904	36 **Kr** 83.80
37 **Rb** 85.468	38 **Sr** 87.62	39 **Y** 88.906	40 **Zr** 91.224	41 **Nb** 92.906	42 **Mo** 95.94	43 **Tc** (97.907)	44 **Ru** 101.07	45 **Rh** 102.906	46 **Pd** 106.42	47 **Ag** 107.868	48 **Cd** 112.411	49 **In** 114.818	50 **Sn** 118.710	51 **Sb** 121.757	52 **Te** 127.60	53 **I** 126.904	54 **Xe** 131.29
55 **Cs** 132.905	56 **Ba** 137.327	57–71	72 **Hf** 178.49	73 **Ta** 180.948	74 **W** 183.84	75 **Re** 186.207	76 **Os** 190.23	77 **Ir** 192.22	78 **Pt** 195.08	79 **Au** 196.967	80 **Hg** 200.59	81 **Tl** 204.383	82 **Pb** 207.2	83 **Bi** 208.980	84 **Po** (208.982)	85 **At** (209.987)	86 **Rn** (222.018)
87 **Fr** (223.020)	88 **Ra** 226.025	89–103	104 **Rf** (261)	105 **Db** (262)	106 **Sg** (266)	107 **Bh** (262)	108 **Hs** (265)	109 **Mt** (266)	110 **–** (271)	111 **–** (272)	112 **–** (277)						

57 **La** 138.906	58 **Ce** 140.115	59 **Pr** 140.908	60 **Nd** 144.24	61 **Pm** (144.917)	62 **Sm** 150.36	63 **Eu** 151.965	64 **Gd** 157.25	65 **Tb** 158.925	66 **Dy** 162.50	67 **Ho** 164.93	68 **Er** 167.26	69 **Tm** 168.934	70 **Yb** 173.04	71 **Lu** 174.967
89 **Ac** 227.028	90 **Th** 232.038	91 **Pa** 231.036	92 **U** 238.029	93 **Np** 237.048	94 **Pu** (244.064)	95 **Am** (243.061)	96 **Cm** (247.070)	97 **Bk** (247.070)	98 **Cf** (251.080)	99 **Es** (252.083)	100 **Fm** (257.095)	101 **Md** (258.10)	102 **No** (259.101)	103 **Lr** (262.11)

Notes: Elements for which the atomic weight is contained within parentheses have no stable nuclides and the weight of the longest-lived known isotope is quot[illegible] three elements Th, Pa, and U do have characteristic terrestrial abundances and these are the values quoted. In cases where the atomic weight is known to bet[illegible] three decimal places, the quoted values are rounded to three decimal places. There is considerable confusion surrounding the Group labels. The top, numeric, [illegible] system (1–18) is the current IUPAC convention. The other two systems are less desirable since they are confusing, but still in common usage. The designation[illegible] are completely arbitrary. The first of these (A left, B right) is based upon older IUPAC recommendations and frequently used in Europe. The last set (main group elements A, transition elements B) was in common use in America. For a discussion of these and other labelling systems see: Fernelius, W.C. and Powell, W.H. (1982). Confusion in the periodic table of the elements, *Journal of Chemical Education*, **59**, 504-508.

Magnetochemistry

A. F. Orchard
Lecturer in Inorganic Chemistry and
Fellow of University College

This book has been printed digitally and produced in a standard specification in order to ensure its continuing availability

OXFORD
UNIVERSITY PRESS

Great Clarendon Street, Oxford OX2 6DP

Oxford University Press is a department of the University of Oxford.
It furthers the University's objective of excellence in research, scholarship,
and education by publishing worldwide in

Oxford New York

Auckland Cape Town Dar es Salaam Hong Kong Karachi
Kuala Lumpur Madrid Melbourne Mexico City Nairobi
New Delhi Shanghai Taipei Toronto
With offices in
Argentina Austria Brazil Chile Czech Republic France Greece
Guatemala Hungary Italy Japan South Korea Poland Portugal
Singapore Switzerland Thailand Turkey Ukraine Vietnam

Oxford is a registered trade mark of Oxford University Press
in the UK and in certain other countries

Published in the United States
by Oxford University Press Inc., New York

Reprinted 2007

ISBN 978-0-19-879278-9

Printed and bound by CPI Antony Rowe, Eastbourne

Series Editor's Foreword

Oxford Chemistry Primers are designed to provide clear and concise introductions to a wide range of topics that may be encountered by chemistry students as they progress from the freshman stage through to graduation. The Physical Chemistry series aims to contain books easily recognised as relating to established fundamental core material that all chemists need to know, as well as books reflecting new directions and research trends in the subject, thereby anticipating (and perhaps encouraging) the evolution of modern undergraduate courses.

In this Physical Chemistry Primer, Tony Orchard presents a clearly written and elegant account of the foundations of Magnetochemistry, building on and developing ideas that undergraduates will have encountered through introductory courses, to give insight and understanding of the magnetic behaviour of atoms and molecules both as isolated species and as bulk materials. This Primer will be of interest to all students of chemistry and their mentors.

Richard G Compton

Physical and Theoretical Chemistry Laboratory

University of Oxford

Preface

The term Magnetochemistry was coined in the 1930s but the subject dates back to Faraday in the early 19th century. In a seminal monograph, Selwood (1943) defined Magnetochemistry as *the application of magnetic susceptibilities and of closely related quantities to the solution of chemical problems*. Nowadays, the "discipline" has naturally a wider canvass, embracing many experimental techniques developed subsequently (most obviously, magnetic resonance). It involves not merely the interpretation of the magnetic properties of compounds but also, where they have technological significance, the optimisation of their useful properties, this with a view always to the design of new materials with improved performance. However, the latter ventures cannot be given adequate attention in a short introductory book. The present text, focussing ultimately on paramagnetism, is devoted more to the role of magnetic measurements, including electron paramagnetic resonance (EPR), purely in the characterisation of molecular electronic structure and thereby the elucidation of chemical bonding.

Magnetism is not an easy subject; it is as difficult conceptually as it is important technologically in its many diverse applications. All magnetic properties are fundamentally quantum-mechanical in origin, and frequently in deeply subtle ways. However, treading a cautious but (I trust) pragmatic path between rigour of exposition and accessibility to the student, considerable use is made here of semi-classical argument. So far as possible, the formal quantum mechanics is relegated to the background. It is hoped that this text will nonetheless provide a useful bridge to more advanced monographs and to the research journals (and perhaps even whet the student's appetite).

I am indebted to many colleagues, too numerous to acknowledge individually, for their willing help and advice. Particular and warm thanks are due to Prof. Allen Hill FRS and Dr. Harry Quiney, who have read and commented constructively on draft versions of the text, and also to *Perseus* (a.k.a. Danny Orchard) for invaluable assistance with many of the illustrations.

For what merit it may have, this book is dedicated with affection and respect to Prof. R J P (Bob) Williams FRS, mentor and former tutor, who introduced me to the subject in my undergraduate days at Wadham College some four decades ago.

AFO *Oxford* January 2003

Contents

1 Introduction to the principles and phenomenology of bulk magnetism

The basic principles that are needed here are developed initially in *classical* terms (section 2). The common forms of magnetism are associated with the presence of *unpaired electrons* and, in the classical theory, the constituent atoms or molecules of a material are endowed with a *permanent magnetic dipole moment* $\mathbf{m}$ analogous to the electric dipole moment $\boldsymbol{\mu}$ possessed by a polar molecule. A brief introductory survey (section 3) of the different forms of bulk magnetism may usefully be conducted on this basis.

However, as will become apparent later, the classical approach has profound, indeed fatal limitations. Most (arguably *all*) magnetic properties of materials are fundamentally quantum mechanical in origin. Much of the material in section 3 is accordingly re-addressed in § 3, and also substantially extended there.

There is a famous theorem due to J H van Leeuwen (1919) and others, according to which the net magnetisation of an ensemble conforming to classical statistics vanishes identically! [See Van Vleck (1932).]

But, first, a brief interlude concerning the history of magnetism.

1.1 Early history of the subject

Pre-historic man must have first become aware of the phenomenon of magnetism through observing that fragments of the mineral *magnetite* Fe_3O_4 could attract each other. Among other names, the ancient Greeks knew this material as "Magnesian stone", this a reference to Magnesia, a district of what is today western Turkey; hence, eventually (from *c.* 1440), the term "magnet" by way of *magneta* in medieval Latin. As a Roman poet put it,

> "The magnet's name the observing Grecians drew
> From the magnetick region where it grew"

(Lucretius Carus: *De Rerum Natura*, 1st century BC; trans. 1714). The name *magnetite* is modern, after the German *magnetit* (1845).

The name of the element magnesium, conferred by its discoverer H Davy in 1808, had a related origin in the names of the minerals *magnesia* (MgO) and *magnesite* ($MgCO_3$).

> A contemporary note: Fe_3O_4 ($\equiv FeO.Fe_2O_3$) can exhibit *permanent* (otherwise termed *remanent*) magnetization of a type called *ferrimagnetism* – to be carefully distinguished from *ferromagnetism*, the form observed in the metals iron, cobalt and nickel.
>
> Fe_3O_4 is a special example of the category *spinel ferrite*. The term *ferrite* embraces all ferrimagnetic materials containing Fe(III). It includes, for example, γ-Fe_2O_3, the everyday magnetic recording medium, as well as many ternary oxides MFe_2O_4 containing various M(II) cations and having the *spinel* ($MgAl_2O_4$) structure. In the case of *magnetite*, M = Fe.

Magnetism naturally had its fascination for the ancient Greeks, who were well aware of what is now called *magnetic induction.* Plato (*c.* 428 – 348 BC) compares divine poetic inspiration to the influence of "that stone which Euripides called Magnesian….. The stone not only attracts iron rings, but also imparts to them a similar power of attracting other rings, so that you may get a long chain of pieces of iron and rings clinging to one another, all of them deriving their power of suspension from the original stone. In the same way, the Muse first of all inspires men herself; and from these inspired persons develop chains of others, who spread the inspiration." (Plato, *Ion* 533d) However, the Greeks understood nothing of, and indeed probably cared little about the underlying nature of magnetism. They were preoccupied with analogy, metaphor and allegory, rather than "science" in the modern sense.

There was not, moreover, any real advance towards an understanding of magnetic forces and magnetization itself for almost two millennia after Plato's time. Magnetism remained for long the preserve of mystics and magicians. However, by the 12th century AD in western Europe, perhaps much earlier in China, the magnetic compass had made its appearance (devised on an heuristic basis, it would seem). A slim piece or sliver of magnetized Fe_3O_4, used as a navigational pointer, became known as a *lodestone* or *loadstone*, literally "way-stone", a name subsequently extended to larger magnetized objects. The magnetic compass, the first magnetic technological invention, was to transform navigation, especially at sea, and in turn stimulated a fresh interest in magnetic phenomena among an emerging scientific community which became a vital part of the Renaissance.

It was William Gilbert (*De Magnete*, 1600) in Elizabethan England who, pulling together the threads of earlier investigations and, conducting further pertinent "magnetical" experiments, first identified the Earth as itself a giant magnet, explaining thereby how the navigational compass works. Gilbert was celebrated by Dryden with

"Gilbert shall live till loadstones cease to draw
Or British fleets the boundless ocean awe."

Peregrinus' account of his investigations (1269 AD) is the earliest known exposition of experimental physics.

As a model for the Earth, Gilbert cited the *terralla*, a lodestone of spherical shape, which had been studied over 300 years before by Pierre Pélerin de Maricourt (*a.k.a.* Petrus Peregrinus).

Comparatively little further progress was made until the connection between electricity and magnetism was established. Ørsted demonstrated in 1819 that an electric current generated a magnetic field. His discovery immediately stimulated further pioneering experiments concerning the magnetic forces between current carrying wires by Ampére (1820–25) and by Biot and Savart (1820). Ampere also demonstrated that, at a distance, a current loop exerted a magnetic influence equivalent to that of a bar magnet.

For an extensive historical survey see Mattis (1981).

In 1831 Michael Faraday discovered *electromagnetic induction* – the induction of a current in a conductor moving in a magnetic field (the "dynamo" effect). He coined the term *magnetic field* in 1845 and demonstrated that all materials are magnetisable in some way.

1.2 The elementary concepts of Magnetism

This text is essentially about Magnetostatics, the properties of materials in time-invariant magnetic fields. For such limited purposes, most of the basic concepts may be established by analogy with those of Electrostatics.

1. The electric dipole moment

The fundamental quantity of Electrostatics is *electric charge* (q), which may be positive or negative. Two electric charges of equal magnitude,[a] but of opposite sign, held (somehow) at a fixed distance apart, constitute an *electric dipole*. This is a *vector* quantity. If the separation of the charges is a, the electric dipole is defined by its moment

$$\boldsymbol{\mu} = q\boldsymbol{a} \tag{1.1}$$

a vector collinear with the vector $\boldsymbol{a}$ – which, by convention, *is directed from the negative to the positive charge* (Fig. 1.1). The major physical realisation in chemistry is the heteronuclear diatomic molecule, such as HF. (In that particular case the dipole is, of course, directed towards the hydrogen atom.)

When subjected to an electric field $\boldsymbol{E}$ (Fig. 1.2), a dipole experiences a couple or *torque* of magnitude $\Gamma = \mu E\sin\theta$, where θ is the inclination with respect to the field, which turns the dipole towards the direction of the applied field. The couple is a *vector* expressible as the "cross-product"

$$\boldsymbol{\Gamma} = \boldsymbol{\mu} \times \boldsymbol{E} \tag{1.2}$$

The potential energy of the dipole is given instead by the *scalar* product

$$U = -\boldsymbol{\mu}.\boldsymbol{E} = -\mu E\cos\theta \tag{1.3}$$

which is minimal when the dipole is parallel to the field. The torque $\boldsymbol{\Gamma}$ turns the dipole towards an orientation of lower potential energy. The *minus* sign in eqn. 1.3 follows from the convention adopted for the direction of the dipole $\boldsymbol{\mu}$.

An electric dipole situated in an electric field *gradient* is subject to a translational force. If the gradient is confined to a particular direction s, the force acting is

$$F_s = \mu_s \frac{dE}{ds}\left(= -\frac{dU}{ds}\right) \tag{1.4}$$

where μ_s is the component of the dipole in the s-direction. It is assumed that the field gradient is constant over the dimension a of the dipole.

Some further properties of electric dipoles, together with other pertinent aspects of Electrostatics, are described in the Appendix (section 5).

2. Magnetic dipoles and magnetic fields

A small bar magnet (*e.g.* a compass needle) has properties analogous to those of an electric dipole, most obviously as regards its behaviour in a magnetic field. Moreover, as detailed in the Appendix [5(2)], the field due to a bar magnet is qualitatively similar to the electric field generated by an electric dipole.[b] The bar magnet is a macroscopic *magnetic dipole*, also a vector and here (as commonly in the Physics literature) denoted $\mathbf{m}$.[c]

The bulk properties of highly magnetisable materials, such as ferromagnetic metals and alloys or ferrimagnetic compounds such as *magnetite* and γ-Fe_2O_3, are attributable to the *concerted* behaviour of magnetic dipole moments existing on the atomic or sub-atomic scale. In classical physics, the model for the source of an atomic or molecular magnetic dipole is the *current loop*. Extensive and varied experiments involving electrical circuits – for

a – strictly *point* charges.
According to what is called the *point charge theorem*, a spherically symmetrical distribution of charge behaves at an external point as though the whole charge resided at the centre.

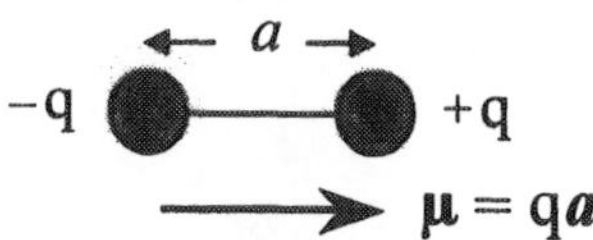

Fig. 1.1 The electric dipole $\boldsymbol{\mu}$
[It has the dimensions C m.]

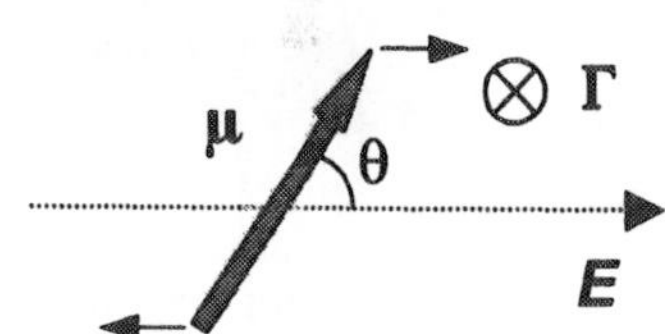

Fig. 1.2 An electric dipole in a uniform electric field $\boldsymbol{E}$.
The torque $\boldsymbol{\Gamma}$ (eqn. 1.2) is directed *into* the page.

b The strength of the field due to an electric dipole and that generated by a magnetic dipole both notably vary as the *inverse cube* of distance r – provided that r is very large in comparison with the dimensions of the dipole.

c The symbol $\boldsymbol{\mu}$ is perhaps more usual in the chemical literature but is potentially confusing here; given its mandatory use in other connections, the symbol μ would also be grossly over-worked.

example, circular coils carrying a constant current – have shown that current loops (like small bar magnets) display magnetic properties that exactly mirror those of electric dipoles in the electrostatic context.

Using current loops of varying size and shape, it is long established that the magnitude m of the magnetic moment is proportional both to the area A and the current I flowing. Thus $m = kIA$, where k is a universal constant. If the unit of magnetic dipole moment is *defined* as that of a single-turn loop of unit area (1 m^2) carrying unit current (1 Ampere, A), then $k = 1$. As pictured in Fig. 1.3, the vector **m** is directed perpendicular to the current loop, *i.e.* parallel to the area vector **A**,[a] and is defined generally by

$$\boxed{\mathbf{m} = I\mathbf{A}} \qquad \text{with the unit } [m] = \text{A m}^2 \tag{1.5}$$

a **A** is in effect an *axial* vector, like the torque **Γ** acting upon a dipole in a field. By convention, its direction is such that circulation of the current is clockwise about it. The direction of **A** may equivalently be defined by the right-hand rule illustrated below.

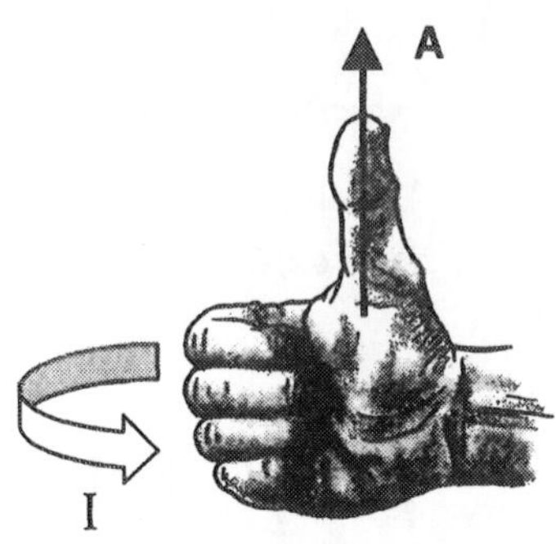

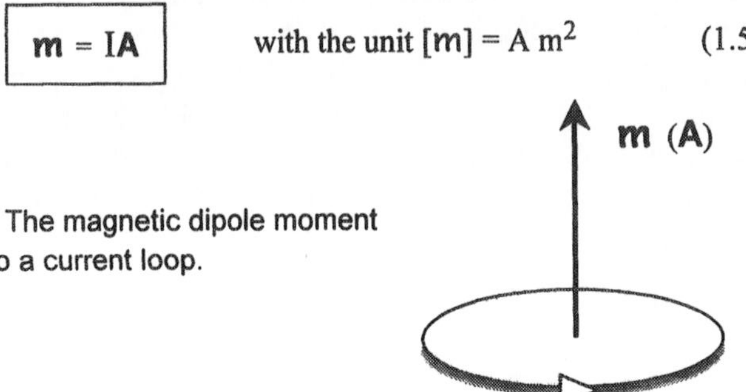

Fig. 1.3 The magnetic dipole moment **m** due to a current loop.

The definition is not confined to *circular* loops. The current loop need not even be symmetrical or planar. If it is not planar – but instead, for example, corrugated – then the vector **A** is directed parallel to the average normal to the loop. But the current I must be constant.

One obvious and important application of (1.5), detailed in § 2.3(1), is in the characterisation of the magnetic dipole moment associated with an orbiting electron.

Among other applications, the relation is also used to evaluate the magnetic moment due to a rotating molecule.

As anticipated, whether it be a current loop or small bar magnet, the interaction of a magnetic dipole with an applied *magnetic field* **B** (Fig. 1.4) is essentially analogous to the electrostatic phenomenon. If the field is uniform the dipole has a potential energy of

$$U = -\mathbf{m}.\mathbf{B} = -mB\cos\theta \tag{1.6}$$

where again θ defines its orientation. At the same time, the magnetic dipole is subject to a torque

$$\mathbf{\Gamma} = \mathbf{m} \times \mathbf{B} = mB\sin\theta \tag{1.7}$$

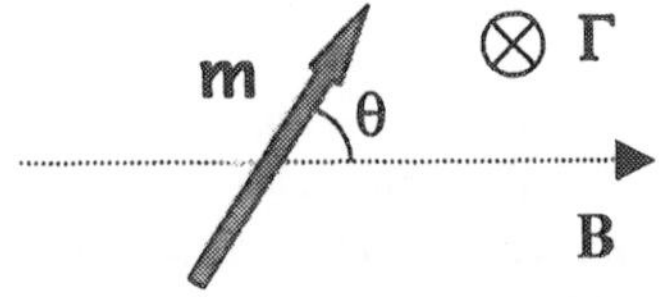

Fig. 1.4 A magnetic dipole in a uniform magnetic field **B**.

which tends to turn the dipole into the magnetic field. This latter effect may be exploited to elucidate what is meant by a *magnetic field*.[b]

The magnetic field **B** – a.k.a. the *magnetic induction* or *magnetic flux density* – is a vector quantity the direction of which at any point is defined as that adopted by a small free current-loop magnetic dipole in stable equilibrium at that point. The magnitude of **B** is defined as the torque or couple exerted on a unit magnetic dipole (1 A m^2) held at right angles to the direction of the field. The magnitude of this couple is given by $\Gamma = m_{\perp}B$ (eqn. 1.7).

b The *electric field* **E** is defined straightforwardly *via* Coulomb's law as the force acting on unit positive charge [Appendix, 5(1)].

Since the unit of torque Γ is the N m (≡ J) and $[m] = \text{A m}^2$, the SI unit for magnetic field is

$$[B] = \text{N A}^{-1}\text{m}^{-1} \equiv \text{J A}^{-1}\text{m}^{-2} \equiv \text{J s C}^{-1}\text{m}^{-2} \equiv \text{V s m}^{-2}$$

for which the collective term is the *tesla* (T). Also occasionally still in use, and equivalent to the tesla, is the SI unit *weber*/m^2 or Wb m^{-2}, where the *weber* (Wb) ≡ V s is the unit of *magnetic flux*.

1 T = 10^4 *gauss* (G), the gauss being the old unit of magnetic induction in the cgs/emu scheme, often retained for practical purposes because the tesla is rather large. The Earth's magnetic field is about 0.5 G at the surface.[a] The field close to an everyday bar magnet ⩽ 50 G, that due to a conventional laboratory (iron-cored) electromagnet ⩽ 1 T. Higher field strengths, up to 30 T or even more, can be obtained with superconducting magnets. The latest NMR spectrometers employ magnet fields in the 5 – 20 T range, and medical magnetic resonance imaging (MRI) equipment fields of up to 10 T.

a To be more precise, the field at the surface varies from about 0.3 G at the equator to about 0.6 G at the poles.

A magnetic dipole also naturally resembles an electric dipole in suffering a translational force when there is a gradient in the field; in place of (1.4), the expression for the force exerted in a particular direction is

$$F_s = m_s \frac{dB}{ds} \tag{1.8}$$

(As before, the dipole is assumed to be sufficiently small, whatever its make-up, that the field gradient is constant within it.) The effect provides the basis of deflection experiments with atomic and molecular beams [§ 4.2(6)];[b] it also underlies certain long-standing methods of measuring the magnetisation (magnetic moment density) of a material [section 4(1)].

b Similarly, the electrostatic force of eqn. 1.4 allows deflection experiments with beams of polar molecules from which electric dipole moments may be determined.

3. Electric Polarisation

(a) The bulk effect

Application of an electric field to a material, solid or fluid, causes what is termed *polarisation.* This is the induction of an electric dipole moment density attributable universally to *distortion* of the charge distribution within the constituent atoms or molecules and also due more importantly, when molecular components have intrinsic dipole moments, to the *orientation* of these dipoles (if only partial) in the direction of the external field. The effect is quantified in the electric polarisation **P**, defined as *the induced dipole moment per unit volume* of material. In the more simple cases – including liquids, gases and many solids – **P** is found to be proportional to the applied field ***E*** (in both scalar and vector terms) and may be formulated as

$$\mathbf{P} = \varepsilon_0 \chi_e \boldsymbol{E} \tag{1.9}$$

where χ_e is known as the electric susceptibility. The factor ε_0 is the ubiquitous *vacuum permittivity*, a vital formal presence in the SI scheme,[c] the units of which leave the (volume) susceptibility χ_e dimensionless. The latter quantity is of course the electric dipole moment per unit volume *induced by unit applied field.*

c $\varepsilon_0 = 8.8542 \times 10^{-12}$ J^{-1} C^2 m^{-1} (or F m^{-1}) = $1/(\mu_0 c^2)$ [see below].

It is important to distinguish the orientation and distortion components of electric polarisation. The former can arise only in materials with inherently polar constituents, when it is the dominant feature.

(b) The polarisability α

In the case of non-polar fluid substances (also the vast majority of *insulating* solids), the polarisation is due solely to the weaker distortion effect, the

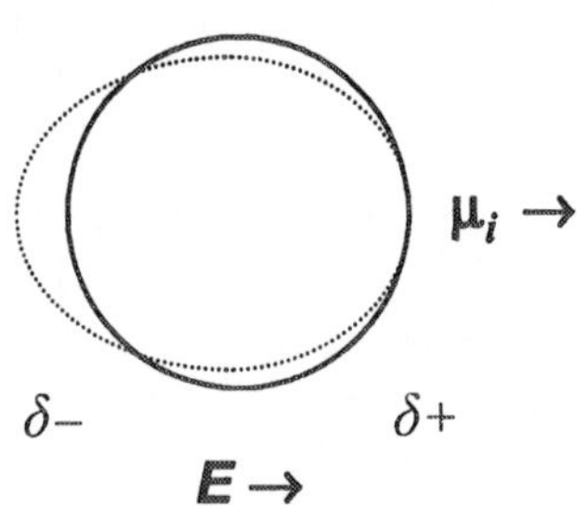

Fig. 1.5 Simple picture of the polarisation of the electron "cloud" of an individual atom.

essential nature of which is illustrated for atoms by Fig. 1.5. The applied electric field causes a relative displacement of nucleus and electron orbits thereby *inducing* an electric dipole moment $\boldsymbol{\mu}_i$ in the field direction. The magnitude of the induced dipole moment proves to be proportional to the field strength, *i.e.*

$$\boldsymbol{\mu}_i = \alpha \boldsymbol{E} \tag{1.10}$$

where α, known as the polarisability, is characteristic of the atom or molecule concerned.[a] There is a concomitant *lowering* in potential energy by $\frac{1}{2}\alpha E^2$ (per atom). [See Appendix, 5(4), for an explanation.]

The bulk *molar* polarisation observed is of magnitude $N_A\mu_i = N_A\alpha E$, where N_A is the Avogadro number, so that (recalling eqn. 1.9) the susceptibility *per mole of substance* may be expressed as

$$\chi_{mol} = \varepsilon_0^{-1} N_A \alpha \tag{1.11}$$

4. Magnetisation

(a) The parameters of bulk magnetism

Magnetisation is a comparable phenomenon to electric polarisation (up to a point), being the induction of a magnetic dipole density in a material, primarily by *orientation* of pre-existing current loops within atoms or molecules, rather than distortion.[b] It is characterised by the magnetisation **M**, defined as the *magnetic dipole moment per unit volume* acquired by a substance when subjected to a magnetic field, and having therefore the units $[M] = A\ m^{-1}$. As with electric polarisation, this is often proportional to the applied field (here **B**),[c] as expressed by

$$\mathbf{M} = \mu_0^{-1}\chi_m \mathbf{B} \tag{1.12}$$

where μ_0 is the *vacuum permeability*, the formal quantity complementary to ε_0 in the SI unit scheme,[d] and χ_m is the magnetic susceptibility, the magnetisation induced by unit applied magnetic field. The units of μ_0 are such that χ_m (like χ_e) is dimensionless.

The magnetic susceptibility *per unit mass*, sometimes denoted K, is just χ_m/ρ, where ρ is the density; this *mass susceptibility* has units $[K] = [\rho^{-1}] = kg^{-1}\ m^3$. Of more general interest in Chemistry is the *molar susceptibility*, the susceptibility per mole of substance, given by

$$\chi_{mol} = K \times M_r/10^3 = \chi_m \times M_r/10^3\rho \quad [m^3\ mol^{-1}] \tag{1.13}$$

where M_r is the relative molecular mass (RMM, in gm).[e]

(b) The magnetic field intensity and the internal field

An auxiliary magnetic field vector **H**, called the magnetic field intensity, may be introduced by expressing the magnetic field *in vacuo* as

$$\mathbf{B}_0 = \mu_0 \mathbf{H} \tag{1.14}$$

This happens to have the same units $[H] = A\ m^{-1}$ as the magnetisation **M**. It is useful here because (as it can be shown) the field within a magnetised material may be written as

$$\mathbf{B} = \mathbf{B}_0 + \mu_0\mathbf{M} = \mu_0(\mathbf{H} + \mathbf{M}) \tag{1.15}$$

[a] Insulators such as diamond, with no local polarity, behave essentially as atomic fluids. So do ionic solids, except that there is additionally a larger contribution to the polarisability α from small relative displacements of cations and anions

[b] The magnetisation does however have *distortion* aspects, notably that producing diamagnetism.

[c] However, the dependency of **M** on **B** is typically more complicated in magnetically ordered materials.

[d] μ_0 is defined to be *exactly* $4\pi \times 10^{-7}\ N\ A^{-2}$. The integration of electrostatic and electromagnetic units is secured by the relation $\varepsilon_0\mu_0 = c^{-2}$, where c is the velocity of light.

[e] Susceptibility data quoted in cgs/-emu terms, *e.g.* in $cm^3\ mol^{-1}$, can be translated into SI quantities ($m^3\ mol^{-1}$) through multiplication by $4\pi \times 10^{-6}$.

– a formulation that will be needed later (§ 3.2). Notice, incidentally, that by the definition (1.12),

$$\chi_m = \mu_0 \mathbf{M}/\mathbf{B} = \mathbf{M}/\mathbf{H} \quad (1.16)$$

The field inside a magnetised material may otherwise be expressed as

$$\mathbf{B} = (1 + \chi_m)\mathbf{B}_0 = \mu_r \mathbf{B}_0 \quad (1.17)$$

where $\mu_r = 1 + \chi_m$ is the relative permeability of the substance. (Hence the term *vacuum permeability* for μ_0.)

In the outmoded cgs/emu system, it was **H** that was known as the *magnetic field* (unit = *oersted*), while **B** was distinguished as the *magnetic induction* (unit = *gauss*). **B** and **H** were however identical numerically (*i.e.* $\mu_0 = 1$ in the old scheme).

The quantity μ_r is analogous to the *relative permittivity* $\varepsilon_r = 1 + \chi_e$ (a.k.a. the *dielectric constant*) in the theory of electric polarisation.

1.3 The different forms of bulk magnetism

As envisaged above, magnetisation is due essentially to the response of elementary current loops to the applied magnetic field, which tends (eqn. 1.7) to orientate their associated magnetic moments in its direction. On this basis the magnetisation **M** is inherently positive, as therefore is χ_m, and the associated field augments the applied field within the sample.

This (positive) magnetisation takes various forms, differing with regard to magnitude and its dependency on temperature and the strength of the applied magnetic field. But, before distinguishing these cases, a brief digression concerning an ubiquitous *negative* magnetisation is called for.

1. Diamagnetism

There is also a universal effect, akin to a *distortion* of the electronic orbital currents[a] within the individual atoms or molecules of the material, whereby a small magnetic moment density directed *against* the external field is generated. Both the magnetisation **M** and the susceptibility χ_m are accordingly *negative*, and the magnetic field within the material is reduced in comparison with the external field [*cf.* (1.15)].[b] This behaviour, *shown by all substances* but having strictly no physical counterpart in electrostatic polarisation, is known as diamagnetism. The susceptibility is usually independent of both temperature and applied field strength for purely diamagnetic materials.

[a] – actually, the induction of small additional currents attributable to the *precession* of electron orbits about the applied magnetic field [§ 3.7].

[b] According to (1.8), a diamagnetic sample is *repelled* by a source of magnetic field (the field gradient being directed towards the source). Samples with positive magnetism are attracted.

That the magnetisation is *negative* follows from Lenz's law of electromagnetic induction. The electron orbits cross the externally applied field, so experiencing a varying magnetic flux and therefore an electromotive force (e.m.f.). According to the Lenz law, the direction of the ensuing current is such as to oppose the change in magnetic flux that causes it; *i.e.* the field due to the induced circulation is directed in opposition to the applied field.

The molar diamagnetic susceptibilities of simple *atomic* materials, *e.g.* an atomic gas or an ionic solid, prove to be roughly -10^{-11} Z $m^3\ mol^{-1}$, where Z is the atomic number [§ 3.7(2)]. Argon gas has $\chi_{mol} = -2.43 \times 10^{-10}\ m^3\ mol^{-1}$ while, for isoelectronic KCl(*s*), $\chi_{mol} = -4.87\ (\pm\ 0.05) \times 10^{-10}\ m^3\ mol^{-1}$ – as might have been expected, almost exactly double that of Ar(*g*).

In the particular case of atoms, their negative values notwithstanding, the diamagnetic susceptibilities resemble electric susceptibilities in that both are determined by innate atomic properties rather than the thermal statistics in the *ensemble*. Recalling eqn. 1.11, the molar diamagnetic susceptibility may be written

$$\chi_{mol} = -\mu_0 N_A \alpha_d \qquad (1.18)$$

where α_d, comparable to the polarisability α, is characteristic of the atom.[a] Both quantities increase with the number of electrons in the atom – *i.e.* with atomic number Z [§ 3.7.2(c)].

[a] The quantity $-\alpha_d$ is a ubiquitous aspect of the *magnetisability* α. Molecules have additionally a paramagnetic (positive) component due to distortion, so also do certain open shell atoms (§ 3.3).

Diamagnetism is a minor effect in the scheme of things and, in absolute terms, considerably smaller than the (positive) forms of magnetism associated with *open-shell* atoms and molecules, *i.e.* species possessing unpaired electrons. It is invariably swamped in such systems.

2. Paramagnetism

The survey of the different types of bulk magnetism continues without the nature of the elementary dipoles being made explicit. It turns out that, in open-shell species, the current loops are due principally to electron *spin* rather than the orbital motion of electrons.

Observed in open-shell species, this involves the simplest form of magnetisation, analogous to orientation polarisation in a polar fluid. In the classical interpretation, each atom or molecule (or ion) is attributed a *permanent magnetic dipole moment* **m**, equivalent to the electric dipole **μ** of a polar molecule. The external magnetic field would orientate these dipoles, but the effect is severely impeded by thermal chaos – in the form of rapidly colliding and tumbling molecules in a fluid substance, *e.g.* oxygen gas or a solution of oxygen in liquid nitrogen (at low temperatures < 77 K). The thermal jostling tends to randomise the orientation of the molecular dipoles and what ensues, therefore, is a magnetisation *varying with temperature*.

The simplest behaviour, as encountered in gaseous O_2, for example, is an inverse dependency on temperature,

$$\chi_m = C/T \qquad (1.19)$$

known as the Curie law (P Curie 1895). The parameter C, termed the *Curie constant* (unit = K), is characteristic of the atomic or molecular species concerned. An analogous expression applies to the electric susceptibility χ_e of polar fluids. The variation as T^{-1} may be explained by classical Maxwell-Boltzmann statistics, provided the interaction between dipoles can safely be neglected[b] and that the Boltzmann energy kT greatly exceeds the magnetic potential energy mB.[c]

[b] Recalling the definition of magnetisation by (1.12), the Curie law (1.19) implies that

$$B/M \propto T$$

A parallel may then be drawn with the ideal gas law, PV = RT, the pressure being replaced by the field B and the volume by M^{-1}. [See Carlin (1986) for development of this analogy.]

[c] See Appendix, 5(5).

The Curie constant appropriate to the *molar* magnetic susceptibility will be shown to be

$$C = \mu_0 N_A m^2/3k \quad [m^3\ mol^{-1}\ K]$$

As explained in § 3.1(2), the magnetic dipole moment so construed is commonly denoted m_{eff}.

The Curie law is not, however, limited to fluid substances: as illustrated by Fig. 1.6, it may apply equally to solids containing a variety of open-shell ions, *e.g.* compounds of Mn(II) and Gd(III). The statistical treatment shows that the constant C is proportional to the *square* of the dipole moment m.[c]

Though invariably strong enough to obscure the underlying diamagnetism, paramagnetism is not a particularly large effect at *ordinary temperatures*. Typically, the (volume) susceptibility χ_m is merely some 10^{-3} and the relative permeability μ_r therefore around 1.001. (The internal field is thus only 0.1 % larger than the applied field.) The molar susceptibility is usually $\gtrsim 10^{-8}$ $m^3\ mol^{-1}$, *i.e.* some two orders of magnitude greater in absolute terms than the (molar) diamagnetic susceptibility.

A different, *weak* form of paramagnetism is observed in *metals*. Known as Pauli paramagnetism, this is essentially independent of temperature (at least in its simplest manifestations). It is attributable to the *itinerant* electrons that promote metallic conductivity [§ 3.8].

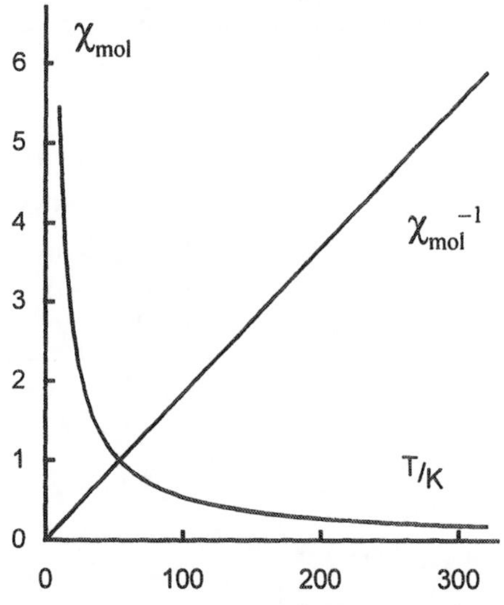

Fig. 1.6 The molar susceptibility of the Mn(II) Tutton salt $(NH_4)_2Mn(SO_4)_2.6H_2O$.

The data (10^{-6} m^3 mol^{-1}) have been corrected for diamagnetism.

The Curie constant C = 54.4 x 10^{-6} m^3 mol^{-1} K. The χ data for the Fe(III) alum $(NH_4)Fe(SO_4)_2$.-$12H_2O$ are practically identical.

Both compounds have *five* unpaired electrons per metal ion.

A precise conformity with the Curie law is also observed for Gd(III) salts and for gaseous O_2 (before it condenses at about 90 K).

3. Cooperative or ordered magnetism

There is a short-range interaction between the magnetic dipoles of individual atoms or molecules formally analogous to that existing between electric dipoles, attributable to the interaction of one dipole with the field generated by the other [Appendix, 5(3)]. However, in materials that display paramagnetism the energy of the interaction happens to be small in comparison with the Boltzmann energy kT at the ambient temperature. In the case of paramagnetic *solids*, this is due to the magnetic centres being spaced well apart. Such materials are referred to as *magnetically dilute*.

The most common form of magnetic ordering in solids, involving antiparallel coupling of each magnetic dipole with its nearest neighbours, gives rise to *antiferromagnetism* – thus named because it is, in a sense, the opposite of *ferromagnetism* (see below).

Antiparallel coupling is the natural expectation, assuming a through-space interaction [5(3)].

(a) Antiferromagnetism

At a sufficiently low temperature, with thermal chaos subsiding, the dipolar coupling will invariably result in a spontaneous magnetic ordering of the magnetic dipoles. It so happens that, in most solids, neighbouring dipoles tend to adopt an *antiparallel* configuration which, repeated throughout the lattice (Fig. 1.7), leads to antiferromagnetism. When a magnetic field of moderate strength is applied, the majority dipoles tend to orientate in the field direction, while the minority are oppositely directed, coupling between the two sets being stronger than interaction of any dipole with the field.

If the magnetic centres are crystallographically equivalent – as *e.g.*, the Mn^{2+} in MnO (rocksalt structure) – so that there are equivalent interpenetrating sub-lattices with oppositely directed dipoles, then the magnetisation ultimately vanishes as the temperature approaches the absolute zero.

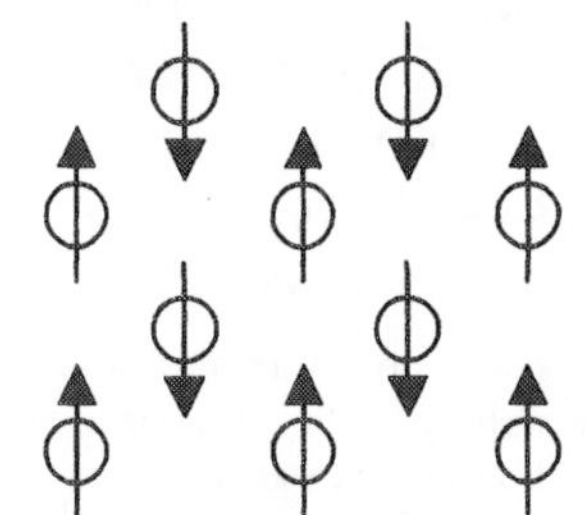

Fig. 1.7 A simple 2-dimensional model for magnetic ordering in an antiferromagnetic solid.

[The arrows represent the magnetic dipoles at individual atomic sites.]

However, as the temperature is allowed to rise, the antiparallel alignment of adjacent magnetic dipoles is increasingly disrupted by thermal fluctations. The susceptibility therefore increases until, eventually, at a temperature characteristic of the substance, it declines sharply as antiferromagnetism gives way to paramagnetism (Fig. 1.8). The transition temperature is known as the *Néel temperature* and is denoted T_N.

There are many compounds of the transition elements that are antiferromagnetic at ordinary temperatures – *e.g.* hexagonal α-Fe_2O_3 and α-Cr_2O_3, which order below 953 K and 308 K, respectively. However, the vast majority of materials that display antiferromagnetism do not do so unless cooled substantially. Thus, while FeF_3 has T_N = 394 K, CrF_3 does not order above T_N = 80 K.

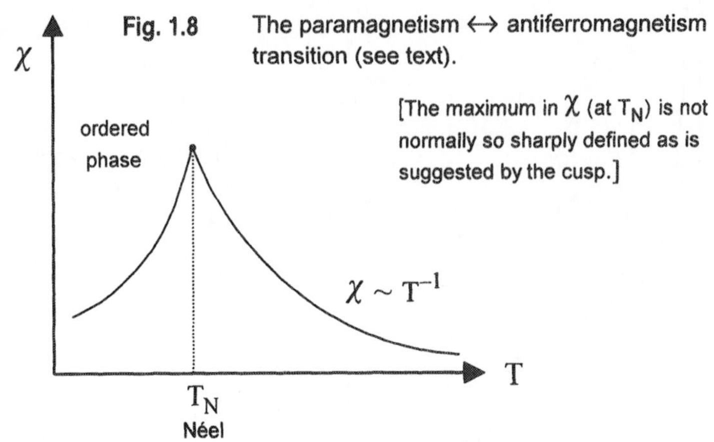

Fig. 1.8 The paramagnetism ↔ antiferromagnetism transition (see text).

[The maximum in χ (at T_N) is not normally so sharply defined as is suggested by the cusp.]

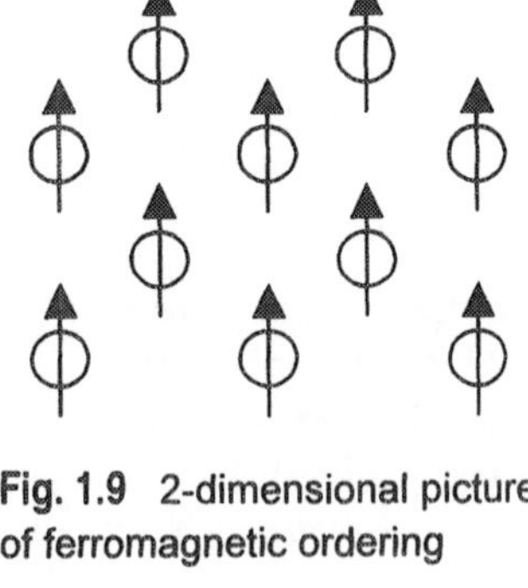

Fig. 1.9 2-dimensional picture of ferromagnetic ordering

(b) Ferromagnetism

Much less common, though better known and very important technologically, is the form of cooperative magnetism due at the atomic level to *parallel* coupling of contiguous dipoles (Fig. 1.9) leading to concerted long-range co-alignment. This is called **ferromagnetism** – originally because it is displayed by metallic iron – and is generally characterised by a huge magnetisation. The (volume) susceptibility χ_m can be as much as 10^5 – *i.e.* some seven, even eight orders of magnitude greater than typical of a paramagnetic solid at ordinary temperatures. The magnetic field inside a sample is correspondingly enhanced.

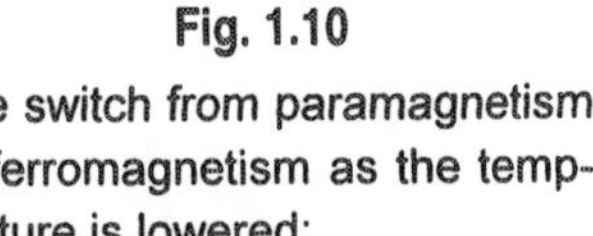

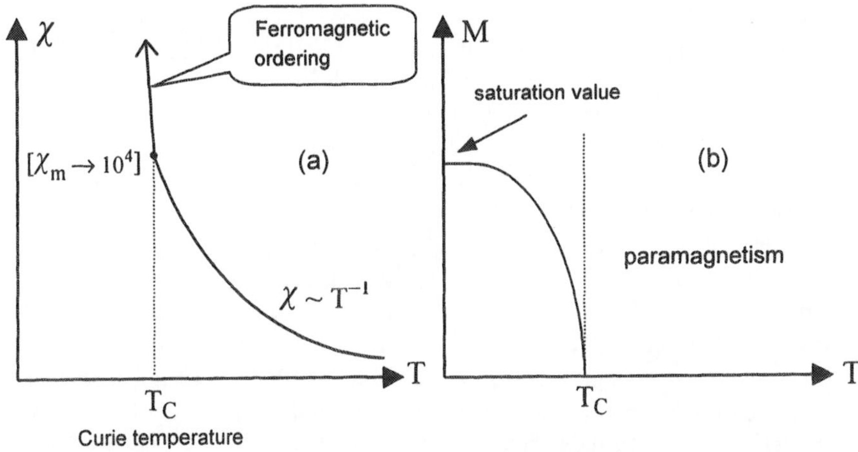

Fig. 1.10
The switch from paramagnetism to ferromagnetism as the temperature is lowered:
(a) the susceptibility χ; and
(b) the magnetisation M.
(The phenomenon is however rare in comparison with antiferromagnetism, illustrated by Fig. 1.8.)

The temperature at which the bulk magnetism of a solid switches from paramagnetic to ferromagnetic type with, concomitantly, a massive increase in the susceptibility (Fig. 1.10), is known as the *Curie temperature*, symbol T_C. In the case of pure (soft) iron $T_C = 1043$ K while CrO_2, a rare example of a *compound* showing ferromagnetism at "room" temperature, has a Curie point $T_C = 393$ K.[a]

As in antiferromagnetism, the dipoles tend to order spontaneously at low temperatures, though again this is resisted by thermal agitation. However, in contrast, a tendency to ferromagnetic ordering is reinforced by an applied magnetic field. As $T \rightarrow 0$ or when, at somewhat higher temperatures, the sample is subjected to a sufficiently strong external field, the magnetisation approaches a limiting value M_{sat} known as the *saturation magnetisation* [Fig. 1.10(b)], the *molar* value of which may be expressed as

$$M_{sat} = N_A m \qquad [\text{unit} = \text{A m}^2\ \text{mol}^{-1}]^b \qquad (1.20)$$

where m is the atomic magnetic moment. In simple cases – so-called *soft* ferromagnetics (below) – the magnetisation is proportional to the strength B of the applied field (recall eqn. 1.12) until close to saturation.

A broad distinction can be made between *soft* and *hard* ferromagnets. The former include iron itself, employed extensively in electromagnets, and a variety of iron-based alloys such as Fe/Si (with up to 4 % Si), used in motors, generators and transformers, and also those involving nickel (*e.g. mumetal* and *permalloy*) used in smaller transformers *etc.* and transducers, and for magnetic screening.

In the ideal *soft* ferromagnetic material the magnetisation disappears as the applied field is reduced to zero. On the other hand, *hard* ferromagnets display *hysteresis*, there being a major retention of magnetisation (or *remanence*) when the magnetic field is switched off. The remanent magnetism of such materials is a familiar property, providing for the construction of *permanent* magnets.[c] A wide variety of ferromagnetic alloys are actually employed, many based on iron – with *e.g.*, Al, Co, Ni & Cu (*Alnico* and *Alcomax*), Cr (5 % steels) and, more recently, Nd & B ($Nd_2Fe_{14}B$, for instance) – but a range of non-ferrous alloys, such as Co_5Sm, are also widely exploited.[d] The materials are invariably polycrystalline, and their technical performance then depends on particle size as well as intrinsic properties.

A ferromagnetic substance may be demagnetised by heating above its Curie temperature T_C (Fig. 1.10). However, a "virgin" sample so prepared does not naturally develop bulk ferromagnetism on cooling below T_C because, even in single crystals, the spontaneous ordering described above is confined to micro-regions or *domains*, the individual magnetisations of which are randomly directed. But application of a magnetic field aligns the local magnetisation vectors, the cooperative magnetism subsequently being largely retained (provided, of course, $T < T_C$) when the field is turned off.

Materials displaying ferromagnetism are generally *metallic*, the dipolar coupling between magnetic centres being partly transmitted by the itinerant conduction electrons.[e] On the other hand, the great majority of metals are diamagnetic, paramagnetic or antiferromagnetic at ordinary temperatures, as can be seen in Fig. 1.12.

[a] Finely dispersed in plastic, CrO_2 is used on some scale as a magnetic recording medium.

[b] The molar M_{sat} is still commonly quoted in cgs/emu (G cm^3 mol^{-1}). To convert to SI units (μ_B mol^{-1}) multiply by $1.791 \times 10^{-4} \equiv (N_A\mu_B)^{-1} \times 10^{-3}$.

[c] – including, of course, the everyday magnets employed in temporary fixing and fastening and also for recreational purposes.

[d] Certain *ferrites* such as $BaFe_{12}O_{19}$ ($\equiv BaO.6Fe_2O_3$) are also commonly used in permanent magnets below).

[e] But there are counter-examples, *e.g.* $LiNiO_2$ ($T_C = 6$ K), $CrBr_3$ (34 K) and EuO (77 K), all electrical insulators or semiconductors. But none is ferromagnetic at ~ 300 K.

(c) Metamagnetism

A crystal lattice may, in general, harbour both ferromagnetic and antiferromagnetic interactions (the latter usually dominant). In $FeCl_2$, which adopts the $CdCl_2$ layer structure, the coupling within each layer of Fe^{2+} ions is *ferro*magnetic, while there is a weak *antiferro*magnetic interaction between the layers. The system orders antiferromagnetically below about 24 K but application of a sufficiently strong magnetic field ($\gtrsim$ 1 T) induces ferromagnetism by overiding the antiferromagnetic coupling between the layers and driving their concerted Fe^{2+} dipoles in the same direction. This is an example of **metamagnetism**, in general a term signifying a magnetic phase transition promoted by an applied field.

(d) Ferrimagnetism

Many oxides containing Fe(III), known collectively as *ferrites*, show cooperative magnetism of a type superficially similar to (hard) ferromagnetism and termed, accordingly, **ferrimagnetism**.[a] *Magnetite* Fe_3O_4 ($\equiv$ $FeO.Fe_2O_3$) and *maghaemite* γ-Fe_2O_3 are prominent, and commercially important examples.

a Ferrimagnetism is also characterised by, in particular, hysteresis and remanent magnetisation.

The temperature below which a substance orders ferrimagnetically is called its *ferrimagnetic Néel temperature*, here denoted T_{fN}.[b] The ferrites Fe_3O_4 and γ-Fe_2O_3 have T_{fN} equal to 858 K and *ca.* 856 K, respectively.

b Sometimes otherwise known as the ferrimagnetic Curie temperature.

As a distinctive common feature, the crystal structures of these compounds have the metal ions arranged in non-equivalent sub-lattices with *different numbers* of ions. An antiparallel dipolar coupling between neighbouring ions belonging to different sub-lattices, stronger than the interaction *within* sub-lattices, leads to oppositely magnetised arrays of ions with a net magnetisation parallel to that of the more populous sub-lattice.

Fig. 1.11

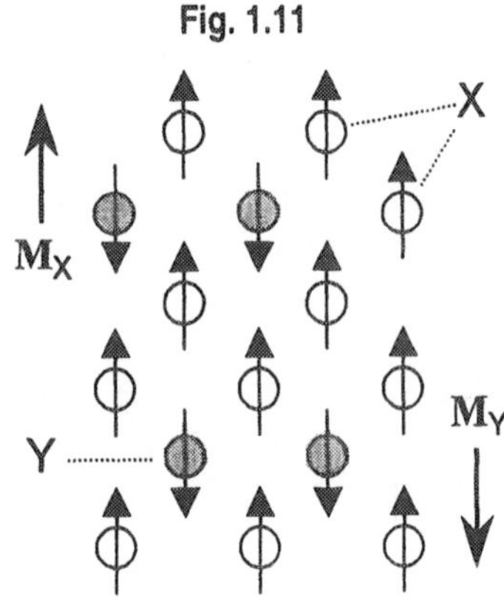

A two-dimensional model for ferrimagnetic ordering.

A simple two-dimensional picture of the phenomenon is provided in Fig. 1.11. The crystal lattice is imagined to contain two distinct kinds of magnetic centre (in practice metal cations), X and Y, and in *unequal* proportion. It is assumed that the dominant interaction happens to be antiferromagnetic coupling between the different ions, this over-riding both the X–X and Y–Y interactions. The outcome is a sub-lattice of type X ions with its magnetisation $\mathbf{M}_X$ in one direction and a sub-lattice of the minority Y ions (shaded) oppositely magnetised. Since the two sub-lattices are inequivalent there is a net magnetisation $\mathbf{M}_X - \mathbf{M}_Y$ in the direction of $\mathbf{M}_X$. The ordering within each sub-lattice is *effectively* ferromagnetic.

In γ-Fe_2O_3 the Fe^{3+} cations are distributed in a particular way among the tetrahedral (A) and octahedral (B) holes in a cubic close-packed oxide lattice.[c] In the usual formulation of the structure, the proportional occupancy of octahedral and tetrahedral holes is 5:3, so that 5/4 of the cations reside in octahedral sites and 3/4 in the tetrahedral. When the compound is fully magnetised, with antiparallel A–B coupling as the dominant interaction, the magnetisation of sub-lattice B therefore exceeds that, oppositely directed, of the sub-lattice A; the saturation moment should be $\frac{1}{2}m$ per formula unit (Fe_2O_3), where m is the dipole moment of an individual Fe^{3+} ion.[d]

c The crystal structure of γ-Fe_2O_3 is closely related to that of Fe_3O_4; it may be formulated as $Fe_{8/3}O_4$ [see § 3.6(4b)].

d The theory of ferrimagnetism is pursued further (on a quantum mechanical basis) in § 3.6(4).

γ-Fe_2O_3 finds wide application in magnetic recording, as do many spinel ferrites such as Fe_3O_4 and $CoFe_2O_4$. Other technologically important ferrites include yttrium iron garnet $Y_3Fe_5O_{12}$ (YIG, T_{fN} = 553 K) and $BaFe_{12}O_{19}$

(barium ferrite, 820 K), both exploited in compressed polycrystalline form as permanent magnet materials. Being essentially electrical insulators, such ferrites are particularly useful for inductance cores, where a metallic material would show large (and undesirable) eddy current losses.

There are applications also in Nature: *magnetotactic* function in biological systems, including the navigational instincts of birds, is provided by organised intra-cellular microcrystals of Fe_3O_4.

(e) Superparamagnetism

Colloidal dispersions of typical ferrites with a particle size of about 10 nm (corresponding to a single domain)[a] are easily saturated magnetically by fields of modest strength, but have zero remanence. At low to medium values of the experimental quantity B/T the colloidal *ensemble* exhibits paramagnetism of the classical Langevin type [Appendix, 5(5)], while being characterised by an unusually large elementary magnetic moment m. At very low B/T the susceptibility conforms to the Curie Law $\chi \propto C/T$.

a – which are known sometimes as *ferrofluids*.

The phenomenon is referred to as **superparamagnetism**. Remarkably, it is displayed also by antiferromgnetic materials if the particle size is comparable with that of the unit cell for, then, the presence of an odd number of layers of cations leaves a significant magnetic moment in relation to volume.[b] Well established instances include NiO and α-Cr_2O_3. Superparamagnetism has also been observed for fine dispersions in very thin surface layers on solids.

b Such a particle may be pictured in terms of Fig. 1.7, if the uppermost or lowest row of ions is deleted.

4. Retrospect

The essential features of the different types of bulk magnetic behaviour are summarised in Table 1.1.

Table 1.1 The characteristics of χ_{mol} for the principal forms of bulk magnetism as observed at ordinary temperatures.

type of behaviour	magnitude (a)	dependence on temperature?	dependence on applied field?
diamagnetism	-10^{-3} Z	X	X
paramagnetism	1	✓	X (β)
antiferromagnetism	$\lesssim 0.1$	✓	✓
ferromagnetism	10^7 (?+)	✓	✓
ferrimagnetism	10^6 (?–)	✓	✓
superparamagnetism	$10 - 10^6$ (γ)	✓	✓
Pauli paramagnetism	$\gtrsim 10^{-2}$	X	X

(a) – referred to a typical paramagnetic susceptibility of about 10^{-8} m^3 mol^{-1} at *ca.* 300 K (when, of course, only certain materials are magnetically ordered).

(β) – unless B/T is extremely large.

(γ) – depending on B.

Except for diamagnetism (and perhaps saturation magnetisation[c]), the susceptibility is a property of the *ensemble* rather than simply an atomic or molecular characteristic.

c – in paramagnets and elementary ferromagnets.

Ferrimagnetism and superparamagnetism aside, the various forms of bulk magnetism are all encountered among the *elements* in their standard states (Fig. 1.12).[d] Only oxygen, occurring as $O_2(g)$, is a perfect Curie paramagnet, though complex paramagnetism is observed in most of the lanthanide (rare earth) metals and in the later actinides.

d The reader may care to ponder the periodicity evident in this figure.

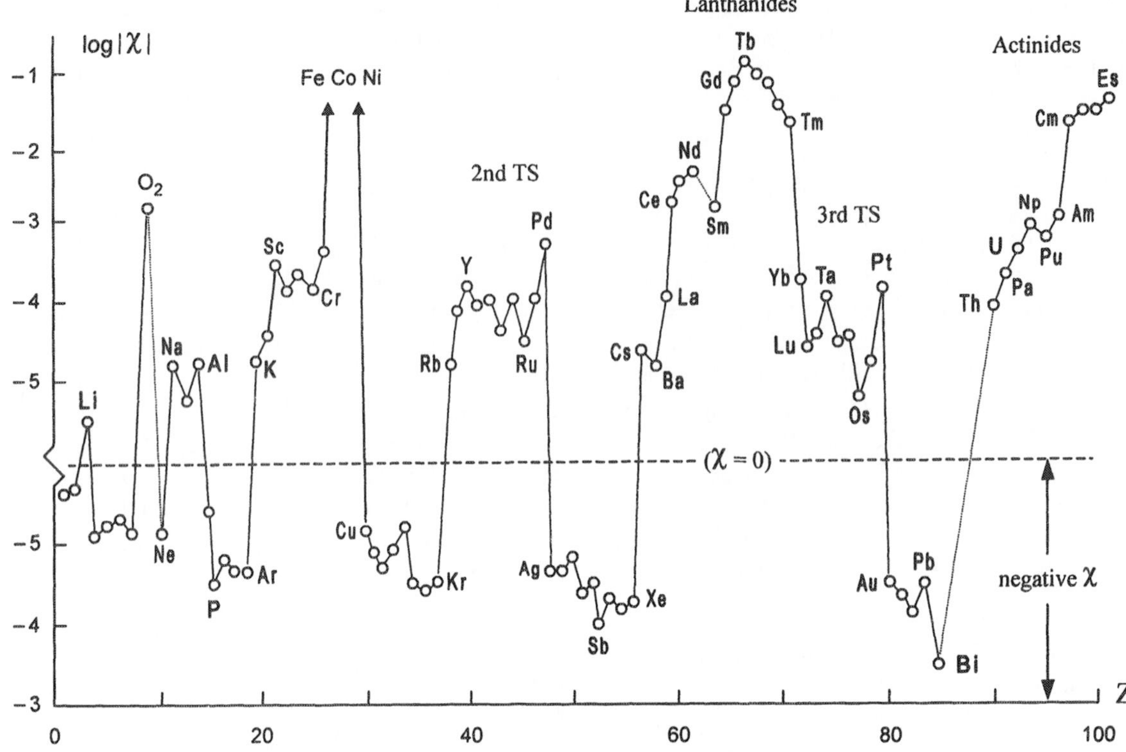

Fig. 1.12 The atomic susceptibilities of the elements at SATP plotted against atomic number Z.
The diagram is based on data expressed (as originally) in cgs/emu (g atom)$^{-1}$.
The entries for carbon and tin relate to the allotropes diamond and α-Sn, respectively.

1.4 The measurement of bulk susceptibilities

A variety of experimental methods is available for determining the bulk susceptibilities of materials,[a] but a detailed survey is not feasible here.

[a] See the bibliography at the end of this chapter.

The most sensitive device is the SQUID magnetometer, which exploits electromagnetic induction. It involves measurement of the electrical effect of the field due to a magnetised specimen as it is translated through superconducting coils. The e.m.f. generated in the coils is proportional to the magnetic moment of the sample. The technique is especially useful for low temperature studies.

The simplest method is that ascribed to Gouy, one of the "classical" techniques based on the force exerted on a magnetised sample when resident in an *inhomogeneous* field. It is particularly suitable for finely divided solids and liquids, though not for single crystals nor for measurements at *very* low temperatures (< 80 K).

1. The Gouy balance

In the Gouy method (Fig. 1.13) the finely powdered sample is evenly compacted in a cylindrical quartz tube – to a height *h* (= XY), say – and suspended from a sensitive balance between the poles of a "horse-shoe" electromagnet or stable permanent magnet. The bottom end of the tube (X) is positioned centrally between the pole pieces where the magnetic field (B_X) is homogeneous[b] while the upper part of the sample (Y) lies in the region of very weak field ($B_Y \sim 0$) beyond the poles. The sample thus resides in a field gradient dB/ds directed downwards in the figure and, therefore (recall eqn. 1.8), a paramagnetic sample experiences an increase in weight. Conversely,

[b] The field is typically ~ 10^3 G in routine measurements.

the weight of a diamagnetic sample is reduced (albeit to a much smaller extent).

Consider an infinitesimal slice of the sample of thickness ds and mass = $(m/h)\mathrm{d}s$, where m is the total mass of the sample. According to (1.12) the magnetic moment of this slice may be expressed as $mK\mathrm{B}\mathrm{d}s/\mu_0 h$, where K is the *mass* susceptibility (eqn. 1.13). Then, invoking (1.8), the downward force on the slice is

$$\mathrm{dF} = (mK\mathrm{B}\mathrm{d}s/\mu_0 h)\,\mathrm{dB}/\mathrm{d}s = mK\mathrm{B}\mathrm{dB}/\mu_0 h$$

and, integrating over the sample, the total force acting is

$$\mathrm{F} = (mK/2\mu_0 h)(\mathrm{B}_\mathrm{X}^2 - \mathrm{B}_\mathrm{Y}^2)$$

The mass susceptibility is therefore

$$K = \gamma\mathrm{F}/m$$

where $\gamma = 2\mu_0 h/(\mathrm{B}_\mathrm{X}^2 - \mathrm{B}_\mathrm{Y}^2)$ is an experimental parameter which may be held constant.

γ may be determined by calibration of the apparatus with a compound of accurately known susceptibility and also temperature dependence. Popular calibrants for paramagnetic susceptibilities include $Hg[Co(NCS)_4]$,[a] containing Co(II) as magnetic centre, and the Mn(II) compound $(NH_4)_2Mn(SO_4)_2\cdot 6H_2O$ which featured in Fig. 1.6. Water is commonly employed as calibrant in measurements on diamagnetic materials.

In accurate work it is necessary to make allowance for the magnetic buoyancy of the air displaced by the sample when comparing the weights of the filled and empty tube.[b]

For more approximate purposes, a small permanent magnet may be used and the susceptibility at "room" temperature instead determined by fixing the sample and measuring the change in weight of the magnet – the *inverted* Gouy method, so called.

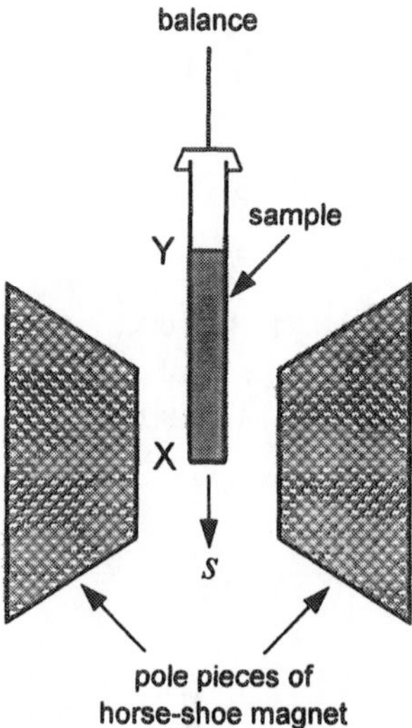

Fig. 1.13 The Gouy method for measuring susceptibilities.

[a] The mass susceptibility of the Co(II) compound is given by

$$K = 62.6/(T + 10) \times 10^{-6}\ \mathrm{m^3\,kg^{-1}}.$$

The susceptibility of water is $-0.905 \times 10^{-8}\ \mathrm{m^3\,kg^{-1}}$ (and independent of T).

[b] To obtain the true paramagnetic susceptibility the observed K must, of course, be corrected for its diamagnetic element.

[See § 3.7(2e)]

2. The NMR method

A quite different technique for the measurement of paramagnetic susceptibilities, applicable to *solutions*, exploits the ubiquity of nuclear magnetic resonance (NMR) spectrometers in modern laboratories. It is based on the shift occasioned in an NMR signal of a co-dissolved compound[c] by the magnetic field due to the paramagnetic species. The magnetic field within the liquid medium being $(1 + \chi_m)\mathrm{B}_0$ (eqn. 1.17), the relative displacement of the reference signal

$$\Delta\mathrm{B}/\mathrm{B} \propto \Delta\chi_m$$

where $\Delta\chi_m$ is the difference in volume susceptibilities of the solutions with, and without the paramagnetic sample. This relation allows useful estimates of the sample χ_m over a wide temperature range.

If paramagnetic, any newly synthesised (and soluble) compound is likely to be identified as such, and its magnetic moment initially determined, by routine NMR spectroscopy.

[c] *e.g.* a ^{1}H resonance of tetramethylsilane (TMS) or *t*-butyl alcohol.

See Deutsch & Poling (1969) for more detail concerning the NMR method.

1.5 Appendix – Some notes on Electrostatics and analogous concepts in Magnetostatics

1. Coulomb's law of force and the electric field *E*

The mutual (repulsive) force acting between point charges q_1 and q_2 is of magnitude[a]

$$F = \frac{q_1 q_2}{4\pi\varepsilon_0 r^2} \qquad \text{(Coulomb's law)} \tag{1.21}$$

[a] The charges are assumed to have dimensions that are small in relation to the distance r between them.

where ε_0 is the *vacuum permittivity*. The electric field $\boldsymbol{E}$ due to a point charge q at a distance r away is defined as

$$\boldsymbol{E} = \frac{q}{4\pi\varepsilon_0 r^3}\boldsymbol{r} \tag{1.22}$$

It is a vector equal to the *force* **F** *acting on unit positive charge* situated at the point concerned, with magnitude $E = q/4\pi\varepsilon_0 r^2$. It radiates equally in all directions.[b]

[b] The *magnetic field* **B** must be defined in a quite different way since there is no such thing as a *magnetic pole* analogous to electric charge.

2. The field due to an electric dipole

The electric dipole $\boldsymbol{\mu}$ was defined in section 2(1). The separation of charges in the dipole ensures that it generates an electric field externally. At distances r large compared with the dimension a of the dipole, the *radial* and *tangential* components, defined in Fig. 1.14, prove to be

$$E_r = \frac{\mu\cos\theta}{2\pi\varepsilon_0 r^3} \quad \text{and} \quad E_t = \frac{\mu\sin\theta}{4\pi\varepsilon_0 r^3} \tag{1.23}$$

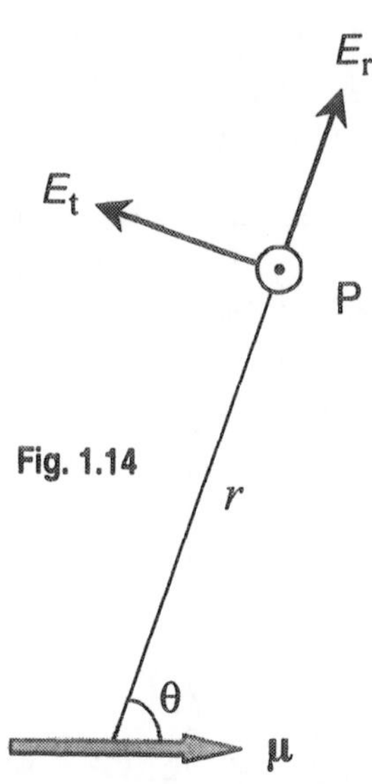

Fig. 1.14

The electric field due to a dipole at some point P.

both varying as r^{-3}. The electric field perpendicular to the page (*i.e.* the plane defined by the dipole and the point P) is zero.

The magnetic field due to a magnetic dipole moment is formally similar, with m replacing μ, and with μ_0 instead of $1/\varepsilon_0$, in the expressions for its components B_r and B_t.

3. The mutual interaction of electric dipoles

It will be sufficient here to treat the simple, particular case of parallel electric dipoles (Fig. 1.15). Dipolar coupling is due to the response of one dipole to the field generated by the other.

The electric field E_1 due to dipole $\boldsymbol{\mu}_1$ at the site of, and *parallel to* dipole $\boldsymbol{\mu}_2$ is of magnitude

$$E_1 = E_r\cos\theta - E_t\sin\theta$$

$$= \frac{\mu_1}{4\pi\varepsilon_0 r^3}(2\cos^2\theta - \sin^2\theta) = \frac{\mu_1}{4\pi\varepsilon_0 r^3}(3\cos^2\theta - 1)$$

(always assuming, of course, that the intervening distance r is much greater than the internal dimensions of the dipoles). Now the energy of interaction of dipole $\boldsymbol{\mu}_2$ with the field E_1 is (by eqn. 1.3) simply $U = -\mu_2 E_1$. The mutual potential energy of the two dipoles is hence

$$U = \frac{\mu_1\mu_2}{4\pi\varepsilon_0 r^3}(1 - 3\cos^2\theta) \tag{1.24}$$

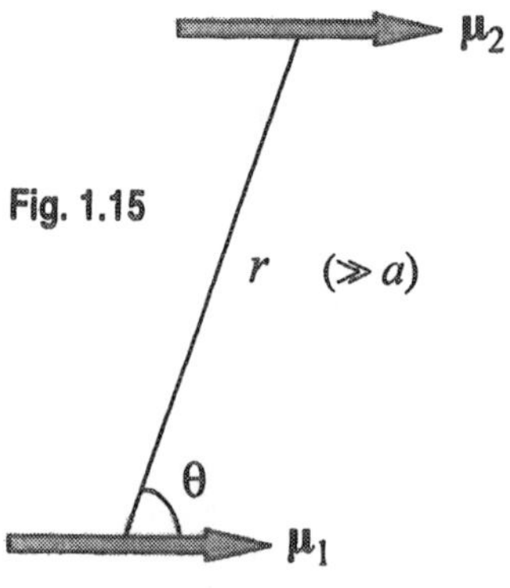

Fig. 1.15

(The same result may naturally be obtained by considering instead the interaction of dipole $\boldsymbol{\mu}_1$ with the electric field E_2 due to dipole $\boldsymbol{\mu}_2$.) The most stable configuration is that with co-linear dipoles ($\rightarrow\ \rightarrow$, $\theta = 0$), when $U = -\mu_1\mu_2/2\pi\varepsilon_0 r^3$; the antiparallel arrangement ($\uparrow\downarrow$, $\theta = \frac{1}{2}\pi$) is also stable, though with a lower potential energy $U = -\mu_1\mu_2/4\pi\varepsilon_0 r^3$.

Magnetic dipoles show an analogous coupling: the expression for the energy of interaction may be deduced immediately from (1.24) by substituting the magnetic moments m_1 and m_2 for the electric dipole moments (while again replacing $1/\varepsilon_0$ by μ_0). Note once more the variation as r^{-3}.

$$U = \frac{\mu_0 m_1 m_2}{4\pi r^3}(1 - 3\cos^2\theta) \tag{1.25}$$

A *direct*, "through-space" interaction between dipoles, as in the electrostatic context, is thereby implied. However, in the solid state, any *substantial* magnetic dipolar coupling has usually some *indirect* mechanism. Thus, the coupling in *metallic* materials is invariably mediated by the conduction electrons. In the case of insulating compounds such as transition metal oxides the interaction occurs through covalent bonding *via* intervening ligands.[a] The direct coupling between magnetic ions is extremely weak in comparison.

a – normally by a mechanism known as *superexchange* [§ 3.6(3)].

In any magnetically ordered material [section 3(3)] the effective field due to a magnetic dipole is clearly very large in relation to the external field B_0.

4. Distortion polarisation

As pointed out earlier, an applied electric field will induce a dipole moment $\boldsymbol{\mu}_i = \alpha \boldsymbol{E}$ (eqn. 1.10) in any atom or molecule. The energy of this polarisation may be calculated by imagining the field $\boldsymbol{E}$ to be generated very gradually from zero. An increment dE induces a dipole moment $d\mu = \alpha dE$ and, by (1.3), the change in potential energy is $dU = -E d\mu = -\alpha E dE$. The potential energy per atom or molecule is therefore

$$U = -\int_0^E \alpha E\,dE = -\tfrac{1}{2}\alpha E^2 \tag{1.26}$$ [c]

b It is worth noting that the electric field E has the dimensions $J\ C^{-1}\ m^{-1}$ and that those of the polarisability α are therefore $J^{-1}\ C^2\ m^2$.

Polarisabilities are often expressed in terms of the *polarisability volume* α/ε_0 with unit = m^3.

5. Orientation polarization – the Langevin-Debye theory

The theory concerns fluid substances composed of polar molecules; the interaction between the molecular dipoles is neglected.

As illustrated by Fig. 1.2, an applied electric field exerts a torque on each individual molecular dipole that tends to orientate it in the field direction. However, the alignment of the dipoles is resisted by thermal agitation of the molecules which tends instead to randomize their orientation. The resulting distribution of molecular dipoles at thermal equilibrium is determined by the

c Conversely, μ_i may be obtained from

$$\mu_E = -dU/dE$$

which follows from (1.3).

classical Maxwell-Boltzmann statistics, whereby the number dn of dipoles aligned at an angle θ to the field direction is just

$$dn = (n/4\pi)\exp(-U_\theta/kT)d\omega \qquad (1.27)$$

$U_\theta = -\mu E\cos\theta$ is the potential energy of the molecule for that particular orientation (eqn. 1.3), $d\omega$ the solid angle corresponding to an increment $d\theta$ in the inclination θ[a] and n the total number of molecules present. The factor $n/4\pi$ is the number of molecular dipoles directed within unit solid angle in the absence of the field.

a $d\omega = 2\pi\sin\theta d\theta$

Each of these dn molecules contributes a dipole moment $\mu\cos\theta$ in the direction of the applied field. Integrating over all possible orientations, the average dipole moment in the field direction is accordingly

$$\langle\mu_E\rangle = \frac{\int \mu\cos\theta dn}{\int dn} \qquad (1.28)$$

It is then not difficult to show that[b]

b Consult, for example, Bleaney and Bleaney (1989).

$$\langle\mu_E\rangle = \mu(\coth x - \frac{1}{x}) = \mu L(x) \qquad (1.29)$$

$$\coth x = \frac{e^x + e^{-x}}{e^x - e^{-x}}$$

where $x = \mu E/kT$. The function $L(x) = \coth x - 1/x$, plotted in Fig. 1.16, is attributed to Langevin.

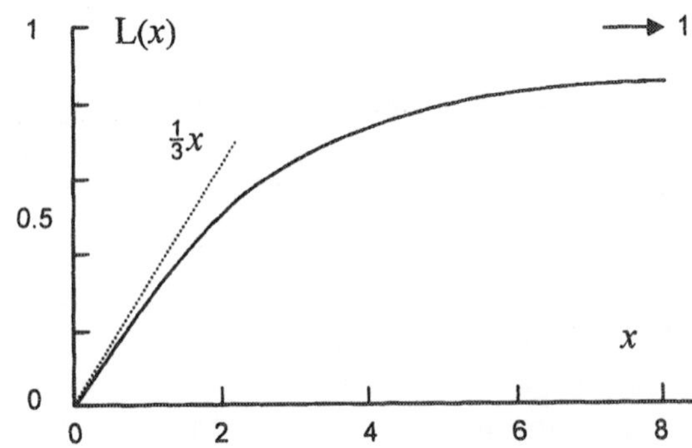

Fig. 1.16
The Langevin function,
$L(x) = \coth x - \frac{1}{x}$

In practice, the energy μE representing the interaction of the dipole with the electric field is always very much less than the Boltzmann energy kT. The parameter x is therefore invariably small, when (it may be shown)[c] the *Langevin function* $L(x)$ reduces to

$$L(x) = x/3 - x^3/45 + \ldots. \qquad (1.30)$$

with the limiting value $x/3 = \mu E/3kT$ as $x \to 0$. In this limit the mean molecular dipole moment in the field direction is therefore

$$\langle\mu_E\rangle = \frac{\mu^2 E}{3kT} \qquad (1.31)$$

c The reader may care to verify eqn. 1.30, remembering that
$e^x = 1 + x + \frac{1}{2}x^2 + \ldots.$
with the general term $= x^n/n!$, and that
$(1+y)^{-1} = 1 - y + y^2 - \ldots.$
with general term $= (-1)^n y^n$, provided $y < 1$ (Maclaurin expansion).

Recalling (1.9), the *molar* susceptibility $\chi_{mol} = N_A\langle\mu_E\rangle/\varepsilon_0 E$. The complete expression for the molar susceptibility, incorporating also the distortion polarisation (eqn. 1.11), is then

$$\chi_{\text{mol}} = \frac{N_A}{\varepsilon_0}\left(\frac{\mu^2}{3kT} + \alpha\right) \quad (1.32)$$

which is known as the *Langevin-Debye equation.*

Langevin (1905) and Debye (1913)

There is an equivalent expression in Magnetostatics for the magnetic susceptibility of an ensemble of non-interacting magnetic dipoles (assuming instead that $mB \ll kT$). Since the temperature independent, distortion term is generally small the above theory rationalises the experimental *Curie law*, $\chi \propto T^{-1}$ (eqn. 1.19).

The corresponding expression for the *paramagnetic* susceptibility is

$$\chi_{\text{mol}} = \mu_0 N_A\left(\frac{m^2}{3kT} + \alpha\right)$$

where α is the *magnetisability*.

The latter is sometimes identically zero so that the susceptibility χ is accurately an inverse function of T (*e.g.* Fig. 1.6).

In the other limit, where $x \to \infty$, L(x) tends to unity and $\langle\mu_E\rangle \to \mu$. This situation, more easily imagined than realised in practice because it would require enormous field strengths and/or exceedingly low temperatures, is termed *saturation*, $\langle\mu_E\rangle = \mu$ being the maximum or *saturation moment.*# Every molecular dipole is driven parallel to the applied electric field, and the (molar) orientation polarisation has its maximum possible value, $N_A\mu$.

While hypothetical as regards the electric polarisation of molecular fluids, saturation can be achieved comparatively easily in the analogous magnetic context – that is, *paramagnetism* [§ 3.1(2c)]. According to the classical treatment above, the saturation magnetic moment, commonly denoted m_{sat}, is to be identified straightforwardly with the magnetic moment m characteristic of the atomic or molecular species concerned. However, it transpires in the quantum theory of paramagnetism that the concept of a *permanent* magnetic dipole moment is not strictly valid, the quantity m varying significantly with the experimental parameter B/T.

\# Saturation effects are however observable in *solids* – for example, many oxides having the *perovskite* ($CaTiO_3$) structure.
[See, for example, West (1984).]

1.6 Bibliography

Further reading:

B I Bleaney and B Bleaney, *Electricity and Magnetism,* Oxford 1976 & 1989

R L Carlin, *Magnetochemistry,* Springer-Verlag, 1986

J Crangle, *The Magnetic Properties of Solids*, Edward Arnold, 1977; *idem, Solid State Magnetism*, Van Nostrand Reinhold (New York), 1991

W J Duffin, *Electricity and Magnetism*, McGraw-Hill, 4th edn. 1990

D C Mattis, *The Theory of Magnetism I* (*Statics and Dynamics*), Springer-Verlag , 1981 & 1988 [for the history of Magnetism, in § 1]

J H Van Vleck, *The Theory of Electric and Magnetic Susceptibilities*, Oxford 1932 [a seminal work, frequently reprinted]

A R West, *Solid State Chemistry and its Applications*, Wiley 1984

Concerning the measurement of bulk susceptibilities:$

D F Evans, *J Chem. Soc.,*1959, 2003; D F Evans and T A James, *J. Chem. Soc., Dalton Trans.*, 1979, 723; J L Deutsch and S M Poling, *J. Chem. Educ.*, 1969, **46**, 167 [NMR method].

W E Hatfield, *Magnetic Measurements*, ch. 4 of *Solid State Chemistry – Techniques*, ed. A K Cheetham and P Day, Clarendon Press, Oxford, 1987

D Jiles, *Introduction to Magnetism and Magnetic Materials*, Chapman and Hall (1991, 2nd edn. 1998), § 3

$ Otherwise see:
R L Carlin (1986), § 11;
J Crangle (1977), § 7, (1991), § 4.

L N Mulay and I L Mulay, *Static Magnetic Techniques and Applications,* in *Techniques of Physical Chemistry* IIIB, 133, 1989.

C J O'Connor, in *Progr. Inorg. Chem.*, ed. S J Lippard, Interscience, 1982, **29**, 203

1.7 Exercises

1. Starting from the Coulomb law (1.21), show that the electric field at a large distance r along the axis of a dipole is

$$E_a = \frac{\mu}{2\pi\varepsilon_0 r^3}$$

while at the same distance along the perpendicular to the dipole, but also parallel to the direction of the dipole, the field is

$$E_b = -\frac{\mu}{4\pi\varepsilon_0 r^3}$$

2. Justify eqn. 1.8 defining the force experienced by a magnetic dipole in a magnetic field gradient.

$\mu_0 = 4\pi \times 10^{-7}$ N A^{-2}

$\mu_B = 9.274 \times 10^{-24}$ A m^2

1 J = 5.035×10^{22} cm^{-1}

3. Referring to section 5(3), estimate the potential energy of a pair of unit magnetic dipoles with antiparallel alignment held 4 Å (= 400 pm) apart, expressing your answer in cm^{-1}; also, explain any assumptions you make. Compare your result with the figure of around 5 cm^{-1} evaluated *via* experiment for many compounds antiferromagnetic at *ca.* 300 K, and comment.

4. Discuss the following information concerning diamagnetism.

(a) The diamagnetic susceptibilities of the noble gases He – Xe decrease (*i.e.* become more negative) with increasing atomic number. So do the susceptibilities of the cations Li^+ – Cs^+, which are uniformly larger.

(b) The diamagnetic susceptibilities of O^{2-}, K^+ and Cu^+ are practically the same despite their different electron configurations.

5. Account for the following observations concerning paramagnetism.

(a) The paramagnetic moment of Gd^{3+} is larger than that of Fe^{3+} and Mn^{2+}.

(b) Dioxygen O_2 is paramagnetic, as is NO_2, but ozone O_3 and the nitrite anion $[NO_2]^-$ are diamagnetic. The paramagnetic moment of O_2 is greater than that of gaseous NO_2, or of any typical organic free radical.

(c) The superoxide anion $[O_2]^-$ (in KO_2, for example) is paramagnetic but also displays a moment that is smaller than that of O_2. The peroxide anion $[O_2]^{2-}$ (in, *e.g.*, Na_2O_2) is however diamagnetic.

(d) Liquid NO_2 is only feebly paramagnetic while the solid (temperature T $< -11^\circ$C) is diamagnetic.

(e) The paramagnetic moment of O_2 is apparently much reduced in the liquid phase (T $< -183^\circ$C) and shows a further, abrupt fall in the solid state ($< -219^\circ$C). However, dilute solutions of O_2 in liquid nitrogen ($< -196^\circ$C) show the same paramagnetic moment as in the gas phase.

[Compare exercise 3(a) of § 3, also Fig. 3.26 (p 81).]

2 Electronic structure of atoms and the implications for magnetism

2.1 The quantum-mechanical description of the electron orbits in atoms: a resumé of the salient features

1. The one-electron atom

According to quantum mechanics, the properties of electron orbits in atoms (and in molecules and solids) are determined through the solution of the appropriate Schrödinger wave equation; so characterised, an electron orbit is termed an *orbital*. The equation can be solved *exactly* for the one-electron, hydrogenic atom – hydrogen H itself, also He^+, Li^{2+} *etc.*

To facilitate the solution of the Schrödinger equation for the hydrogen atom, the position of the electron in relation to the nucleus (proton) is defined in terms of *polar coordinates* (r, θ, φ) rather than cartesian (x, y, z). These alternative coordinate systems (Fig. 2.1) are related by

$$\begin{aligned} z &= r\cos\theta \\ y &= r\sin\theta\sin\varphi \\ x &= r\sin\theta\cos\varphi \end{aligned} \tag{2.1}$$

Fig. 2.1

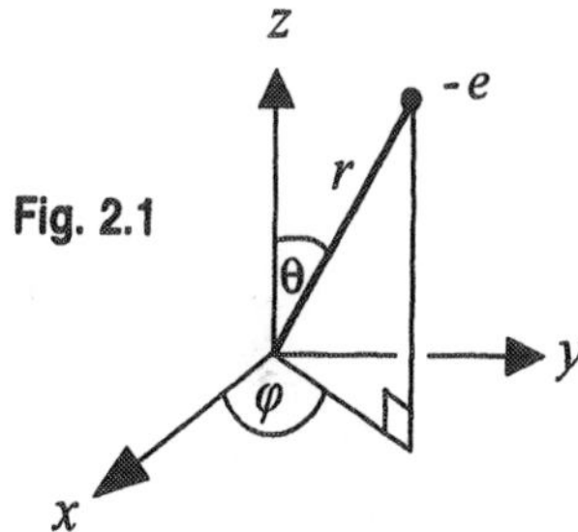

Spherical polar coordinates

r is the distance from the origin, while θ, the *polar angle* (or *co-latitude*), defines the inclination of the radius vector $\boldsymbol{r}$ with respect to the z-axis. The angle φ, known as the *azimuth* (or *latitude*), is that between the x-axis and the projection of the vector $\boldsymbol{r}$ onto the xy-plane.

Note that $r^2 = x^2 + y^2 + z^2$.

When expressed in terms of polar coordinates the Schrödinger equation may be separated into two parts, one dependent solely on the radial coordinate r, and the other only on the angular coordinates (θ, φ). The wave-function characterising an electron orbit accordingly proves to be of the product form

$$\psi = \mathrm{R}_{n,l}(r).\mathrm{Y}_{l,m_l}(\theta, \varphi) \tag{2.2}$$

The component R(r), the *radial wave-function*, depends upon both the principal quantum number n and the orbital quantum number l, while the *angular wave-function* Y(θ, φ) changes only with the two orbital quantum numbers, l and m_l. The algebraic form of the radial wave-functions R(r) is unique to the one-electron atom, reflecting the simple coulomb potential $\Phi(r) \propto r^{-1}$ in which the electron moves – as does the well-known quantisation of the electronic orbital energy $\in$ in the form $\in \propto -1/n^2$.

$$\Phi = +\frac{Ze}{4\pi\varepsilon_0 r}$$

$$\in = -\frac{m_e Z^2 e^4}{8h^2\varepsilon_0^2 n^2}$$

(Z = atomic number)

The angular wave-functions $\mathrm{Y}_{l,m_l}(\theta, \varphi)$ are of more general significance.

2. Many-electron atoms

The Schrödinger equation cannot be solved exactly for atoms containing two or more electrons because of the coulomb repulsion between electrons, an additional feature that causes intractable algebraic difficulties. A separation into radial and angular equations is only possible by use of the *orbital approximation*, according to which each electron is assumed to move in an

The use of (one-electron) orbitals is an "approximation" because a rigorous description of each electron orbit should involve the inter-electronic coordinates, which is not generally viable.

average (repulsive) potential $\Phi(r)$ due to the other electrons; moreover, the radial equation can only be solved by numerical methods.

Even if it were possible to characterise the individual electron orbits analytically, it is inconceivable that the principal quantum number n could have a simple general meaning since, taking into account the inter-electronic repulsions, the effective potential experienced by any electron would be a more complicated function of r than $\Phi(r) \propto r^{-1}$ and, moreover, its form would in general vary from one electron to another. But n remains useful as a label for the principal quantum shells.

The spherical harmonics $Y(\theta, \varphi)$ are the wave-functions for the *particle on a sphere.*

The description of the angular aspect of the electron orbits is unchanged, however, because the shape of the wave-functions $Y(\theta, \varphi)$ and the role played by the orbital quantum numbers, l and m_l, is independent of the precise form of the average potential $\Phi(r)$. The angular functions $Y(\theta, \varphi)$ are quintessentially a property of spherical symmetry and are known accordingly as *spherical harmonics*. The simplest context in which they arise, in quantum theory, is as solutions to the Schrödinger equation for the model system of a particle confined to the surface of a sphere. The potential Φ is constant in this model, which is tantamount to freezing the purely radial motion in an atomic orbital.

3. Specification of the orbital angular momentum

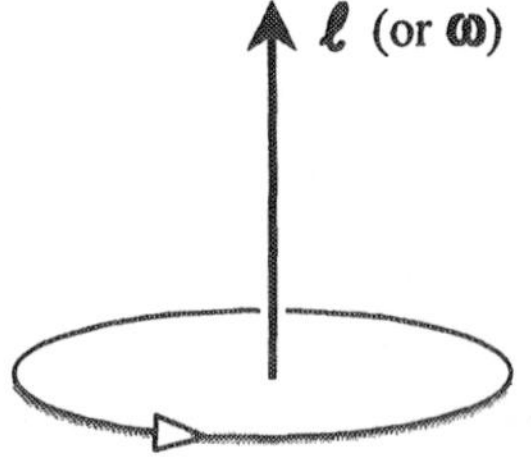

By convention, the direction of the ℓ vector is such that, when looking down towards the arrow-head, the circulation is *clockwise.*

The direction of ℓ may otherwise be pictured in terms of the *right-hand rule.*

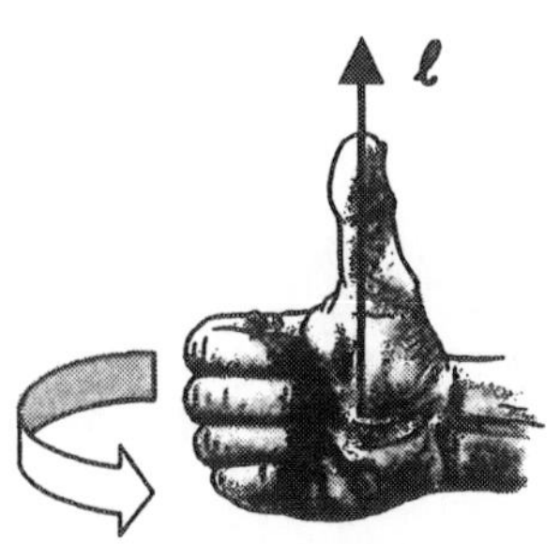

The motion of a particle in a circular orbit with a velocity v is naturally described in terms of its *angular velocity*, $\omega = v/r$ (with unit rad s^{-1}), where r is the radius. It may be characterised also by its *orbital angular momentum*, here denoted $\boldsymbol{\ell}$, the moment of the (tangential) linear momentum about the centre – that is, the linear momentum p = mv multiplied by the distance r from the centre (unit = kg m^2 s^{-1}). Both $\boldsymbol{\ell}$ and $\boldsymbol{\omega}$ are *vectors*, directed perpendicular to the plane of the orbit. In the case of a circular orbit, the angular momentum $\boldsymbol{\ell}$ has the magnitude

$$|\boldsymbol{\ell}| = mvr = mr^2\omega \tag{2.3}$$

More formally, and to embrace also *elliptical* orbits, the angular momentum is defined as the *vector product* $\boldsymbol{\ell} = \boldsymbol{r} \times \mathbf{p}$, illustrated beneath.

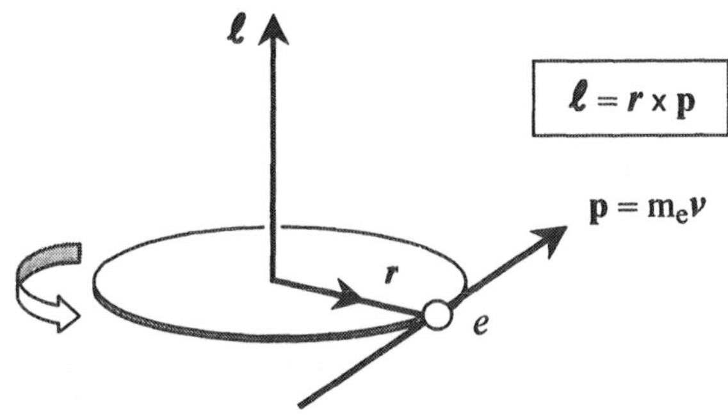

Fig. 2.2 Classical picture of the orbital angular momentum $\boldsymbol{\ell}$ of an electron in an atom.

When the classical orbit is an ellipse, both the radius r and the velocity v vary, but such that $|\boldsymbol{\ell}|$ is constant. The orbit is stable – that is, $\boldsymbol{\ell}$ is invariant in both magnitude *and* direction – unless subjected to an external torque.

Solution of the Schrödinger equation defines both the angular functions $Y(\theta,\varphi)$ and the purpose of the quantum numbers l and m_l. For a given quantum shell, l may assume any integer value from 0 to $n-1$ and specifies the magnitude of the vector $\boldsymbol{\ell}$ through

$$|\boldsymbol{\ell}| = [l(l+1)]^{\frac{1}{2}}\hbar \qquad (2.4)$$

where $\hbar = h/2\pi$ (h being Planck's constant). Remarkably, there is a category of orbital, the s-orbital (having $l = 0$), which has no observable angular momentum. An s-orbital has the angular wave-function $Y(\theta,\varphi)$ = constant ($1/2\sqrt{\pi}$, to be exact), and the motion is purely radial.

The auxiliary quantum number m_l, often known as the (orbital) *magnetic* or *axial* quantum number, defines the component of the vector $\boldsymbol{\ell}$ in a chosen direction. By convention, the direction selected is always referred to as the z-axis, and commonly termed the *axis of quantization*. The projection of $\boldsymbol{\ell}$ onto this axis is given by

$$\ell_z = m_l\hbar \qquad (2.5)$$

The quantum number m_l is allowed the integer values $\pm l, \pm(l-1), \ldots.\ 0$, *i.e.* $2l+1$ different values altogether; an orbital <u>sub</u>shell (n,l) therefore has a degeneracy of $2l+1$. Limiting m_l to the range $-l \leq m_l \leq +l$ ensures, of course, that the component ℓ_z cannot exceed $\boldsymbol{\ell}$ in absolute magnitude.

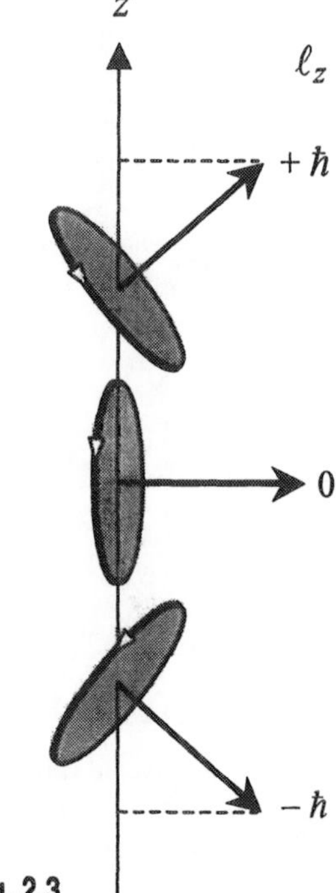

Fig. 2.3

Classical interpretation of the space quantization of the p-orbitals – for which $l = 1$, and $m_l = \pm 1, 0$.

No other orientation of the $\boldsymbol{\ell}$ vector is permitted.

The constraint on the alignment of the orbital vector $\boldsymbol{\ell}$, described as *space quantization*, is illustrated by Fig. 2.3. In classical mechanics, of course, an angular momentum vector may in principle assume any orientation in space whatsoever. Notice moreover that, according to eqns. 2.5 and 2.4, $|\ell_z|$ can never be equal to $|\boldsymbol{\ell}|$ – that is, the vector $\boldsymbol{\ell}$ can never align with the chosen axis of quantization. If it could the vector $\boldsymbol{\ell}$ would be specified completely, which is contrary to the celebrated *uncertainty principle*.

Only *one* component of $\boldsymbol{\ell}$ can be specified. If this is ℓ_z then the perpendicular components, ℓ_x and ℓ_y, are indeterminate, a situation which may be pictured in terms of a conical surface that would be traced out were the vector $\boldsymbol{\ell}$ to precess about the z-direction (Fig. 2.4). It is known only that, for each discrete orientation, $\boldsymbol{\ell}$ lies *somewhere* on such a conical surface; the latter is often referred to, therefore, as the *cone of uncertainty*.

In practice, the choice of z-axis is not arbitrary because any experimental measurement will somehow physically impose an axis of quantization. An experiment might, for example, involve the application of an electric or magnetic field, the direction of which pre-empts the selection of the z-axis.

When it is a *magnetic* field that is applied the orbital angular momentum vector $\boldsymbol{\ell}$ is driven *actually* to precess about the z-axis (and, as it transpires, in a clockwise direction with an angular velocity ω_L proportional to the field strength). This is known as the *Larmor precession* [see § 3.7(1)]; it is of great importance in various connections, not least the theory of *diamagnetism*.

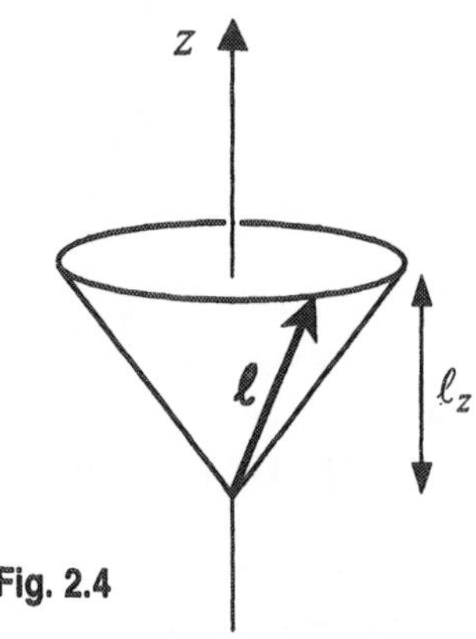

Fig. 2.4

The *cone of uncertainty*

This is the locus of possible orientations of the vector $\boldsymbol{\ell}$ for a given value of ℓ_z.

4. The nature of the angular wave-functions

Though polar coordinates must be used in order to solve the wave equation for the angular motion, the wave-functions are easier to visualise when formulated in terms of cartesian coordinates. (The relationship between the two sets of coordinates was defined earlier by eqn. 2.1). The cartesian forms of the angular wave-functions for p and d-orbitals are given in the table below.

The angular wave-functions $Y(\theta,\varphi)$ may alternatively be expressed in cartesian form using as a basis the functions

$$Y_l \sim x^a y^b z^c / r^l$$

where $a + b + c = l$.

Thus p orbitals, with $l = 1$, are of the angular form x/r *etc.*

The d orbitals, having instead $l = 2$, possess angular functions built from xy/r^2, x^2/r^2, y^2/r^2, *etc.*

Note, however, that the combination $(x^2 + y^2 + z^2)/r^2$ is actually an s-orbital in disguise, with $Y_0 \sim 1$.

That leaves *five* possible, independent combinations.

Table 2.1 Angular wave-functions for p- and d-orbitals expressed in cartesian coordinates

$$p_z = z/r \qquad d_{xy} = \surd 3\,xy/r^2$$
$$p_y = y/r \qquad d_{xz} = \surd 3\,xz/r^2$$
$$p_x = x/r \qquad d_{yz} = \surd 3\,yz/r^2$$
$$d_{3z^2-r^2} = (3z^2 - r^2)/2r^2 \quad \text{(a.k.a. } d_{z^2}\text{)}$$
$$d_{x^2-y^2} = \surd 3\,(x^2 - y^2)/2r^2$$

Normalising factors of $\surd(3/4\pi)$ and $\surd(5/4\pi)$, for the p and d sets, respectively, have been omitted.

Actually $p_z = Y_{1,0}$ and $d_{z^2} = Y_{2,0}$. The $Y_{l,m_l}(\theta, \varphi)$ are in general complex, however.

(s-orbitals are spherically symmetrical with $Y_{0,0} = 1/2\surd\pi$.)

The complete orbital wave-functions are obtained on multiplication of these angular functions by the relevant radial wave-functions, $R_{n,l}(r)$.

The angular dependence of the p and d-orbitals is illustrated by Figs. 2.5 and 2.6; in both the z-axis is directed out of the page. Each sketch defines a *contour of constant electron probability density* ψ^2 in the xy-plane – the locus of points, viewed in all directions from the nucleus, with the same value of ψ^2. (The relative phases of the amplitude ψ are indicated.)

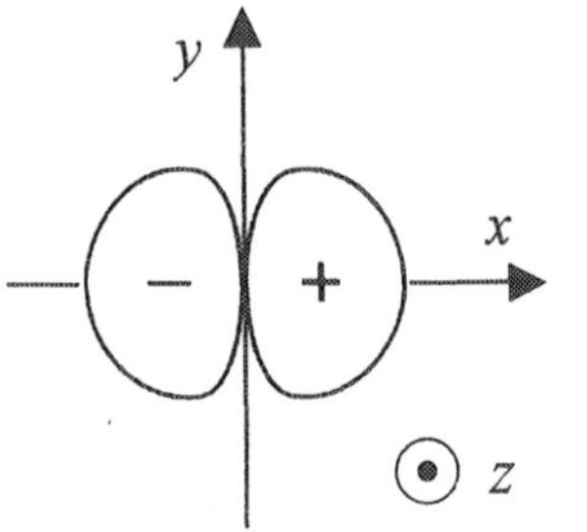

Fig. 2.5 The p_x orbital – one of three equivalent orbitals (having together $l = 1$).

The other two orbitals, p_y and p_z, are generated by rotation through 90° about the perpendicular z and y-axes, respectively.

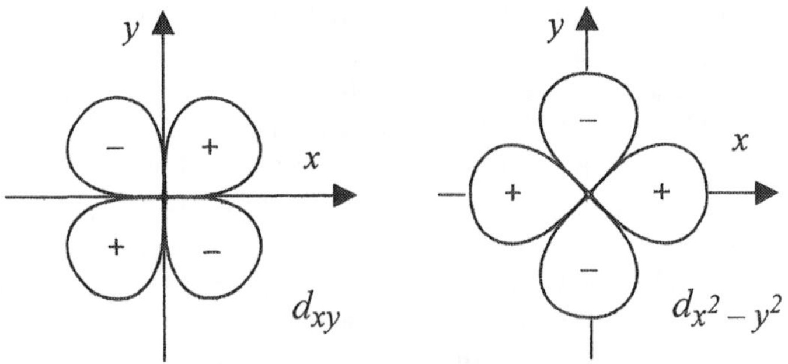

Fig. 2.6 Two d-orbitals of representative shape. (Note that they are related by rotation through 45°.)

The figures are equivalent to *polar representations* in which ψ is plotted as a function of the angular coordinate φ. Thus Fig. 2.5 might be interpreted instead as a plot of $p_x \sim \cos\varphi$ and those in Fig. 2.6 as plots of $d_{xy} \sim \frac{1}{2}\sin 2\varphi$ and $d_{x^2-y^2} \sim \cos 2\varphi$, respectively.

In the contour representation the size of an orbital may be conveyed by choosing that contour which encloses a certain arbitrary proportion (90 %, say) of the electron density ψ^2. A three-dimensional picture of an orbital constructed on this basis is called a *boundary surface*. The boundary surface of an *s*-orbital is simply a sphere.

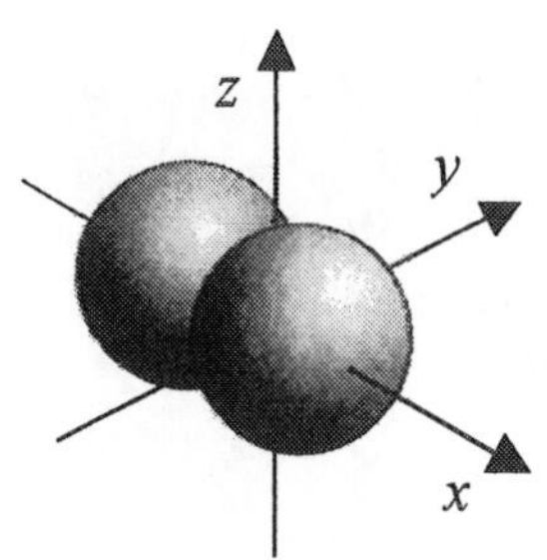

Boundary surface of the p_x orbital.

The lobal structure of the orbitals reflects the presence of *angular nodes*, commonly nodal planes – that is, surfaces on which the wave-function ψ, and therefore the probability density ψ^2 of the electron, is zero. The number of such nodes, which contain the nucleus, is equal to the orbital quantum number l. Each of the *p*-orbitals has therefore just *one* nodal plane. Thus, the p_x orbital (Fig. 2.5), with angular wave-function $\sim x/r$, has the nodal plane yz (perpendicular to the page), corresponding to $x = 0$. The *d*-orbitals have an additional nodal surface: for example, the d_{xy} orbital has the nodal plane xz as well as yz (Fig. 2.6). The $d_{x^2-y^2}$ orbital has instead the two nodal planes $y = \pm x$, both also containing the *z*-axis.

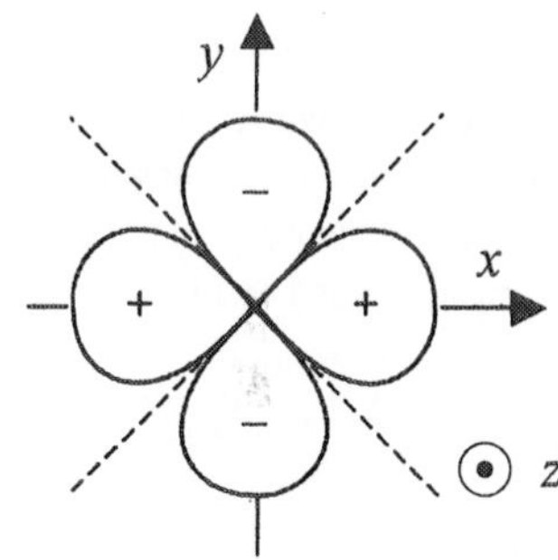

The intersection of the nodal planes $y = \pm x$ with the xy–plane in the $d_{x^2-y^2}$ orbital.

There are altogether *five* *d*-orbitals (since $l = 2$ and $2l + 1 = 5$). In addition to the two pictured in Fig. 2.6, there are d_{yz} and d_{xz} orbitals analogous to d_{xy}, but orientated in the yz and xz planes, respectively. These three orbitals are related by rotations through 90° and, in that respect, resemble the set of *p*-orbitals (*cf.* Figure 2.5). Thus d_{yz} may be obtained by rotating d_{xy} about the *y*-axis.

Some freedom of choice exists regarding the definition of the *complete* set of cartesian *d*-orbitals. The orbitals d_{xy}, d_{yz} and d_{zx} may be natural members of the set but, if $d_{x^2-y^2}$ is also to be included, why not $d_{y^2-z^2}$ and $d_{z^2-x^2}$? The latter are equivalent orbitals concentrated instead in the yz and xz planes. Only *five* functions are needed, however, so a simple course of action, if $d_{x^2-y^2}$ is indeed to be preferred, is to take a linear combination of $d_{y^2-z^2}$ and $d_{z^2-x^2}$, giving the two functions equal weight. The particular combination

The figures show the phase of each wave-function for its different lobes. There is always a change of sign on traversing a nodal surface.

$$(d_{z^2-x^2} + d_{z^2-y^2})/\sqrt{3} = d_{3z^2-r^2} \qquad (2.6)$$

– known for short as d_{z^2} – ensures that there is no redundancy in the formulation of the d orbital set. (Note that $d_{z^2-y^2} = -d_{y^2-z^2}$, *etc.*)

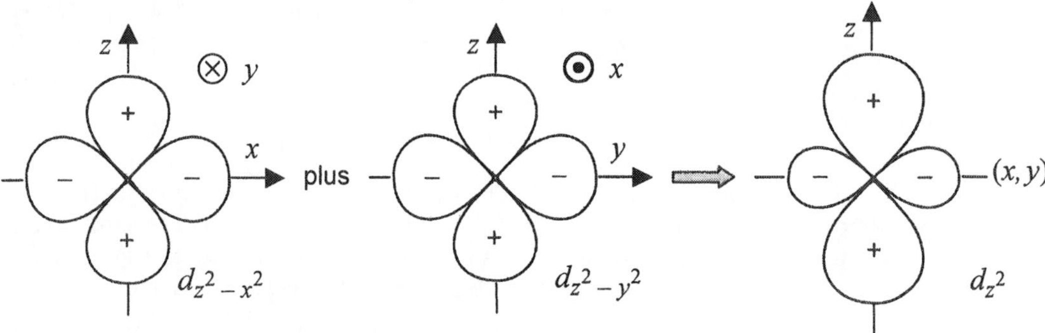

The reader may care to verify the expression for $d_{z^2} \equiv d_{3z^2-r^2}$ by referring to the definitions of the relevant angular functions in Table 2.1.

As is the case for the p orbital set, none of the chosen d functions can be expressed as a linear combination of the others. The orbitals are said to be *linearly independent*.

The choice of $d_{x^2-y^2}$ together with d_{z^2} to complete the set of d-orbitals is a matter of convention. It would be equally acceptable in principle to use instead the orbitals $d_{z^2-y^2}$ and $d_{x^2} \equiv d_{3x^2-r^2}$, or the pair $d_{z^2-x^2}$ and $d_{y^2} \equiv d_{3y^2-r^2}$. But by convention, as with space quantization, the z-axis is given a certain arbitrary priority.

The d_{z^2} orbital, like each of the three p-orbitals, is *axially symmetric*. The boundary surface is then akin to the figure of revolution of the contour (or polar) representation.

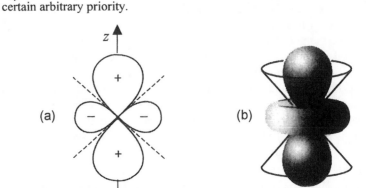

Fig. 2.7 Representations of the $d_{z^2} \equiv d_{3z^2-r^2}$ orbital showing the conical nodal surfaces, $r = \pm\sqrt{3}\,z$: (a) a section in any plane containing the z-axis, and (b) a perspective view of the boundary surface.

Observe, in the sketches of the d_{z^2} orbital above, the characteristic "tyre" of electron density aligned in the xy-plane, wrapped around the waist of the main double-lobed structure. The unique shape of the orbital derives from the presence of two intersecting nodal surfaces, here *conical* rather than planar, between which there is again a change of phase.

5. The angular momentum within particular orbital sets

As described above the p and d orbitals do not *individually* display angular momentum. The electron distribution in each orbital is *stationary*, the (real) wave-function being analogous to a standing wave in classical mechanics, and there is no net circulation about the z-axis.

The orbitals having well-defined angular momentum (eqn. 2.5) are the spherical harmonics $Y_{l,m_l}(\theta, \varphi)$ which, reversing the transformations, are *complex* linear combinations of the real orbitals in Table 2.1. For a d subshell ($l = 2$), the orbitals with $\ell_z = \pm 2\hbar$ and $\pm\hbar$ are

The absolute value of m_l is equal to the number of nodal planes containing the z-axis in the real orbitals.

$$d_{\pm 2} = (d_{x^2-y^2} \pm i d_{xy})/\surd 2 \equiv Y_{2,\pm 2} \tag{2.7a}$$

$$d_{\pm 1} = (d_{xz} \pm i d_{yz})/\surd 2 \equiv Y_{2,\pm 1} \tag{2.7b}$$

where $i = \surd(-1)$. Such orbitals correspond to stable *travelling* (or *running*) waves in classical mechanics. Thus d_{+2} and d_{-2}, for example, represent equivalent angular momenta about the z-axis of, respectively, $+2\hbar$ (clockwise) and $-2\hbar$ (anticlockwise). Being of complex form, these functions are not visualisable individually, but (as pictured in Fig. 2.8) the corresponding probability densities $|\psi|^2 = \psi^*\psi$ can be pictured in the same way as ψ^2 for a real orbital.

The p-orbitals displaying well-defined angular momentum are (*cf.* Fig. 2.3):

$$p_{\pm 1} = (p_x \pm i p_y)/\surd 2 \equiv Y_{1,\pm 1}$$
$$p_0 = p_z \equiv Y_{1,0}$$

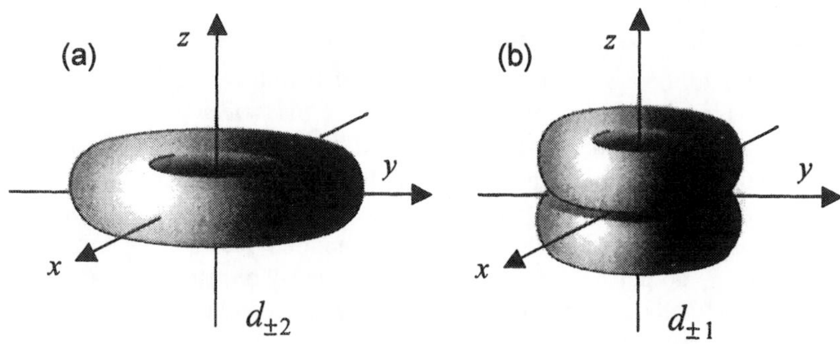

Fig. 2.8 Boundary surfaces for *d*-orbitals having well-defined angular momentum (eqn 2.7).

(a) $d_{\pm 2}$, resembling a tyre (or "doughnut") with a vanishing aperture, like the d_{z^2} feature – *cf.* Fig. 2.7(b).

(b) $d_{\pm 1}$ with, on both the top and underside, concavities that merge at the nucleus. There is a node in the *xy*-plane.

The *d*-orbital corresponding to $\ell_z = 0$ is simply

$$d_0 = d_{z^2} \equiv \mathrm{Y}_{2,0} \tag{2.7c}$$

with a *real* wave-function (like $p_0 = p_z$).

The description of angular momentum outlined above may be recast in simple terms, shorn of mathematical formalities. Recognising the "fixed" nature of the electron density in the *real* d-orbitals (Table 2.1), it might innocently be asked: How can the electron circulate about the *z*-axis ?# One answer is apparent on looking back at Fig. 2.6, which portrays the d_{xy} and $d_{x^2-y^2}$ orbitals.

\# This is reminiscent of the question: How does an electron travel from one side of a node to the other?

Dirac, the principal architect of *relativistic* quantum theory, is said to have remarked that "it spins across".

The motion of the electron about the *z*-axis (with $\pm 2\hbar$ units of angular momentum, as it happens) may be understood in terms of its continuous transfer between the two orbitals, these being *identical* save for rotation about the *z*-axis. If the electron were placed in either of these orbitals then, in quantum theory, nothing could prevent it from migrating to the other orbital and thereby circulating smoothly. According to this interpretation, the occurrence of equivalent clockwise and anticlockwise motion about a particular axis, and hence the existence of a well-defined (*i.e.* observable) orbital angular momentum, depends crucially upon the rotational equivalence of a particular pair of *conjugate* orbitals. In a *d*-subshell, the conjugate orbitals allowing circulation about the *z*-axis with $\pm\hbar$ units of angular momentum are d_{xz} and d_{yz}. The corresponding pair of *p*-orbitals consists of p_x and p_y.

A simple but general rule may be formulated (which will prove useful later):

> An electron displays orbital angular momentum about a specified axis only when its orbital has a degenerate partner related by a rotation about that axis.

 NB (A)

Any non-degenerate orbital has $\ell_z = 0$ and also therefore a null vector $\boldsymbol{\ell} = 0$. (An *s*-orbital would be an example, if a trivial one.)

[Statements of special importance will be emphasised as above, and labelled for future reference, throughout the chapter.]

It should be emphasised, in respect of the rule given above, that degeneracy is not of itself a sufficient condition; also essential is the rotational equivalence of the orbitals.

2.2 Electron spin

1. The intrinsic angular momentum of an electron

Spin is a fundamental property of a particle, like mass or charge.

The *proton* and the *neutron* also possess spin – and so do, therefore, the nuclei of many atoms.

In addition to the angular momentum associated with the orbital motion (in orbitals having $l \geqslant 1$), the electron possesses an *intrinsic* angular momentum called ***spin*** and denoted $\mathbf{s}$. It was so named because, originally, the angular momentum was imagined to arise from the electron, pictured as a small sphere, rotating about a diagonal axis. Such an interpretation is now known to be inappropriate and misleading; the spin motion occurs somehow *inside* the electron or, put another way, in an additional dimension of space. It cannot be visualised except as a vector analogous in certain vital respects to the orbital angular momentum $\boldsymbol{\ell}$.

The existence of an intrinsic angular momentum (spin), with the quantum number $m_s = \pm\frac{1}{2}$, follows from the Dirac equation, an extension of the Schrödinger equation incorporating the theory of special relativity.

It is well established, both experimentally and by relativistic quantum mechanics, that the magnitude of the spin angular momentum of an electron is given by

$$|\mathbf{s}| = [s(s+1)]^{\frac{1}{2}}\hbar \qquad (2.8)$$

analogous to eqn 2.4 specifying $|\boldsymbol{\ell}|$. However, the spin quantum number s is confined to the single, invariant value $\frac{1}{2}$, and so $|\mathbf{s}| = \frac{1}{2}\sqrt{3}\,\hbar$ always. The component of the spin in a chosen direction is defined as

$$s_z = m_s\hbar \qquad (2.9)$$

where m_s, with alternative values $\pm\frac{1}{2}$, is the auxiliary (a.k.a. *magnetic* or *axial*) spin quantum number. Thus the spin vector $\mathbf{s}$ is space quantized, like $\boldsymbol{\ell}$, but has only *two* orientations with respect to the axis of quantization (z). (Note that $2s+1 = 2$.) There are only two spin states for a single electron, crudely but conveniently distinguishable as "up" (↑) or "down" (↓) (or, in other connections, as the α and β states). The up-state displays a clockwise angular momentum of $\frac{1}{2}\hbar$ about the chosen axis, the down-state an equal angular momentum anticlockwise.

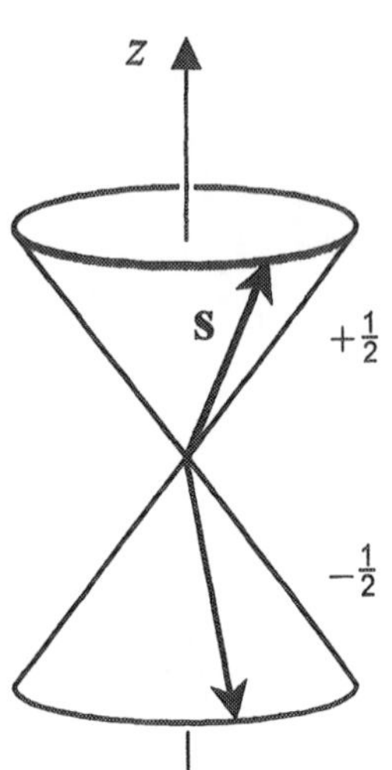

Fig. 2.9 Space quantization of the electron spin

– an attempt to visualise the two possible alignments, $m_s = \pm\frac{1}{2}$ or "up" and "down", of the spin of a solitary electron (*cf.* Figs. 2.3 and 2.4). The vector $\mathbf{s}$, of length $\frac{1}{2}\sqrt{3}\,\hbar$, is drawn with its cone of uncertainty (the perpendicular components s_x and s_y being indeterminate) for the alternative orientations.

This description of the quantization of electron spin is of general significance; as an intrinsic motion, not occurring in cartesian space, the spin of an electron is impervious to electric fields, and so displays essentially the same behaviour in atoms and in molecules and, when localised, in the solid state. (The same is far from true of the orbital angular momentum, as will be explained in section 3(4).)

The spin property of the electron, otherwise inscrutable from a classical viewpoint, is of crucial importance because it affects in a fundamental way how electrons are arranged among the available orbitals in a many-electron system, thereby being a major determinant of the structure of the Periodic Table. The spin configuration also influences the repulsive interactions between electrons.

2. Two or more electrons

The electron spins in a many-electron atom or molecule may be combined as a vector sum

$$\mathbf{S} = \Sigma_i \, \mathbf{s}(i) \qquad (2.10)$$

where $\mathbf{s}(i)$ denotes the spin of the ith electron. The total spin **S** is characterised by quantum numbers S and M_S, the former defining its magnitude as

A comparable sum $\mathbf{L} = \Sigma_i \boldsymbol{\ell}(i)$ of the orbital angular momenta may of course be usefully constructed. [See eqn. 2.20 later.]

$$|\mathbf{S}| = [S(S+1)]^{\frac{1}{2}} \hbar \qquad (2.11)$$

– a generalisation of eqn 2.8. Correspondingly, the component of **S** in the (arbitrary) *z*-direction is given by

$$S_z = M_S \hbar \qquad (2.12)$$

(*cf.* eqn 2.9) where the subsidiary (magnetic) quantum number $M_S = \Sigma_i \, m_s(i)$ has $2S+1$ distinct values ranging from S to $-S$ in integral steps. The degeneracy $2S+1$ is often termed the (spin) *multiplicity*.

There being two spin states for an individual electron, an *N*-electron system will have 2^N states altogether. The values of S that arise range from $\frac{1}{2}N$ through $\frac{1}{2}N-1$, *etc.* to either 0 or $\frac{1}{2}$, according to whether *N* is even or odd (and are such that $\Sigma_S(2S+1) = 2^N$). For each distinct state S the variation of the quantum number M_S embraces $M_S = 0$ if S is integral; but when S is half-integral (odd *N*) M_S ranges from $\pm S$ to $\pm\frac{1}{2}$.

If there are just *two* electrons then S = 1 or 0. The latter state is a *singlet* ($M_S = 0$ only) corresponding to the *pairing* of the two electron spins – *i.e.* the spins are aligned antiparallel (denoted ↑↓) with consequent cancellation of their angular momenta. (The combined spin angular momentum is then the null vector **S** = 0.) On the other hand, the S = 1 state is a *triplet* having space quantized components $M_S = \pm 1, 0$ (Fig. 2.10) like the orbital angular momentum of a *p*-subshell (Fig. 2.3). The state involves a parallel orientation (↑↑) of the electron spins.

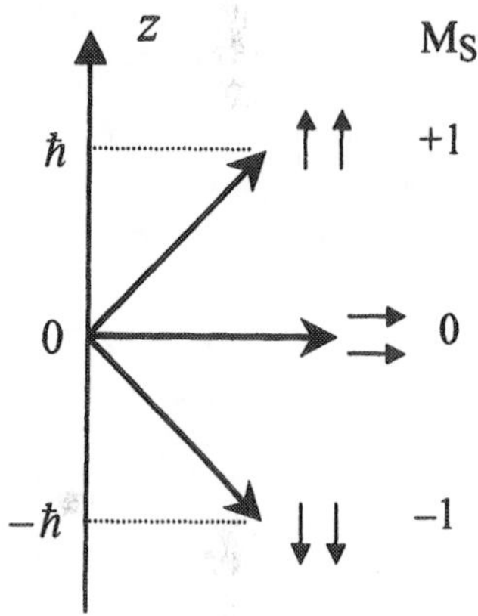

Fig. 2.10 A vector picture of the three components of a spin triplet state.

3. The Pauli principle and its consequences

Electron spin exerts its influence on the preferred orbital configurations in many-electron atoms, and on the form of the Periodic Table, through the operation of the *Pauli principle*, an independent postulate of quantum mechanics additional to the protocols of Schrödinger, Born and others. It is most familiar in a limited version, the *exclusion principle*, which may be stated as follows:

> A single orbital may accommodate no more than two electrons and, if it is doubly occupied, the electron spins must be paired.

 NB (B)

Therefore the occupancy of an *l*-subshell (having a degeneracy $2l+1$) cannot exceed $4l+2$ electrons. When a subshell is full – *e.g.* s^2, p^6 *etc.* – it is described as *closed*.

In its general form the Pauli principle has a further important implication for electronic structure in general, known as the *exchange effect*. Electrons naturally adjust their relative orbital motions to offset so far as possible their mutual repulsion; this is known as *charge correlation*. It transpires, essentially because any two electrons are indistinguishable by experiment, that the

There is no experiment that can tell two electrons apart.
They do not carry identity tags.

repulsion between them depends on their spin configuration. If the electrons reside in separate orbitals then, remarkably, they tend to avoid each other to a greater degree when their spins are parallel than when their spins are opposed. The effect is described as *spin correlation* (and it is additional to charge correlation). It follows that

NB (C) ➡

> The coulomb repulsion between two electrons occupying different orbitals is significantly less when their spins are aligned parallel than when they are antiparallel.

Since the effect is associated with the indistinguishability of electrons it is often interpreted in terms of *exchange* between the electrons, as if they can somehow swap identities. The relative stabilisation of a two-electron system when the spins are co-directional is accordingly attributed to an *exchange interaction.*

The energy of repulsion in a two-electron system, involving different orbitals a and b, may be expressed in the form

If the electrons pair up in a single orbital – a say – then $E_{ee} = J_{aa}$.

E, rather than U, is henceforth used as the symbol for energy (as is the more usual in the context of electronic structure).

$$E_{ee} = J_{ab} \mp K_{ab} \tag{2.13}$$

where the minus sign relates to the triplet state (↑↑). Both quantities J and K are inherently positive and $J \gg K$ always. The energy difference between the triplet and singlet spin states is just 2K.

The quantity J is known as the *coulomb integral*; it is analogous to the electrostatic repulsion between two extended charge distributions in classical physics. The other quantity K, called the *exchange integral*, has no classical counterpart, but may be interpreted (perhaps not altogether helpfully) as the repulsion of the *overlap* density of the two charge clouds with itself. While K is small in relation to J it is far from negligible: for example, in the transition metal ion V^{3+}, with electron configuration (Ar)$3d^2$, J_{dd} is about 142,000 cm^{-1} (17.5 eV) while $K_{dd} \approx 6{,}000$ cm^{-1} (0.75 eV). In other cases, K can be as much as 10% of J.

4. Hund's first rule

Spin correlation applies to every pair of electrons in a multi-electron atom or molecule and has the simple consequence in any *open*-shell system, *i.e.* one having a partly filled subshell, that:

NB (D) ➡

> The ground electronic state is that with the maximum possible value of the spin quantum number S.

This is known as *Hund's first rule*, otherwise as *Hund's rule of maximum multiplicity*. The maximum value of S (or *multiplicity* 2S + 1) derives from maximal parallel alignment of the electron spins, and therefore corresponds to the minimum interelectronic repulsion. Thus the ground states of carbon (He) $2s^22p^2$ and nitrogen (He) $2s^22p^3$ have, respectively, S = 1 (a triplet) and $S = \frac{3}{2}$ (a quartet).

m_l	+1	0	−1		+1	0	−1
	↑	↑	—		↑	↑	↑
	(a) p^2 S = 1				(b) p^3 S = $\frac{3}{2}$		

Fig. 2.11 The ground states of the configurations p^2 and p^3 pictured in terms of the maximal value of $M_S = S$ (& $M_L = L$).

The p^3 electron configuration of nitrogen is an example of a *half-filled subshell*, when the effects of spin correlation are greatest.

2.3 The electronic magnetic moment

1. The magnetic moment due to the orbital motion

The electron being charged, its circulation about the nucleus produces a magnetic dipole moment $\mathbf{m}^o$ which, pictured classically, is a vector directed perpendicular to the plane of the orbit (Fig. 2.12). It may be characterised by means of the current loop formula, $\mathbf{m} = I\mathbf{A}$ (eqn 1.5).

Consider, for simplicity, a circular electron orbit of radius r. The current I $= -ev/2\pi r$, where v is the velocity of the electron, and the area of the orbit is A $= \pi r^2$. The negative value of I reflects the charge on the electron. The orbital magnetic moment is therefore of *magnitude* $|\mathbf{m}^o| = \frac{1}{2}evr$ which, since $|\boldsymbol{\ell}| = m_e vr$, may be recast as $|\mathbf{m}^o| = (e/2m_e)|\boldsymbol{\ell}|$.

Recall that the *atomic unit* of angular momentum is $\hbar = h/2\pi$. An atomic unit of magnetic moment, known as the *Bohr magneton* and denoted μ_B (in the older literature β or BM), may accordingly be defined:

$$\mu_B = \frac{eh}{4\pi m_e} = \frac{e\hbar}{2m_e} = 9.273 \times 10^{-24}\ \mathrm{A\ m^2} \qquad (2.14)$$

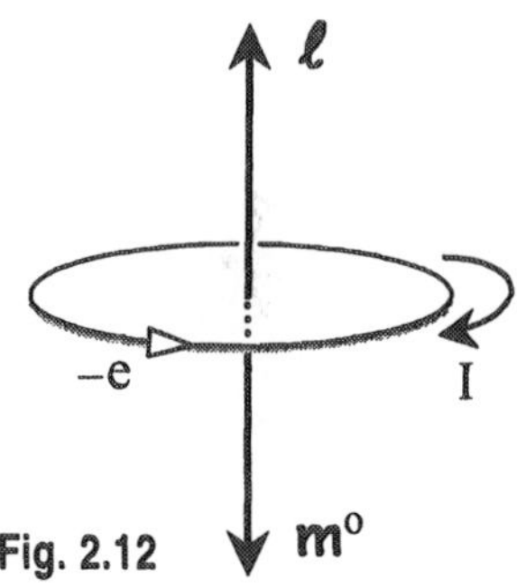

Fig. 2.12 The orbital magnetic dipole moment $\mathbf{m}^o$.

Expressed in vector terms, the orbital magnetic moment is therefore

$$\boxed{\mathbf{m}^o = -\frac{e}{2m_e}\boldsymbol{\ell} = -\left(\frac{\mu_B}{\hbar}\right)\boldsymbol{\ell}} \qquad (2.15)$$

The proportionality constant in eqn. 2.15 is known as the *gyromagnetic ratio* and denoted γ_o – that is,

$\mathbf{m}^o = \gamma_o\boldsymbol{\ell}$ where $\gamma_o = -e/2m_e$.

(γ_o is also known as the *magnetogyric* ratio.)

Because of the negative charge on the electron *the orbital magnetic moment is directed in opposition to the angular momentum.*

It may be shown that eqn. 2.15 applies also to elliptical orbits. The quantum mechanical value of $\mathbf{m}^o$ is obtained by substituting the quantised values of $|\boldsymbol{\ell}|$ (eqn. 2.4) or, to specify instead the z-component of the magnetic moment, a well-defined value of ℓ_z (eqn. 2.5).

2. The spin magnetic moment

The intrinsic, *spin* motion of an electron also has an associated magnetic moment, which is given by

$$\boxed{\mathbf{m}^s = -g_e\frac{e}{2m_e}\mathbf{s} = -g_e\left(\frac{\mu_B}{\hbar}\right)\mathbf{s}} \qquad (2.16)$$

The gyromagnetic ratio for electron spin is $\gamma_s = -g_e e/2m_e \approx 2\gamma_o$.

It has been said that $g_e \approx 2$ because space is curved (relativistic quantum mechanics). The deviation of g_e from the exact value of 2 is due to space having also a slight "wobble" (relativistic quantum electrodynamics).

equivalent to (2.15), apart from the factor g_e – a constant, equal to 2.002319, known as the (electron) *spin g-factor*. This extremely important formula is well authenticated experimentally, and may be explained by relativistic quantum electrodynamics. Again, *the magnetic moment is directed against the angular momentum*, reflecting the negative charge on the electron. The vector **s** is characterised by eqns. 2.8 and 2.9 of the previous section.

In sum, the magnetic moment due to a single electron may be written as

$$\mathbf{m} = -\mu_B(\boldsymbol{\ell} + g_e\mathbf{s})/\hbar \qquad (2.17)$$

3. Many-electron atoms

The magnetic moment of a complex, open-shell atom is the vector sum of the individual electronic dipole moments. The contribution due to spin may be formulated as

$$\boxed{\mathbf{m}^s = \Sigma_i\, \mathbf{m}^s(i) = -\mu_B g_e \mathbf{S}/\hbar} \qquad (2.18)$$

(a generalisation of eqn. 2.16). The expression applies also to molecules and to ions in crystals, *e.g.* transition metal cations.

The orbital motions make a similar contribution,

$$\mathbf{m}^o = \Sigma_i\, \mathbf{m}^o(i) = -\mu_B \mathbf{L}/\hbar \qquad (2.19)$$

where **L** is the total orbital angular momentum, the vector sum of the individual angular momenta,

$$\mathbf{L} = \Sigma_i\, \boldsymbol{\ell}(i) \qquad (2.20)$$

The vector **L** is characterised by two quantum numbers, L and M_L, comparable to the spin quantum numbers except that they are exclusively *integral*. The magnitude of the total angular momentum is given by

$$|\mathbf{L}| = [L(L+1)]^{\frac{1}{2}}\hbar \qquad (2.21)$$

with space-quantized components,

$$L_z = M_L\hbar \qquad (2.22)$$

The auxiliary (magnetic) quantum number $M_L = \Sigma_i\, m_l(i)$. It has $2L + 1$ values ranging from $\pm L$ to 0, *i.e.* each distinct state L has a degeneracy of $2L + 1$.

The nature of these states is elucidated further in § 2.4.

In light atoms the orbital and spin magnetic moments contribute independently to the observed magnetism. But in heavier atoms – including transition metal cations, for example – there is an internal magnetic interaction between **S** and **L**, called the *spin-orbit coupling*, which complicates matters. The spin and orbital aspects of the magnetism become confused. The subject is taken up later in section 5.

While the spin of an electron is not directly affected by its physical location, the orbital angular momentum is severely perturbed by the internal electric fields within a molecule or crystal, and may then contribute comparatively little, or even nothing to the observed magnetism.

4. The quenching of orbital angular momentum

Within any molecule the electrostatic force due to one nucleus or atomic core disturbs the electron orbits of another, neighbouring atom, and the effect is reciprocal; similar interactions naturally exist in extended solids. A particular consequence is, of course, covalency; otherwise, it is the polarization of localised orbitals. Either way, the orbital angular momentum of electrons is significantly affected.

The internal electric field of a molecule or crystal can be enormous, vastly greater in magnitude than any static field that can be generated in a controlled manner in the laboratory, and it will, moreover, vary substantially with location. It may also be highly anisotropic. A major consequence, as regards magnetic properties, is the disruption of orbital motion such that, in general, smooth circulation with a well-defined angular momentum about a particular axis is not possible. The turbulent electrostatic environment largely dissipates the orbital angular momentum – or, as more commonly stated, the angular momentum is ***quenched***. For the most part at least, it fails to contribute to the measured magnetic moment of an open-shell system.

Electrons have a rather bumpy ride in molecules and crystals and commonly fail therefore to display orbital angular momentum about any particular axis.

> The observed magnetic moment of an open-shell molecule, or ion in a crystal lattice, is often due mainly or entirely to electron spin.

⇐ NB (E)

As for the quantum mechanical description of quenching, it should be recalled that the quantization of $\boldsymbol{\ell}$, outlined in section 1(2), is a reflection of spherical symmetry and therefore cannot normally be sustained when the environment is non-spherical (whatever the physical factors involved). More specifically, the mechanism of quenching is simply the loss of the degeneracy of conjugate orbitals, those that *together* carry well-defined angular momentum – the d_{xy} and $d_{x^2-y^2}$ orbitals in Fig. 2.6, for example. If such orbitals differ in energy, smooth circulation about the *z*-axis, as shown by the $d_{\pm 2}$ orbitals pictured in Fig. 2.8(a), is not viable. The vector sum $\mathbf{L} = \Sigma_i\, \boldsymbol{\ell}(i)$ will naturally reflect what happens to the individual $\boldsymbol{\ell}(i)$.

Recall NB (A) of section 1(5).

However, that the quenching of $\boldsymbol{\ell}$ is frequently *incomplete*. For instance, the *doubly*-degenerate subshells (label *e*) encountered in *axially symmetric* molecules,[a] such as the *metallocenes* $(\eta^5\text{-}C_5H_5)_2M$ (with a 5-fold symmetry axis), generally display angular momentum, though it is quantised in integer form only in *linear* molecules – *e.g.* NO and O_2. The quenching is also only partial in molecules of *cubic* symmetry, notably octahedral and tetrahedral complexes of transition metal ions, where the *triply*-degenerate (or *t*) orbital sets retain some angular momentum in a form not dissimilar to that of an atomic *p*-subshell.[b]

[a] An axially-symmetric molecule is one having a rotational symmetry axis of order 3 or higher.

Linear molecules occupy a special place in the scheme of things since the component of $\boldsymbol{\ell}$ about the figure axis is well-defined – much like $\boldsymbol{\ell}_z$ in atoms.

The partial resolution of the degeneracy of the components of an atomic *l* subshell by an electric field can be observed in the atomic spectra of many-electron atoms as the splitting of emission lines when a strong electric field is applied externally. This is known as the *Stark effect*. The quenching of orbital angular momentum in molecules or crystals may therefore be ascribed to an *internal* Stark effect.

[b] The labels *e* and *t* are part of the *Mulliken notation*. Non-degenerate states are denoted *a* or *b*, depending on the rotational symmetry of the wave function.

2.4 Many-electron states: the Russell-Saunders scheme

1. The orbital aspect

In general, an orbital configuration in a multi-electron atom gives rise to more than one separately observable electronic state, these being distinguishable by reference to their total orbital angular momentum L, introduced in

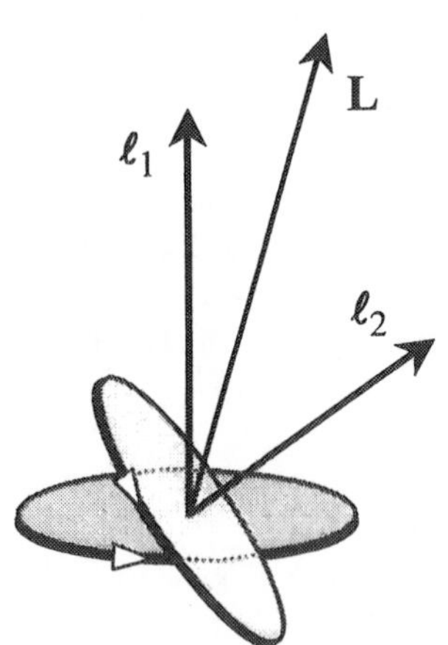

Fig. 2.13 The combination of two angular momentum vectors to form one of their resultants.

section 3(3). The latter is defined by the quantum number L, together with $M_L = \Sigma_i\, m_l(i)$ which specifies the 2L+1 space components. The possible values of L, which are necessarily *integral*, can conveniently be characterised by a semi-classical vector model, due to Russell and Saunders (1925), incorporating *ad hoc* subsequent developments in quantum mechanics.

A given configuration may have a large number of so-called *microstates*, these differing in the set of quantum numbers $m_l(i)$ allocated to the individual electrons. The grouping of these states according to the quantum number L identifies the degeneracies that exist simply by virtue of spherical symmetry. *States differing in their* L*-parentage will in general have different energies due to electron-electron repulsion.* The different values of L – *viz.* 0, 1, 2, 3 *etc.* – are denoted by capital letters S, P, D, F *etc.*, following the notation for orbitals.

For present purposes, the essential features of the Russell-Saunders scheme are brought out adequately by the two-electron case (Fig. 2.13). It is assumed, first of all, that the electrons are *inequivalent* – that is to say, they reside in different subshells, and l_1 may well be different from l_2. The angular momenta, $\boldsymbol{\ell}_1$ and $\boldsymbol{\ell}_2$, are permitted only certain discrete orientations with respect to each other, these relating to values of the orbital quantum number L ranging in *integral* steps from $l_1 + l_2$ to $|l_1 - l_2|$. When $L = l_1 + l_2$ the individual $\boldsymbol{\ell}$ vectors are aligned as close to parallel as possible; the state $L = |l_1 - l_2|$, with the minimum value of L, involves instead the equivalent antiparallel orientation.#

\# The figure depicts a mutual orientation corresponding to a high value of the L quantum number.

Thus the configuration p^1d^1, with $l_1 = 1$ and $l_2 = 2$, yields states with L = 3, 2 and 1 – *i.e.* F, D and P states. In this case, only one intermediate, relative orientation (L = 2) is allowed.

The number of distinct L states realised is $2l_1 + 1$ or $2l_2 + 1$, whichever is the *smaller*. Suppose $l_2 > l_1$: then the vector $\boldsymbol{\ell}_1$ has $2l_1 + 1$ different orientations with respect to $\boldsymbol{\ell}_2$. Moreover, since $\boldsymbol{\ell}_2$ itself has $2l_2 + 1$ space quantized components with respect to some other axis,

$$\Sigma_L (2L + 1) = (2l_1 + 1)(2l_2 + 1) \tag{2.23}$$

– *i.e.* the total number of microstates. In the case of the p^1d^1 configuration, $\Sigma_L (2L + 1) = 7 + 5 + 3 = 15$, correctly equal to 3 x 5.

Compare the expression for the total spin states

$$\Sigma_S (2S + 1) = 2^N$$

described in section 2(2).

As remarked earlier, the discrimination of many-electron states by the L quantum number is important because sets of microstates that must be degenerate are thereby identified. The microstates may differ in energy through the inter-electronic repulsion since the interaction between any two electrons depends usually on the orbitals they happen to occupy. This is obvious when the various shapes of orbitals, described in section 1(4), are contemplated. It can also be understood in classical terms through vector pictures such as Fig. 2.13, in which it is evident that the inter-electronic repulsion must in general depend on the relative orientations of the planes of the electron orbits. It is equally clear that the repulsion does not vary within an L group, the 2L + 1 components of which differ merely with regard to space quantization of the coupled vector **L**.

In summary (leaving aside the question of spin configuration):

States having the same value of L are degenerate; those with differing values of L have different energies due to the interelectronic repulsion.

NB (F)

The quantum number L differentiates the electron-electron repulsion, like the spin quantum number S – though little of *general* import can be said about how exactly the interelectronic repulsion depends on L, save that there is apparent in the observations of atomic spectroscopy a marginal tendency for the states of higher L to be lower in energy. Exactly why is unclear.

Such considerations are largely superfluous when the orbital configuration gives rise to only *one* state, as is the case for *closed*-shell atoms or those having just one electron in an open subshell (or one *hole* in an otherwise closed subshell). Any closed subshell $(l)^{4l+2}$, *e.g.* p^6 or d^{10}, has simply L = 0 (an S state) because $M_L = \Sigma_i m_l(i) = 0$, and is non-degenerate. If a subshell contains only one electron – configuration $(l)^1$ – then there is again a unique state, but with L = l. The very same, unique state is obtained for the corresponding *hole* configuration, $(l)^{4l+1}$, *e.g.* p^5 or d^9, as opposed to p^1 or d^1, both either P states or D states, respectively.

2. Russell-Saunders terms

Incorporating also the spin **S** into the description, many-electron states are usually defined through the notation $^{2S+1}(L)$ and referred to as *terms*. Thus the unique states of p^N mentioned above are denoted ^{1}S (p^6) and ^{2}P (p^1 or p^5). They are spoken of as "singlet S", "doublet P" *etc.* A Russell-Saunders term has a degeneracy of (2S + 1)(2L + 1).

"Term" is synonymous with many-electron state. It dates back to the early days of atomic spectroscopy, as does the notation $^{2S+1}(L)$.

The spin and orbital aspects of a configuration may be freely combined when the electrons are inequivalent and so p^1d^1 displays spin singlet and triplet states for each of the allowed L values; altogether six terms ^{3}F, ^{1}F, ^{3}D, ^{1}D, ^{3}P & ^{1}P, are observed.

[The reader may care to check that the sum of the degeneracies of the states of p^1d^1 amounts correctly to 6 x 10 = 60.]

However, configurations that involve *equivalent* electrons, occupying the same degenerate subshell (*e.g.* p^N with N = 2, 3 or 4), are less straightforward because of the exclusion principle.

3. Configurations with equivalent electrons

In the case of equivalent electrons, the Pauli principle prohibits many (often roughly half) of the states that might otherwise be anticipated. The p^2 configuration, for example, allows L = 2, 1 & 0 – *i.e.* D, P & S terms – but none has *both* singlet and triplet spin states. The terms that occur, it transpires, are ^{1}D, ^{3}P & ^{1}S. These have a combined degeneracy of 5 + 9 + 1 = 15, which is readily understood: allowing for spin, a p-subshell offers *six* places for electrons, so the total number of microstates should be just (6 x 5)/2 = 15.

i.e. the binomial coefficient $\binom{6}{2}$

An atomic D state (L = 2) must have components with $M_L = \pm 2$, which can be contrived only by allocating both electrons of the p^2 configuration to *either* the p_{+1} *or* the p_{-1} orbital; the spins must then be paired, with the unavoidable outcome $M_S = 0$ and S = 0. The hypothetical state ^{3}D is thus disallowed by the Pauli principle. It is possible, however, to construct a spin triplet state (S = 1) by placing the electrons in *different* orbitals, when $M_L = \pm 1$ or 0, and L = 1 – *i.e.* a P state. This is illustrated in Fig. 2.11(a).

It is unnecessary to explore the term structure for equivalent electrons in great detail here. In an introductory account of magnetic properties, only the *ground* electronic states are of immediate concern and these can always be identified by means of two rules due to Hund (1927). The first rule, concerning the dominant# influence of the spin configuration, has already been discussed in section 2(4). In a restricted form convenient for present purposes, *Hund's second rule* asserts that:

While the interelectronic repulsion varies according to orbital placement, the effect is less important than spin correlation.

NB (G) ➡

> Where a configuration $(l)^N$ gives rise to more than one state of maximum S, the ground state is that with the largest value of L.

While *Hund's first rule* enjoys the status of a theorem in applied quantum mechanics, the second rule has a purely empirical basis; it is not properly understood how it works, though it certainly does for *equivalent* electrons.

The determination of electronic ground states may be further expedited by the *hole theorem*, touched upon in section (1). With again configurations of equivalent electrons in mind, it may be stated:

NB (H) ➡

> The terms of an atomic configuration $(l)^N$ are the same as those of the corresponding hole configuration $(l)^{4l+2-N}$.

The hole theorem applies also to molecular electron configurations.

The application of these rules is illustrated below by reference to the d^N valence electron configurations encountered in transition metal ions.

4. The ground states of d^N configurations

The configurations d^1 and d^9 (the hole equivalent) have the single state ^{2}D. Each of the other configurations, with $2 \leqslant N \leqslant 8$, gives rise to several terms. Their *ground* states can be characterised by identifying, first, the spin configuration of maximum S and, secondly, the maximum compatible value of L. The latter may be equated with the largest value of $M_L = \Sigma_i\, m_l(i)$ that can be realised.

In the d^2 case, S = 1 and the maximum value of M_L is then +3, which must belong to a state with L = 3. Therefore, the ground state has to be ^{3}F. (The argument is illustrated by the figure beneath.)

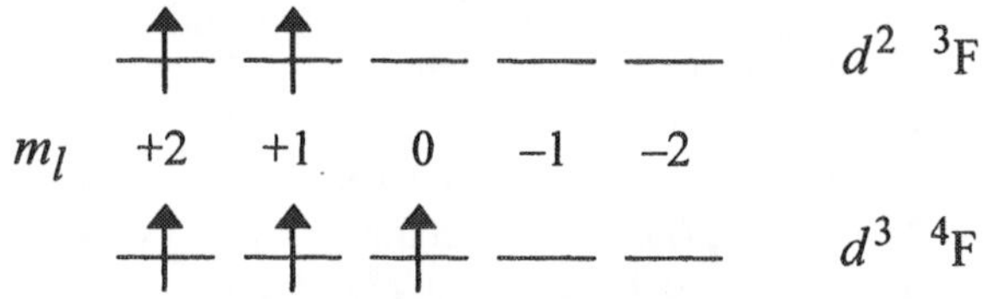

	↑	↑	—	—	—	d^2 ^{3}F
m_l	+2	+1	0	−1	−2	
	↑	↑	↑	—	—	d^3 ^{4}F

Fig. 2.14 The ground states of the d^2 and d^3 configurations, viewed in terms of the maximum values of M_S = S and M_L = L.

The configuration d^3 has instead a maximum S = $\frac{3}{2}$, but the same maximum M_L = +3, indicating a ^{4}F ground state. By the same reasoning, the ground state for d^4 must be ^{5}D. It may be noted that, by the very method of their derivation, these ground states are unique. Given a parallel spin configuration, the states of maximal M_L can be constructed in only one way.

The ground terms of the related configurations – d^8, d^7 & d^6 – may be deduced from the *hole theorem*.

The half-filled subshell d^5 is a special case because, as may be seen in Fig. 2.15(a), the maximal spin S = 5/2 allows only one orbital arrangement, this with $M_L = 0$, and thereby determines the non-degenerate ground term ^6S ($L = 0$). The cations Mn^{2+} and Fe^{3+} are prominent examples.

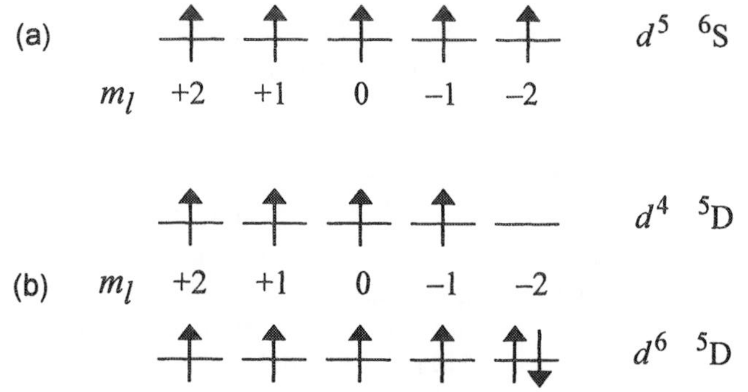

(a) ↑ ↑ ↑ ↑ ↑ d^5 ^6S

m_l +2 +1 0 −1 −2

↑ ↑ ↑ ↑ — d^4 ^5D

(b) m_l +2 +1 0 −1 −2

↑ ↑ ↑ ↑ ↑↓ d^6 ^5D

Fig. 2.15 Ground states of the half-filled subshell d^5 and also the configurations d^4 and d^6.

(a) The $M_S = 5/2$ component of ^6S (d^5) with $M_L = 0$ ($L = 0$);

(b) Components of maximum M_S and M_L for ground states of d^4 and d^6.

Evidently the ground state of a half-filled subshell will always be an S state: for p^3 it is ^4S ($S = 3/2$), pictured earlier in Fig. 2.11(b), while for f^7, encountered among the lanthanide elements in the (isoelectronic) ions Eu^{2+}, Gd^{3+} and Tb^{4+}, it is ^8S ($S = 7/2$).

The half-filled subshell is also special in that exchange stabilisation (or spin correlation) is then maximised for that subshell – *cf.* section 2(3).

It is also worth noting that the ground states of d^4 and d^6 might have been identified expeditiously by reference to that of the half-filled subshell d^5. The ground states may be deduced by removing an electron (making a d hole in the S state) or by adding a d electron, respectively – either way generating a ^5D term. By this device, the ground states of both Eu^{3+} (f^6) and Tb^{4+} (f^8) may be seen immediately to be ^7F.

2.5 The spin-orbit interaction

1. The elementary effect (one-electron atom)

The spin of an electron interacts magnetically with its own orbital angular momentum. This spin-orbit "coupling" may be understood by examining the effect at the electron of the relative motion of the nucleus, of charge Ze (where Z is the atomic number).

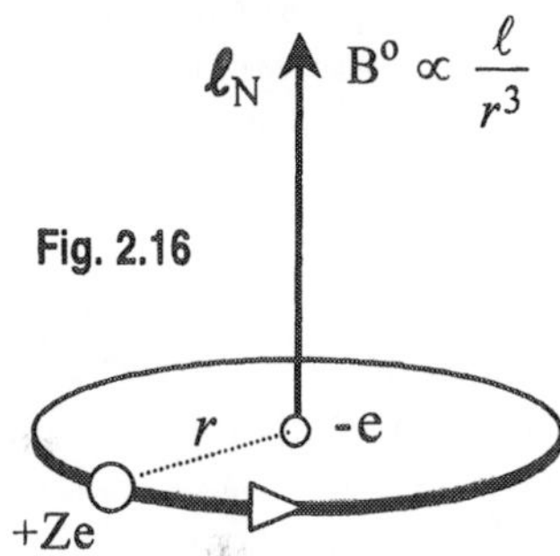

Fig. 2.16 The magnetic field at an electron due to the relative motion of the nucleus.

To an observer located at the electron, the nucleus will appear to circulate about the electron with the same angular momentum $\boldsymbol{\ell}$ as the electron orbiting the nucleus (Fig. 2.16). This relative motion of the nucleus produces a current of magnitude $I = Ze\,|\boldsymbol{\ell}|/2\pi m_e r^2$ (assuming again a circular orbit).

Now, as shown elsewhere# (but originally by Ampére), such a current loop generates at its centre a magnetic field given by

$$B^o = \frac{\mu_0 I}{2r} \tag{2.24a}$$

directed perpendicular to its plane which, on substituting for I, becomes

$$B^o = \frac{Ze\mu_0}{4\pi m_e r^3}|\ell| \tag{2.24b}$$

co-directional with $\boldsymbol{\ell}$. Note the essential dependency on $\boldsymbol{\ell}/r^3$.

The spin magnetic moment $\mathbf{m}^s$ (eqn 2.16) resident "within" the electron interacts with the field $\mathbf{B}^o$. Invoking eqn. 1.6, its potential energy may be expressed as $E_{so} = -\,\mathbf{m}^s.\mathbf{B}^o$, that is

e.g., Duffin (1990), § 7.3.

The field at a distance R along the axis of the loop is given by

$$B_R = \frac{\mu_0 I r^2}{2(R^2 + r^2)^{3/2}}$$

This reduces to (2.24a) when $R = 0$. If instead $R >> r$ then

$$B_R = \frac{\mu_0 I r^2}{2R^3} = \frac{\mu_0 m}{2\pi R^3}$$

as expected from § 1.5(2) (though r is used there rather than R).

$$E_{so} = \frac{\mu_0 Z e^2 g_e}{8\pi m_e^2 r^3}\, \mathbf{s}.\boldsymbol{\ell} \tag{2.25}$$

where $\mathbf{s}.\boldsymbol{\ell}$ is the scalar product of the spin and orbital angular momenta. More commonly, this *spin-orbit interaction energy* is written simply as

The true value of $\xi(r)$, and therefore of the constant ζ, is actually one-half of that suggested by eqn. 2.25. This is due to a relativistic effect called the *Thomas precession*, which thus essentially cancels the factor g_e.

$$E_{so} = \xi(r)\, \mathbf{s}.\boldsymbol{\ell} \tag{2.26}$$

where $\xi(r) = ZA/r^{-3}$, A being the universal constant implicit in eqn. 2.25. This semi-classical formula for E_{so} is valid also for elliptical orbits.

The quantum mechanical expression for the spin-orbit interaction energy is obtained by taking the average of E_{so} for the electron orbit. The scalar product $\mathbf{s}.\boldsymbol{\ell}$ has an exact value [eqn. 2.31 later], because both $\mathbf{s}$ and $\boldsymbol{\ell}$ are well-defined, leaving simply the need to average $\langle r^{-3}\rangle$ over the electronic charge cloud, when

[ζ is the Greek letter *zeta* while ξ is the letter *xi*, both lower case.]

$$\boxed{E_{so} = \zeta_{nl}\, \mathbf{s}.\boldsymbol{\ell}/\hbar^2} \tag{2.27}$$

The quantity $\zeta_{nl} = \hbar^2\langle\xi(r)\rangle = \hbar^2 ZA\langle r^{-3}\rangle$, which is characteristic of the both the orbital and atom concerned, is known as the (one-electron) *spin-orbit coupling constant*. It is inherently positive and as defined in (2.27) has the dimensions of energy.[a]

[a] However, ζ_{nl} is more commonly expressed in wave-number terms, like the illustrative data provided in sub-section 3 below.

The electron spin $\mathbf{s}$ has just two distinct orientations (*i.e.* up ↑ or down ↓) with respect to the orbital field $\mathbf{B}^0$ or the collinear orbital vector $\boldsymbol{\ell}$. The scalar product $\mathbf{s}.\boldsymbol{\ell}$ has therefore two different values and, consequently, the orbital energy level concerned shows a splitting[b] (of magnitude Δ_{so}) which is proportional to ζ_{nl}. Since $E_{so} \propto \mathbf{s}.\boldsymbol{\ell}$, the more stable state must be that in which $\boldsymbol{\ell}$ and $\mathbf{s}$ are *opposed* (Fig. 2.17).

[b] Such a splitting is responsible for the doublet structure of the so-called D-line in the yellow region of the emission spectrum of atomic sodium. This line is due to the transition 3p ⟶ 3s in the excited neutral atom.

This *doublet* splitting is not observed in the case of *s*-orbitals which have $l = 0$ and carry no net orbital angular momentum. There is no internal magnetic field $\mathbf{B}^0$ nor, accordingly, a spin-orbit interaction, for an *s*-electron.

2. The total one-electron angular momentum j

It proves useful to define a total angular momentum $\mathbf{j}$, the vector sum of the spin and orbital angular momenta of an electron,

$$\mathbf{j} = \boldsymbol{\ell} + \mathbf{s} \tag{2.28}$$

The vector $\mathbf{j}$ is characterised by quantum numbers j and m_j, with functions analogous to those of l and m_l, and s and m_s. Thus its magnitude is

The quantum number m_j has $2j + 1$ values with the range $\pm j$, $\pm(j-1)$, ... $\pm\frac{1}{2}$. The microstates are distinguished by $m_j = m_l + m_s$.

$$|\mathbf{j}| = [j(j+1)]^{\frac{1}{2}}\hbar \tag{2.29}$$

while the component of the (space quantized) $\mathbf{j}$ in a selected direction is

$$j_z = m_j\hbar \tag{2.30}$$

The quantum number j has alternative values, $l \pm \frac{1}{2}$, which correspond, roughly speaking, to parallel and antiparallel alignments of the coupled vectors $\boldsymbol{\ell}$ and $\mathbf{s}$ (Fig. 2.17), and therefore have different values of the spin orbit interaction energy E_{so}. These states, with well-defined $\mathbf{j}$, are known as *spin-orbitals*. A p electron ($l = 1$) has spin-orbitals denoted $p_{1/2}$ and $p_{3/2}$.

A spin-orbital (or j state) has a degeneracy of $2j + 1$, the constituent states being discriminated by the auxiliary quantum number $m_j = m_l + m_s$. The degeneracies of the two states $j = l \pm \frac{1}{2}$ are respectively $2l + 2$ and $2l$, and their sum $= 4l + 2$ correctly reproduces the number of microstates, $2 \times (2l + 1)$, identifiable at the outset. In *many-electron* atoms, a spin-orbital may accommodate up to, but no more than $2j + 1$ electrons.

An orbital subshell has a degeneracy of $2l+1$, while electron spin has a two-fold degeneracy. An electron has therefore $4l+2$ distinct microstates, these being differentiated by the spin-orbit interaction.

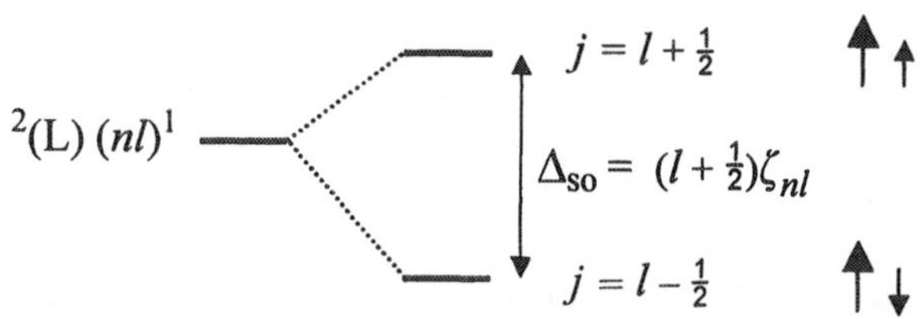

Fig. 2.17 The splitting of a one-electron energy level $(nl)^1$ under the spin-orbit interaction.

Relative to the unperturbed state the energies of the two states are

$$E_{so}(l + \tfrac{1}{2}) = +\tfrac{1}{2}l\zeta_{nl}$$
$$E_{so}(l - \tfrac{1}{2}) = -\tfrac{1}{2}(l+1)\zeta_{nl}$$

The doublet splitting Δ_{so} illustrated above may be calculated as follows. The square $\mathbf{j}^2 = \mathbf{j.j}$ of $\mathbf{j} = \boldsymbol{\ell} + \mathbf{s}$ can be expanded to give

$$\mathbf{j}^2 = \boldsymbol{\ell}^2 + \mathbf{s}^2 + 2\mathbf{s}.\boldsymbol{\ell}$$

so that the scalar product $\mathbf{s}.\boldsymbol{\ell} = \boldsymbol{\ell}.\mathbf{s}$ may be written as

$$\mathbf{s}.\boldsymbol{\ell} = \tfrac{1}{2}[\mathbf{j}^2 - \boldsymbol{\ell}^2 - \mathbf{s}^2] \qquad (2.31a)$$

Entering the well-defined values of the angular momenta, the expression becomes

$$\mathbf{s}.\boldsymbol{\ell} = \tfrac{1}{2}[j(j+1) - l(l+1) - s(s+1)]\hbar^2$$
$$= \tfrac{1}{2}[j(j+1) - l(l+1) - \tfrac{3}{4}]\hbar^2 \qquad (2.31b)$$

The energy of the spin-orbit interaction (eqn. 2.27) is therefore

$$E_{so} = \tfrac{1}{2}\zeta_{nl}[j(j+1) - l(l+1) - \tfrac{3}{4}] \qquad (2.32)$$

from which it follows that the spin-orbital $j = l - \frac{1}{2}$ is the more stable. The doublet splitting, $\Delta_{so} = E_{so}(l + \frac{1}{2}) - E_{so}(l - \frac{1}{2})$, is easily shown to be

$$\boxed{\Delta_{so} = (l + \tfrac{1}{2})\zeta_{nl}} \qquad (2.33)$$

The strength of the internal field B^o (eqn. 2.24) can be estimated *via*

$$\Delta_{so} \simeq 2\mu_B B^o$$

In the case of the excited sodium atom, (Ne)$(3p)^1$, the emission spectrum gives $\Delta_{so} = 17.2\ cm^{-1}$, indicating $B^o \simeq 18$ T. Still larger internal magnetic fields are found in heavier atoms: thus, the "outer" electron of Ce^{3+} (Xe)$(4f)^1$, with $\Delta_{so} = 2251\ cm^{-1}$, has $B^o \simeq 2410$ T.

3. Variation of the spin-orbit coupling constant ζ_{nl}

The one-electron parameter ζ_{nl}, defined through eqn. 2.27, has the make-up

$$\zeta_{nl} = \hbar^2 ZA\langle r^{-3}\rangle \qquad (2.34)$$

where $A = \dfrac{\mu_0 e^2 g_e}{16\pi m_e^2}$ and, as before, Z is the atomic number.

In the case of *hydrogenic* atoms, where the radial wave functions are known in analytical form, the average value of r^{-3} for an orbital can be specified exactly as

$$\langle r^{-3}\rangle = \frac{2Z^3}{n^3 l(l+1)(2l+1)a_0^3} \qquad (2.35)$$

$\langle r^{-3}\rangle$ is expressed in terms of the atomic unit a_0, the Bohr radius.

Note that $\langle r^{-3}\rangle$ is indeterminate for s-orbitals ($l = 0$) for which $E_{so} = 0$.

Using this result, eqn. 2.34 accurately reproduces the observed values of ζ_{nl} in the excited states of a one-electron atom – whether H, or He^+, Li^{2+}, *etc.*

It can be seen that $\zeta_{nl} \propto Z^4$ in one-electron atoms. However, this strong dependency on Z is not maintained in many-electron atoms, although

NB (I) ➡

> ζ_{nl} is found to increase both *across rows* and also, more sharply, *down subgroups* of the periodic table.

The radial wave functions in many-electron atoms are considerably distorted in relation to hydrogenic form [*cf.* section 1(2)], such that $\langle r^{-3} \rangle$ cannot reliably be evaluated by eqn. 2.35.# In heavier atoms, the electron orbitals are further "deformed" by the effects of special relativity. There is, moreover, a contribution to the spin-orbit coupling, negligible in light atoms but increasing steadily with atomic number Z, which arises from the interaction of the spin magnetic moment of an electron with the magnetic fields generated by the orbital motion of the *other* electrons.

\# This remains the case when Z is replaced by an *effective nuclear charge* Z_{eff} that allows for the partial shielding from the nuclear charge due to the other electrons present.

But eqn. 2.34 remains a useful approximation, and $\langle r^{-3} \rangle$ may be determined *ab initio* through numerical solutions to the Schrödinger equation (or its relativistic elaboration). The figures below show that eqn. 2.34 provides a good account of the variation of experimental ζ_{nl} for valence electrons in both *horizontal* and *vertical* atomic sequences.

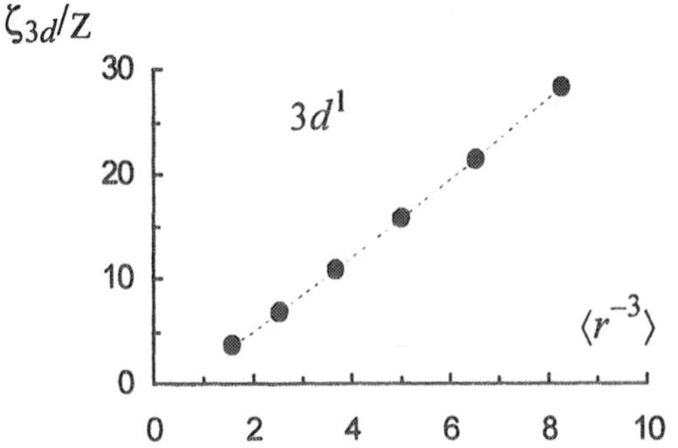

Fig. 2.18 The $3d^1$ isoelectronic series Sc^{2+} – Fe^{7+}.

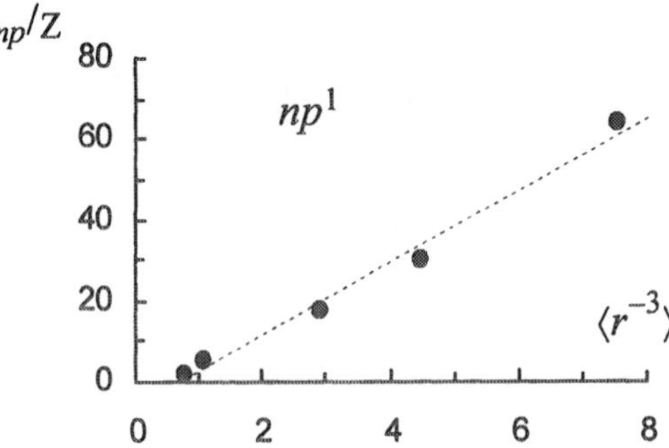

Fig. 2.19 Ground states np^1 of the atoms B – Tl.

The ζ_{nl} values in these diagrams are expressed in the wave-number unit cm^{-1}, *i.e.* as $10^{-2}\,\zeta_{nl}/hc$. The data have been obtained by analysis of atomic emission spectra.

ζ_{nl} also increases significantly with the external charge on an atom.

In both the series chosen the atoms have a solitary electron outside closed subshells, the configurations being $(Ar)3d^1$ and $(core)ns^2np^1$ ($n = 2 - 6$), respectively. The $\langle r^{-3} \rangle$ values come from non-relativistic calculations.

The trend in the data plotted in Fig. 2.18 is easily explained: proceeding from $Sc^{2+} \rightarrow Fe^{7+}$, the $3d$ orbital contracts progressively and $\langle r^{-3} \rangle$ naturally increases in step; so, therefore, do both ζ_{3d}/Z and also ζ_{3d} itself. Sequences such as Sc^{2+} $(3d^1) \rightarrow Cu^{2+}$ $(3d^9)$ show a similar trend.

In group 13 (or IIIB) the spin-orbit coupling constant increases from $\zeta_{2p} = 11\ cm^{-1}$ at boron to $\zeta_{6p} = 5195\ cm^{-1}$ at thallium.

The vertical plot in Fig. 2.19 has a more subtle basis. The increase in both ζ_{np} and ζ_{np}/Z observed on descending the boron subgroup, B → Tl, reflects a monotonic increase in $\langle r^{-3} \rangle$ due to the build-up of an increasing proportion of the valence electron density at very short distances r from the nucleus.

Somewhat paradoxically, this occurs as the outer orbital grows more diffuse, as judged by the average quantities $\langle r \rangle$ and $\langle r^2 \rangle$, common measures of the "size" of an orbital. In other words, the electron density simultaneously spills out and contracts as Z increases down the subgroup. From the hydrogenic standpoint, this can be understood in terms of the outer orbital acquiring additional *spherical* nodal surfaces and hence an increasingly flexible form permitting the probability density to contract towards the increasing nuclear charge while, at the same time, expanding more freely in response to centrifugal tension in the orbital.

Hydrogenic radial wave functions have $n-1$ spherical nodes.

The tendency for ζ_{np}/Z to increase down a subgroup (due to increasing $\langle r^{-3} \rangle$) is part of a general phenomenon. Thus a comparable growth in the parameter ζ_{nd}/Z is observed in the *transition-metal* subgroups, as illustrated by the following data for the neutral atoms of group 6 (VIA):

		Z	ζ_{nd}	ζ_{nd}/Z
Cr	$3d^4 4s^2$	24	227	9.46
Mo	$4d^4 5s^2$	42	589	14.02
W	$5d^4 6s^2$	74	3347	45.23

In the case of both Cr and Mo, d^4s^2 is actually an excited configuration. The ground states are $3d^5 4s^1$ and $4d^5 5s^1$ (both ^7S), respectively.

(The ζ_{nd} are again expressed in cm^{-1}.) The enormous increase in ζ_{nd} on going from molybdenum to tungsten, a typical development in the transition-metal triads, is in large part due to the poor shielding provided by the $4f^{14}$ outer "peel" of the lutetium core. In the third transition series, the $5d$ orbital accordingly shows a particularly large increase in $\langle r^{-3} \rangle$.

The valence orbitals are said to become more *penetrating* on descent of a subgroup and, when $l \geqslant 1$, experimental values of ζ_{nl}/Z ($\propto \langle r^{-3} \rangle$ theoretically) provide an empirical measure of their *penetrating character*.

4. The spin-orbit interaction in many-electron states

When the open-shell aspect of an atomic configuration involves two or more electrons, and therefore several Russell-Saunders terms $^{2S+1}(L)$ [*cf.* section 4(2)], the energy of the spin-orbit interaction is usually formulated as

$$\boxed{E_{so} = \lambda \mathbf{S}.\mathbf{L}/\hbar^2} \qquad (2.36)$$

[λ = Greek *lambda* (lower case)]

where **S** and **L** are the spin and orbital angular momenta, respectively, and λ is a *many-electron* spin-orbit coupling constant analogous to ζ_{nl} in eqn. 2.27. The above expression for the spin-orbit coupling energy represents a sum of one-electron terms of the form $\xi(r)\,\mathbf{s}.\boldsymbol{\ell}$ (eqn. 2.26). For configurations of *equivalent* electrons [section 4(3)], the principal concern here, the energy may be written simply as

In general, λ depends on the configuration type and is a function of the spin and orbital quantum numbers,

$$\lambda = f(S,L).$$

$$E_{so} = \zeta_{nl} \Sigma_i \mathbf{s}(i).\boldsymbol{\ell}(i) \qquad (2.37)$$

since the different orbitals of the degenerate (n, l) subshell involved have ζ_{nl} as a common value of the spin-orbit coupling constant. For any configuration with a less than half-filled subshell, the parameter λ is *positive*, like the one-electron quantity ζ_{nl}.

It is expedient, following the treatment of the one-electron problem [in section 5(2)], to construct a grand total angular momentum

[Eqns. 2.38 – 2.40 are, of course, generalisations of the eqns. 2.28 – 2.30 earlier.]

J has an associated quantum no. J, analogous to L and S. A J state has a degeneracy of 2J +1.

In the jargon of atomic spectroscopy a J state is known as a *multiplet.*

$$\mathbf{J} = \mathbf{L} + \mathbf{S} = \Sigma_i\, \mathbf{j}(i) \tag{2.38}$$

of magnitude

$$|\mathbf{J}| = [\mathrm{J}(\mathrm{J}+1)]^{\frac{1}{2}}\hbar \tag{2.39}$$

The vector **J** has the space-quantized components,

$$\mathrm{J}_z = \mathrm{M_J}\hbar \tag{2.40a}$$

where $$\mathrm{M_J} = \mathrm{M_L} + \mathrm{M_S} = \Sigma_i\, m_j(i) \tag{2.40b}$$

Invoking again the vector coupling model [introduced in section 4(1)], the total angular momentum quantum number J is allowed the values J = L + S, L + S − 1, ... |L − S|, which are integral or half-integral according as S is integral or half-integral. J is permitted *either* 2L+1 *or* 2S+1 different values, depending on which is the *smaller* (invariably S), and each distinct J state harbours a degeneracy of 2J + 1.

Thus, a ^{3}P (p^2) state may have J = 2, 1 and 0 (as shown in Fig. 2.20), while a ^{3}F (d^2) state is allowed J = 4, 3 and 2. The reader may care to confirm that, in both these examples,

$$\Sigma_\mathrm{J}\,(2\mathrm{J}+1) = (2\mathrm{S}+1)(2\mathrm{L}+1) \tag{2.41}$$

as must always be the case [*cf.* eqn. 2.23].

The quantum no. J discriminates the energy of the spin-orbit interaction.

Each distinct value of J defines, for its parent atomic term, one of the permitted mutual orientations of the vectors **S** and **L** and therefore, by eqn. 2.36, a particular value of the spin-orbit interaction energy. The term is consequently split into two or more energy levels differentiated by the J quantum number (*e.g.* Fig. 2.20). This is often described as *multiplet splitting.*

The spin-orbit coupling may be interpreted in terms of an *internal field* $\mathrm{B_{int}}$, co-directional with **J**, to which both **S** and **L** are coupled, and sometimes very strongly. [See § 4.1(1).]

Retracing the arguments earlier that culminated in eqn. 2.31, it may easily be shown that

$$\mathrm{E_{so}}(\mathrm{J}) = \tfrac{1}{2}\lambda[\mathrm{J}(\mathrm{J}+1) - \mathrm{L}(\mathrm{L}+1) - \mathrm{S}(\mathrm{S}+1)] \tag{2.42}$$

The figure below, which might represent, say, the ground state of carbon (He)$2s^2 2p^2$, has been constructed on the basis of this formula.

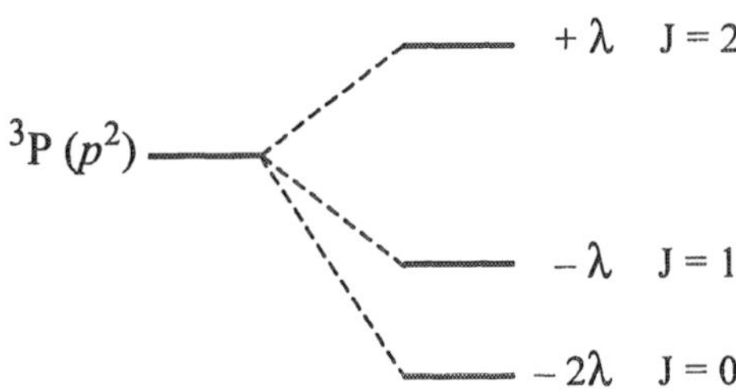

Fig. 2.20

Multiplet splitting of the ^{3}P ground state of the p^2 configuration.

[The non-degenerate J = 0 ground multiplet may be viewed in terms of double occupancy of the $j = \frac{1}{2}$ spin-orbital.]

Observe that the ground multiplet is that with the lowest possible value of J, as in the one-electron case. The figure illustrates also the celebrated *Landé interval rule* (1923), following straightforwardly from eqn. 2.42, that the energy separation of adjacent J states, known as the *multiplet interval*, is given by#

$$\mathrm{E_{so}}(\mathrm{J}+1) - \mathrm{E_{so}}(\mathrm{J}) = (\mathrm{J}+1)\lambda \tag{2.43}$$

The rudimentary case (eqn. 2.33), in which $\mathrm{J} = l - \frac{1}{2}$, is encompassed by eqn. 2.43.

Another quantity of interest is the *multiplet width,* $W_{so} = |E_{so}(L + S) - E_{so}(L - S)|$, the overall energy spread of the multiplets of a term. Assuming that L > S, this proves to be

Landé observed that the separation between any two adjacent multiplets is proportional to the higher J value.

$$W_{so} = \lambda S(2L + 1) \qquad (2.44)$$

The Landé rule antedates the theory that is needed to justify it, like the Russell-Saunders vector model.

In the case of ^3P (p^2), for example, $W_{so} = 3\lambda$, as evident in Fig. 2.20.

The many-electron parameter λ can be shown to have a simple general form for the *ground* terms of equivalent electron configurations, namely

$$\lambda = \pm \zeta / 2S \qquad (2.45)$$

where the *minus* sign applies if the subshell is *more* than half-occupied. The ^3P ground state of p^2 thus has $\lambda = \frac{1}{2}\zeta_{np}$ (and so $W_{so} = \frac{3}{2}\zeta_{np}$).

The case of the half-filled subshell is excluded since L = 0 and $E_{so} = 0$.

The change in sign of λ for occupancies N in excess of $2l + 1$ can be understood in terms of the hole formalism [section 4(3)]. The spin-orbit interaction energy in a configuration $(l)^N$ may be expressed as that in the filled subshell $(l)^{4l+2}$ minus that in the hole configuration $(l)^{4l+2-N}$. But the interaction is zero for the completed subshell since S = L = 0. Therefore,

$$E_{so}\{(l)^N\} = -E_{so}\{(l)^{4l+2-N}\}$$

and the energy of the spin-orbit interaction changes sign in the hole configuration. Accordingly, when applying eqn. 2.42, it must be recognised that the parameter λ is negative when the subshell is more than half-occupied, *i.e.* when $N > 2l + 1$.

As an obvious consequence there is a reversal of the energy ordering of the multiplets J in the ground state of a configuration with a more than half-filled subshell. For example, the ^3P ground state of the p^4 configuration – *e.g.* oxygen (He)$2s^2 2p^4$ – has the ground multiplet J = 2, rather than J = 0 as in Fig. 2.20. This phenomenon of *multiplet inversion*, as it is called, is the subject of *Hund's third rule*:

[However, multiplet inversion had been spotted earlier by Landé.]

> The most stable multiplet of the ground term of a configuration $(l)^N$ is that with the minimum value of J possible when the subshell is less than half-occupied but, instead, that with the maximum possible value of J if the subshell is more than half-filled.

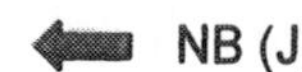 NB (J)

5. Breakdown of Russell-Saunders coupling

Implicit in the foregoing account of spin-orbit coupling is the assumption that *the interaction is small in relation to the inter-electronic repulsion* that distinguishes the Russell-Saunders terms; it does not then interfere significantly with the pre-existing vector couplings. This is a good approximation for light atoms, up to argon (say), and remains tolerable with increasing atomic number up to and including the lanthanide elements. The approach is unrealistic beyond this point, however, as the multiplet widths W_{so} become comparable in magnitude to the term separations. The theory of section 5(4) is not remotely applicable in the third transition series, nor to the elements that follow, including the Actinide elements.#

The failure of the Russell-Saunders scheme makes the interpretation of magnetism in compounds of the third row transition metals, and also the Actinide elements, a very complicated matter.

The theory for the one-electron atom also collapses at high Z (*e.g.* U^{91+}), but this is of limited chemical import.

In the full quantum mechanical theory, the spin-orbit interaction not only splits the term energy levels but *also*, in an important secondary effect, mixes together multiplet states with a common J value stemming from different

Russell-Saunders terms. The strength of this "blending" is the greater the closer are the multiplets in energy. As it develops (with increasing atomic number), the effect is revealed by the *compression* of the multiplets of the ground term and the breakdown of the Landé interval rule.

A strong spin-orbit interaction tends to "scramble" the spin and orbital angular momenta such that S and L cease to be valid quantum numbers.

As its distinctive feature, a Russell-Saunders term has well-defined values of both **L** and **S**. But, when the mixing of multiplets between terms becomes substantial, the quantum numbers L and S lose their identities. **L** and **S** can no longer be separately characterised, and only $\mathbf{J} = \Sigma_i\, \mathbf{j}(i)$ remains well-defined.

When the spin-orbit coupling is very strong, such that the multiplet widths would exceed the term separations, it becomes more realistic to base the analysis of electronic structure on an initial assignment of the electrons to the spin-orbitals of the atom, defining a configuration pre-organised as regards the spin-orbit interaction. This is known as **j–j** coupling, the inter-electronic repulsions being introduced *after* the spin-orbit coupling. In this alternative approach, the ground state of the lead atom, Pb (Pt)$6s^2 6p^2$, may be approximated reasonably well as (Pt)$(j_0)^2(j_{\frac{1}{2}})^2$ with J = 0 (*cf.* Fig. 2.20). Inclusion of the electron-electron repulsions results in a certain mixing of the J states. Where this mixing is substantial the coupling scheme is described as *intermediate.* The term **intermediate coupling** is applied equally to the approach based on Russell-Saunders terms interacting considerably under the spin-orbit coupling; indeed, the alternative treatments amount to the same thing when rigorously applied.

The spin-orbit coupling constant ζ in the $6p$ subshell of the lead atom is almost 10,000 cm^{-1}. At carbon, ζ_{2p} is a mere 29 cm^{-1}.

2.6 Postscript: perturbation theory and the mixing of states

1. Quantum mechanical mixing

It is frequently expedient to characterise a complex system by refinement of the description of a closely related, but simpler system which can be analysed exactly (or nearly so). Often this is the only viable approach. If the system of interest differs but slightly from another, by a *minor* additional energy term, then its states may be satisfactorily approximated by reference to those of the less complicated system by means of what is known as **perturbation theory**. The small extra energy term is described as the *perturbation*, here denoted *P*.

A full discussion of perturbation theory is beyond the scope of an elementary text. For the details consult, *e.g.*, Green (1997 & 1998) and Atkins and Friedman (1996),

Applied in *first order* (as the jargon goes), this simply involves adjusting the energies by evaluating the perturbation *P* for the states of the reference, or "unperturbed" system; should the latter have degeneracies, then these are likely to be lifted, at least in part. This is the essence of the development in section 5(4) where, in effect, the spin-orbit interaction was introduced as a perturbation $P = \lambda \mathbf{S.L}/\hbar^2$ (eqn. 2.36) of the Russell-Saunders terms. In other words, eqn. 2.42 is correct to first order in perturbation theory.

In *second order* perturbation theory, the wave functions of the system are adjusted through the mixing together of those states that are indistinguishable under the perturbation *P*. This is illustrated, albeit in a limited way,# by Fig. 2.21, where for simplicity it is assumed that the separation of the two states, $E_b - E_a = \Delta$, is large in comparison with the energy π typically associated with the perturbation.

A more general formulation is needed if $\pi \sim \Delta$.

Notice in particular that the second-order effects depend *inversely* on the energy spacing Δ. The states are mutually admixed to an extent $\kappa^2 \propto (\pi/\Delta)^2$; in addition, there is a further minor displacement in energy by an amount $\delta \propto \pi^2/\Delta$, the lower state (with wave function ψ_1) being stabilised.

[κ = Greek *kappa* (lower case)]

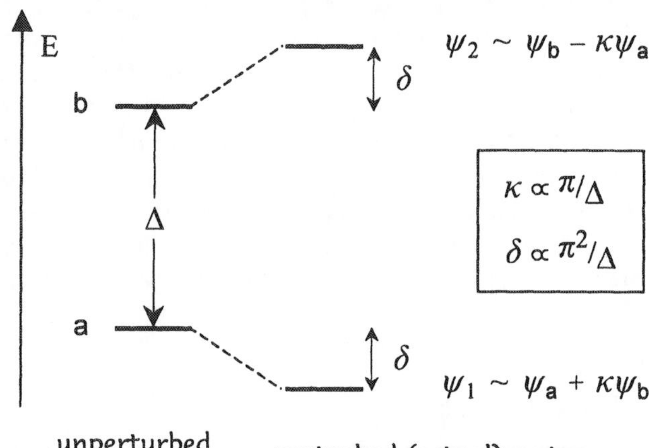

Fig. 2.21 Quantum mechanical mixing in a two-level system. The perturbation is of magnitude $|P| \sim \pi$ (assumed $\ll \Delta$).

In the theory of multiplet structure, the spin-orbit interaction mixes states having the same value of J, these necessarily belonging to different terms. If two such multiplets belong to the same configuration then the characteristic energy of the perturbation (Fig. 2.21) is $\pi = \zeta$, otherwise $\pi = \sqrt{|\zeta_a\zeta_b|}$. The mixing becomes stronger on proceeding down a periodic subgroup due, not only to the increase in ζ [section 4(3)] but also to the fact that term separations diminish as the valence orbitals become more diffuse and the interelectronic repulsions decrease correspondingly. The *intermediate coupling* of outer electrons in elements of the third transition series, and also the Actinide elements, requires second order perturbation theory applied without further approximation.

This topic will be revived later, especially in connection with the paramagnetic properties of transition-metal compounds. As foreshadowed by section 3(4), the orbital contribution to the electronic magnetic moment m is largely quenched by the ligand field. However, it can never be *entirely* suppressed, even when the degeneracy of the *d* subshell is completely razed, because of the second-order effects of the spin-orbit coupling. By mixing excited multiplets into the ground state the spin-orbit interaction may, in effect, restore some of the quenched orbital angular momentum. Suppose the two states pictured in Fig. 2.21 represent spin-orbitals that involve conjugate orbitals such as (say) $d_{x^2-y^2}$ and d_{xy}, discriminated by the internal electric field: then the spin-orbit interaction can "resuscitate" the quenched orbital angular momentum ($\pm 2\hbar$) to an extent of the order $\kappa\hbar = (\zeta_d/\Delta)\hbar$ by partially "re-mixing" the separated *d* orbitals.

2. The polarizability of an atom or non-polar molecule

Among numerous other important applications, perturbation theory may be used to describe the phenomenon of *distortion polarization* through which [recall § 1.2(3b) & 5(4)] an electric dipole moment $\boldsymbol{\mu}_i = \alpha \boldsymbol{E}$ is induced by an applied electric field $\boldsymbol{E}$. In this case, the perturbation is

$$P = eE\Sigma_i z(i) \tag{2.46}$$

it being assumed that the electric field $\boldsymbol{E}$ is applied in the z-direction. P is of the form $-\mu_E E$, where μ_E is the component of the electronic dipole moment parallel to the electric field, defined by $\mu_E = -e\Sigma_i z(i)$, the summation embracing all electrons in the atom.

In a not dissimilar application, perturbation theory is used to describe the distortion effect of a magnetic field. This leads, in closed-shell molecules, to a weak paramagnetism independent of temperature (*a.k.a.* TIP – § 3.3.)

μ_E is, of course, zero for an unperturbed atom or non-polar molecule.
In the case of polar molecules there will be naturally a much stronger *orientation* effect in the ensemble [§ 1.5(5)].

Referring back to Fig. 2.21, the lower state (a) might be the ground state of the unperturbed atom, and the upper state (b) a representative excited state. The second order stabilisation of the ground state would then be of the form $\delta = cE^2/\Delta$, where c is a constant characteristic of that excited state. A sum over all excited states yields a total stabilisation $\Sigma\,\delta$ which may be expressed as $\frac{1}{2}\alpha E^2$, where α is the *polarizability* of the atom. Hence eqn. 1.26 earlier.

α has the make-up $2\Sigma_k c_k E^2/\Delta_k$, where k indexes the excited states.

2.7 Bibliography

Suggestions for further reading:

[OCP = Oxford Chemistry Primer, published by Oxford Univ. Press.]

P A Cox, *Introduction to Quantum Theory and Atomic Structure* (OCP 37), Oxford 1996

T P Softley, *Atomic Spectra* (OCP 19), Oxford 1996

R McWeeny, *Coulson's Valence*, Oxford 1979 (3rd edn.)

P W Atkins, *Physical Chemistry*, Oxford 2001 (7th edn.)

Regarding perturbation theory in particular:

P W Atkins and R S Friedman. *Molecular Quantum Mechanics*, Oxford 1996 (3rd edn.)

N J B Green, *Quantum Mechanics 1*, Oxford 1997; *Quantum Mechanics 2*, Oxford 1998 (OCP 48 & 65)

2.8 Exercises

1. Confirm the following ground state assignments: Cu^{2+} $^2D_{5/2}$, Co^{2+} $^4F_{9/2}$, Sm^{3+} $^6H_{5/2}$, and Ho^{3+} $^5I_8$.

2. Comment on the following values (in cm^{-1}) of the average exchange integral K_{dd} [section 2(3)] deduced from atomic energy level data.

Cr^{2+}	5275	Co^{2+}	6584	Mo^{2+}	3839
Cr^{3+}	5979	Co^{3+}	7266	Mo^{3+}	4324

3. Explain why the ground state multiplet width observed for Tm^{3+} is practically twice that observed for Pr^{3+}.

4. Discuss the following empirical ζ_{nd} data (in cm^{-1}): Cr^{2+} 230, Cr^{3+} 270 (ζ_{3d}); Mo^{2+} 749, Mo^{3+} 818 (ζ_{4d}).

5. Comment on the observation that, in the first transition series, empirical values of ζ_{3d} for neutral atom configurations $3d^n4s^2$ are closely similar to those of the $3d^n$ cations M^{2+}.#

How are the exchange integrals K_{dd} likely to compare in the two cases?

e.g. Cr^0 $3d^44s^2$ and Cr^{2+} $3d^4$ have ζ_{3d} of 227 cm^{-1} and 230 cm^{-1}, respectively.

3 The major aspects of bulk magnetism – additional considerations

The various forms of bulk magnetism were surveyed from the classical viewpoint in § 1.3 (albeit briefly). This chapter extends the review in certain areas, incorporating the quantum mechanical concepts set out in the previous chapter, while dwelling on systems of more obvious chemical interest, *e.g.* *compounds* of the metallic elements rather than the elementary metals, and mainly those that are electrical insulators (or semiconductors).

3.1 Spin paramagnetism

It was emphasised in § 2.3(4) that the common forms of magnetism often arise principally or exclusively from the *spin* motion of unpaired electrons, the orbital contribution being quenched by internal electric fields.[a]

a Circumstances in which the orbital motions contribute importantly will be dealt with in subsequent chapters. [See, however, section 4(2).]

1. The Zeeman effect

The spin magnetic moment of an open-shell atom or molecule is given by $\mathbf{m}^S = -g_e(\mu_B/\hbar)\mathbf{S}$ (eqn. 2.18) where $\mathbf{S}$ is space quantised according to $S_z = M_S\hbar$ (eqn. 2.12). The z-component of the magnetic moment is thus

$$m_z^S = -g_e M_S \mu_B \qquad (3.1)$$

Now the applied magnetic field imposes the axis of quantisation, *i.e.* z is the field direction. The energy in the magnetic field, given (recalling eqn. 1.6) by $E = -\mathbf{m}^S.\mathbf{B}$, is therefore

$$E(M_S) = -m_z^S B = +g_e M_S \mu_B B \qquad (3.2)$$

The (2S + 1)-fold degeneracy of the spin state is accordingly lifted, as shown in the figure below. This is the Zeeman effect.[b]

b Strictly, the *first order* Zeeman effect.

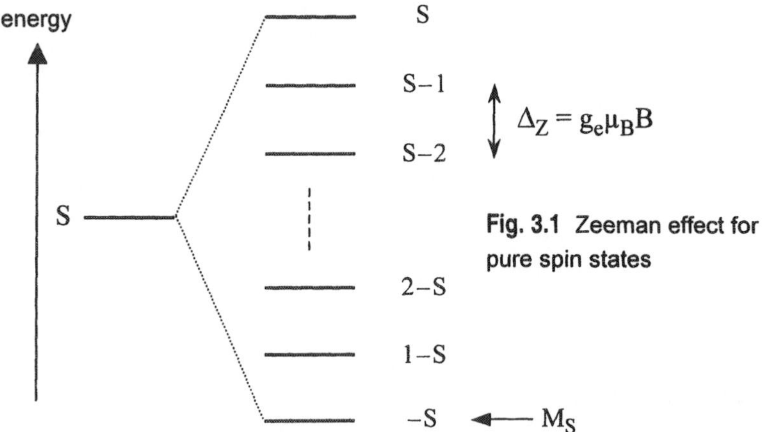

Fig. 3.1 Zeeman effect for pure spin states

The term *Zeeman effect* derives from atomic spectroscopy. The first order effect is the splitting of certain lines when a strong magnetic field is applied to an emitting atom (Zeeman 1897, Preston 1898). The splitting is proportional to the field strength.

The *second order* or *quadratic* effect, requiring very strong fields, is the shift in the position of a line by an amount ∝ B^2.

Other forms of angular momenta, *e.g* **J**, show analogous splittings, though with different g-factors.

For pure spin states the resonance condition is $h\nu = g_e\mu_B B$.

ESR may be contrived either by holding the field constant and varying the microwave frequency ν, or *vice versa*. In practice it is more convenient to sweep the magnetic field, ν being held constant. The technique is enormously sensitive.

Notice that the most stable of the "Zeeman levels" is that with $M_S = -S$, and also that the overall splitting is symmetrical about the unperturbed level. The energy difference between adjacent levels – the *Zeeman interval* – is a constant $\Delta_Z = g_e\mu_B B$, *i.e.* $2\mu_B B$ for ordinary purposes. The splitting is extremely small in relation to the separation of different electronic states.

In an applied magnetic field of 1 T (10^4 G) the Zeeman interval amounts to *ca.* 2×10^{-23} J, equivalent to 1 cm^{-1} in *wave-number* units, and electromagnetic radiation of wave-length *ca.* 1 cm, belonging to the microwave region, can induce transitions between adjacent Zeeman levels. This is the basis of electron spin resonance spectroscopy (ESR), a technique of fundamental importance in the study of compounds having unpaired electrons.

2. The limiting Curie susceptibility

Consider an ensemble of atoms or molecules which possess only spin angular momentum in their ground states and do not interact magnetically. This might be a fluid material or a crystal containing well separated open-shell ions. It is assumed that no *excited* electronic state is thermally accessible.

(a) The Curie limit ($kT \gg \mu_B B$)

In routine measurements in the chemical laboratory, the temperature T is comparatively high (commonly ≥ 80 K) and the applied magnetic field of modest strength ($\lesssim 1$ tesla) such that the Boltzmann energy kT exceeds the magnetic energy $\mu_B B$ by at least an order of magnitude. At thermal equilibrium the populations of the different Zeeman levels (Fig. 3.1) are then practically equalised and the Zeeman splitting is irrelevant as regards the properties of the *ensemble*; in other words, the system behaves classically, like a polar fluid in the limit $kT \gg \mu E$ [§ 1.5(5)].

[The Boltzmann energy kT corresponds to 0.6952 T cm^{-1}.]

However, the elementary magnetic dipole moment [m of § 1.3(2)] is of quantum mechanical nature, with the magnitude

$$\mathsf{m}_{eff} = g_e(\mu_B/\hbar)\,|\mathbf{S}| = g_e[S(S+1)]^{\frac{1}{2}}\mu_B \qquad (3.3)$$

commonly termed the **effective magnetic moment**.[a] The magnetic moment is instead often characterised through the *effective Bohr magneton number* $\mathsf{n}_{eff} = \mathsf{m}_{eff}/\mu_B$ (a dimensionless quantity, of course).

[a] The qualifying "effective" is meant to emphasise that m_{eff} is the magnetic moment particular to the Curie limit.

If $kT/\mu_B B$ is indeed very large then, by analogy with the Langevin-Debye equation (1.32), the molar paramagnetic susceptibility may be expressed in *semi-classical* terms as[b]

$$\boxed{\chi_{mol} = \frac{\mu_0 N_A}{3kT}\mathsf{m}_{eff}^2} \equiv \frac{\mu_0 N_A}{3kT} g_e^2 S(S+1)\mu_B^2 \qquad (3.4)$$

[b] – a formula attributed to F Hund (1925), among others.

There is no T-independent term – equivalent to that involving the polarisability α in the Langevin-Debye equation – for pure spin states.

The Hund formula may otherwise be deduced from the general quantum mechanical treatment of the susceptibility in § 4. See also section 10(2).

The Curie constant defined by (1.19) is thus

$$C = \frac{\mu_0 N_A}{3k}\mathsf{m}_{eff}^2 \equiv \frac{\mu_0 N_A}{3k} g_e^2 S(S+1)\mu_B^2 \qquad (3.5a)$$

or, more conveniently,

$$C = \frac{\mu_0 N_A \mu_B^2}{3k} n_{eff}^2 \equiv C' n_{eff}^2 \qquad (3.5b)$$

where the new constant $C' = 1.571 \times 10^{-6}$ m^3 mol^{-1} K.

The Hund equation (3.4) explains the susceptibility data plotted in Fig. 1.6 for a salt containing Mn^{2+} $3d^5$. The cation has a total spin S = 5/2 and hence, by (3.3), m_{eff} should be 5.92 μ_B; the experimental value is 5.89 μ_B.

(b) A postscript concerning the averaging of spin angular momentum

The susceptibility depends on the square of m_{eff} and therefore on $|\mathbf{S}^2|$. If the Zeeman levels of Fig. 3.1 are equally populated then, for each atom or molecule,#

$$\langle S_z^2 \rangle = \frac{\hbar^2}{2S+1} \sum_{-S}^{S} M_S^2 = \tfrac{1}{3} S(S+1)\hbar^2 \qquad (3.6)$$

The system being isotropic, the *x*- and *y*-components of $\mathbf{S}^2$ must be the same. Therefore $|\mathbf{S}^2| = S(S+1)\hbar^2 = |\mathbf{S}|^2$, equivalent to (2.11).

The algebraic result,

$$\sum_1^n r^2 = \tfrac{1}{6} n(n+1)(2n+1)$$

is assumed. This is derived in almost any intermediate level Maths text.

3. Paramagnetic saturation

The converse of the Curie limit, when $\mu_B B \gg kT$, is known as the *saturation limit*; it is usually reached by a combination of high field strength and low temperature. At sufficiently large values of B/T the magnetization **M** becomes saturated: the open-shell species are all induced to occupy the ground Zeeman level, with $M_S = -S$, and the average moment $\langle m_B \rangle$ reduces to a common value,

$$\langle m_B \rangle = g_e \mu_B S = m_{sat} \qquad (3.7)$$

The value of m_{sat} – known as the saturation magnetic moment – *invalidates the classical assumption of a permanent atomic or molecular moment.* A semi-classical treatment would predict instead a saturation moment equal to $g_e\sqrt{[S(S+1)]}\mu_B$. The quantum mechanical result, $m_{sat} = g_e S \mu_B$, is a simple manifestation of the fact that an individual spin vector S, and therefore its associated magnetic moment, can never lie along the axis of quantization (defined here by the applied field).

Unlike the electric dipole moment $\boldsymbol{\mu}$ of a polar molecule, the magnetic dipole moment of an open-shell species is inextricably an *ensemble* property, though with limiting values m_{eff} and m_{sat} that may be characteristic of that species. The saturation moment m_{sat}, in particular, is essentially an atomic or molecular property, in the same sense as the diamagnetic moment.

The general case, where $kT \sim \mu_B B$, is analysed in the Appendix (section 10) and the semi-classical and quantum theories compared further.

The concept of *saturation* in orientation polarisation or magnetisation was introduced at the end of § 1.5(5).

The classical concept of a permanent magnetic moment **m** is untenable in quantum theory.

The Mn(II) Tutton salt recalled above has $m_{sat} = g_e(5/2)\mu_B = 5.01\ \mu_B$.

The molecular property $\boldsymbol{\mu}$ exists in the absence of an external electric field (though philosophers might well quarrel with such a statement).

However, a magnetic moment comes into being only on application of a magnetic field.

4. The lessons

The material of the preceding sections is extremely important. So far as everyday chemical applications are concerned, the most significant result is the expression (3.4) for the spin susceptibility, together with the definition of the associated magnetic moment m_{eff}. This has considerable scope because the orbital angular momentum is so very often quenched [*cf.* § 2.3(4)], both in free-radical molecules and in open-shell ions situated in crystals. The spin state is then readily determined by measurement of the bulk susceptibility, an essential first step in the electronic characterisation of a newly synthesised compound found to be paramagnetic.

The theory of the spin susceptibility is important also as prototypal for the treatment of paramagnetism when either the orbital angular momentum **L**, or the total angular momentum **J**, is well-defined (see § 4.2) – or, indeed, for the description of nuclear spin paramagnetism (see section 9). Furthermore, the introduction to magnetic saturation above paves the way for the further consideration of magnetic ordering in solids (section 6).

3.2 The Weiss field and the Curie-Weiss law

1. The Curie-Weiss law

Paramagnetic solids having susceptibilities that obey the Curie law (1.19) accurately are actually somewhat rare. Far more common is a variation with temperature of the form[a]

$$\chi = C/(T - \theta) \tag{3.8}$$

where θ, known as the *Weiss temperature*, is characteristic of a substance. It is typically positive for materials that order ferromagnetically at a sufficiently low temperature, but usually *negative* in the case of those that instead display antiferromagnetism. If C is constant then, as shown in Fig. 3.2, graphs of χ^{-1} against temperature are straight lines intercepting the abscissa at $T = \theta$.

[a] The reader should beware that the law is often expressed with instead the denominator $(T + \theta)$.

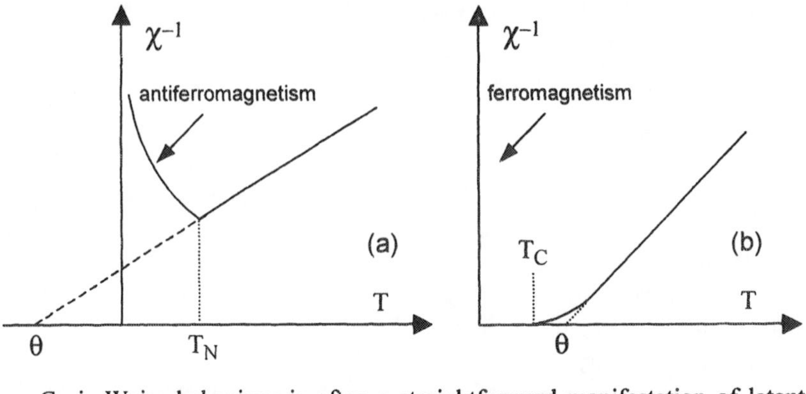

Fig. 3.2
Plots of the inverse susceptibility *vs.* temperature for
(a) an antiferromagnetic and
(b) a ferromagnetic material,
illustrating the Curie-Weiss law.
Note the negative value of θ in the antiferromagnetic case (a).

Curie-Weiss behaviour is often a straightforward manifestation of latent magnetic dipolar coupling within the solid. Associated with the interactions between the magnetic centres is an internal field $\mathbf{B}_W$, called the *Weiss field*, which is defined to be proportional to the magnetization,[b]

$$\mathbf{B}_W = \lambda \mathbf{M} \tag{3.9}$$

where the parameter λ, referred to as the *Weiss constant* (or *molecular field constant*), mirrors the strength of the coupling between the centres, as shown in section 6(2) below. A simple lattice with equivalent sites is assumed.

[b] The concept of the Weiss field (P Weiss 1907) is the basis of what is termed, rather opaquely, *molecular field theory*. Note that λ has the dimensions N A^{-2}, like μ_0 [*cf.* (1.15)]. This is another, unfortunately overworked symbol.

2. Ferromagnetic substances

In the paramagnetic phase existing above the Curie temperature T_C the total internal field is of magnitude

$$B = B_0 + B_W \qquad (3.10)$$

and the inherent volume susceptibility $\chi_m = \mu_0 M/B$ therefore

$$\chi_m = \mu_0 M/(B_0 + \lambda M) = C/T$$

(assuming the Curie law). Rearranging, the *experimental* susceptibility is

$$\chi_m = \mu_0 M/B_0 = C/(T - \lambda C/\mu_0) \qquad (3.11)$$

which has the Curie-Weiss form (3.7) with

$$\theta = \mu_0^{-1} C\lambda \qquad (3.12)$$

Note that θ increases with both the Curie constant C ($\propto m_{eff}^2$) and λ. The (positive) Weiss field naturally facilitates the magnetization.

Note also that, according to (3.11), $\mu_0 M(T - \theta) = CB_0$. Now, because of the internal field, M does not vanish when the temperature is lowered to T_C and the applied magnetic field turned off; therefore θ should be identifiable with the ferromagnetic Curie temperature T_C. However, as implied by Fig. 3.2(b), this simple expectation is not exactly borne out. Due to complicating factors, T_C and θ commonly differ by 10 – 20 K or more, with $\theta > T_C$.

The Weiss field is here extremely high in relation to static magnetic fields that are controllable in the laboratory.

Above T_C thermal chaos overcomes the ferromagnetic coupling. An estimate of the strength of the Weiss field B_W may therefore be obtained from

$$m_{sat} B_W \approx kT_C$$

In the case of CrO_2, for example, T_C is about 400 K while $m_{sat} \approx 2\mu_B$; as the reader may care to verify, B_W is hence *ca.* 300 T. In metallic iron B_W exceeds 700 T.

3. Inherently antiferromagnetic materials

Consider the elementary case of a solid with two geometrically equivalent but oppositely magnetised sub-lattices (each behaving like a ferromagnetically ordered system).[a] Assume for simplicity that the dipoles of one sub-lattice A interact only with those of the other sub-lattice B, and *vice versa.* Then the net magnetic fields acting at the two sites may be expressed as

$$B_A = B_0 - \lambda M_B$$

$$B_B = B_0 - \lambda M_A$$

These are much more common than materials that are intrinsically prone to ferromagnetic ordering.

a MnO is one case in point [section 6(3a)].

The local Weiss fields B_W are here negative owing to the reversed sign of the dipolar coupling. For temperatures above the Néel point T_N, again invoking the Curie law,

$$M_A = \tfrac{1}{2}C(B_0 - \lambda M_B)/\mu_0 T$$

$$M_B = \tfrac{1}{2}C(B_0 - \lambda M_A)/\mu_0 T$$

where the factor of $\tfrac{1}{2}$ is needed because each sub-lattice contains just half of the magnetic sites. The measured susceptibility is then[b]

$$\chi_m = \mu_0(M_A + M_B)/B_0 = C/(T + \lambda C/2\mu_0) \qquad (3.13)$$

with

$$\theta = -\tfrac{1}{2}\mu_0^{-1} C\lambda \qquad (3.14)$$

and such that the susceptibility is *decreased* as a result of the antiferromagnetic coupling.[c]

b A more realistic theory allowing for interactions *within* the individual sub-lattices naturally leads to a more complex expression for χ. [See the discussion of ferrimagnetic ordering in section 6(4a).]

c If this simple model is pursued further, θ is predicted to equal T_N; however, this conclusion proves unreliable in general.

4. Other instances of Curie-Weiss behaviour

A very large number of paramagnetic compounds display susceptibilities that conform closely to a Curie-Weiss law, often with substantial values of $|\theta|$. Unfortunately, despite the foregoing rationalisation, this is not necessarily a sure indicator of strong incipient ordering; nor can the nature of the latent dipolar coupling in a solid safely be deduced from the sign of θ.

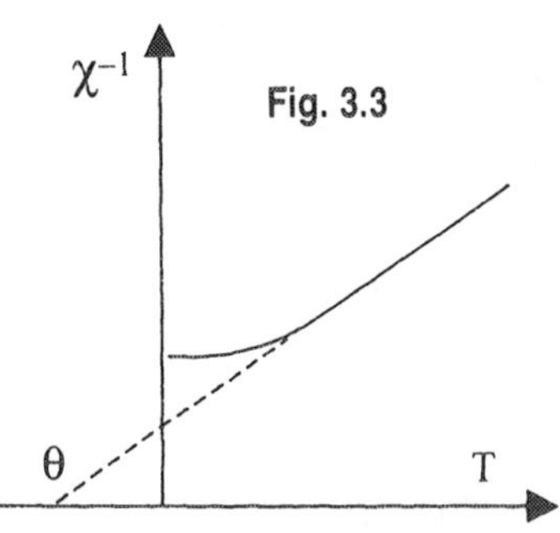

A typical plot of the inverse susceptibility *vs.* T for an octahedral $3d^1$ complex (of *e.g.*, Ti^{3+} or V^{4+}).

The system is assumed to be magnetically dilute (*i.e.* $T_N \sim 0$).

There are innumerable compounds having electronic ground states which exhibit fine structure such that there is a significant thermal distribution between the ground level and low-lying excited levels; in these cases, the T-dependence of the susceptibility often happens to be of essentially Curie-Weiss form when kT is comparable to, or somewhat larger than the ground state splitting. Well-known instances include octahedral complexes of $3d^1$ transition metal ions, such as Ti^{3+} [Fig. 3.3; see § 5.6(5d)], and very many compounds containing lanthanide ions [§ 4.3(5)]. The $C/(T - \theta)$ simulation obscures the fact that the deviation from simple Curie behaviour is due principally to variation with temperature of the parameter C (reflecting a variation in m_{eff}). The Weiss constant θ is thus of no immediate physical significance in these cases.

The Curie-Weiss form of T-dependence is frequently forced upon susceptibility data for no apparent reason, and is then always potentially misleading. Due caution must be exercised in attributing non-zero values of θ to any particular effect.#

Van Vleck, the architect of much of the quantum theory of paramagnetism, described the Curie-Weiss law as "the most overworked formula in the history of paramagnetism" (1973).
[Van Vleck shared the Nobel Prize for Physics in 1977.]

One powerful means of exploring the origin of θ in a paramagnetic phase is to investigate the compound progressively diluted in an isomorphous, diamagnetic host lattice. If |θ| decreases markedly with dilution then it is likely that the Curie-Weiss law genuinely reflects a tendency to magnetic ordering. This same technique of magnetic dilution is often necessary to obtain satisfactory EPR spectra.

3.3 The second order Zeeman effect and its consequences

1. The mixing of electronic states by the external magnetic field

The concept of quantum mechanical mixing in (second order) perturbation theory was introduced in § 2.6(1), albeit in somewhat hand-waving terms. It was subsequently illustrated [§ 2.6(2)] by application to the distortion polarisation of an atom or molecule subjected to an electric field. An analogous, also universal *magnetic* effect occurs in molecules and heavy atoms (Fig. 3.4) – the second order Zeeman effect, whereby the electronic energy levels are displaced by amounts proportional to the square of the strength of the applied magnetic field. This proves the sole source of paramagnetism in closed shell molecules (see below).

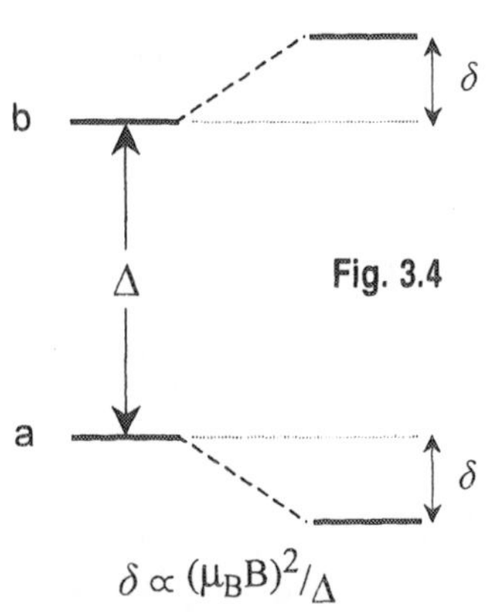

The mixing of molecular electronic states by an applied magnetic field.

(a) might be the ground state and (b) a representative excited state.
(Compare Fig. 2.21.)

Recalling (1.6), the perturbation *P* is here of the form $-m_B B$, where m_B is the projection of the magnetic moment onto the field, and its characteristic energy $\pi \sim \mu_B B$. More specifically, following § 2.3(3),

$$P = -(\mu_B/\hbar)(L_z + g_e S_z)B \tag{3.15}$$

where *z* is the field direction. This reduces to $P = -(\mu_B/\hbar)L_z B$ if Russell-Sanders coupling is applicable because, as it may be shown, there is no mixing between electronic states when S_z is well-defined (quantised). By the same token, the perturbation vanishes if L_z is also well-defined; therefore *light* atoms are not susceptible to the second order Zeeman effect.

However, *heavy* atoms – with the distinction between the spin and orbital angular momentum blurred – are prone to second order mixing of electronic states [§ 4.1(3)] and may therefore in principle display TIP.

2. Temperature-independent paramagnetism (TIP)

(a) Closed-shell species

The ground electronic states of closed-shell atoms and molecules are non-degenerate and there is no scope therefore for a (*first* order) Zeeman splitting and ordinary paramagnetism. But, as a result of the field-induced mixing, the energy of the ground state (initially U^0) acquires the form

$$E = E^0 - \tfrac{1}{2}\alpha B^2 \qquad (3.16)$$

where α is termed the *magnetisability*, equivalent to the polarisability in the analogous electrostatic situation (eqn. 2.47), and also characteristic of the system. The stabilisation $\frac{1}{2}\alpha B^2$ corresponds to δ in Fig. 3.4 (taking (a) to be the ground state), so that $\alpha \equiv 2\delta/B^2$.

α has the dimensions $m^3\ N^{-1}\ A^2$.

It is also sometimes known as the *Van Vleck coefficient.*

Developing (1.6), the magnetic moment induced in the field direction in each molecule may be obtained from

$$m_B = -dE/dB \qquad (3.17)$$

Hence, $$m_B = \alpha B \qquad (3.18)$$

which is inherently *positive.* It may be attributed to distortion of the electron density such as to force a circulation about the field direction equivalent to a (positive) current co-directional with the field; this is sometimes called the *paramagnetic current.* The susceptibility, which is *independent of temperature* if no excited states are thermally accessible, has the molar value

$$\chi_{TIP} = \mu_0 N_A \alpha \qquad (3.19)$$

where the suffix stands for temperature-independent paramagnetism.[a]

[a] – originally described by Van Vleck as *high frequency paramagnetism.* [See § 4.1(1).]

It may be noted that, like diamagnetism, TIP is a *molecular* characteristic rather than a property of the *ensemble* reflecting equilibrium thermal statistics.

TIP is a property of all *molecules* but is not, in general, displayed by atoms; only *heavy* atoms, in which the spin-orbit interaction leaves J_z alone well-defined, may exhibit the effect [§ 4.1(3b)]. The phenomenon is, moreover, entirely absent in closed-shell atoms so that

Closed-shell atoms are *purely* diamagnetic.

As a second order effect, TIP is generally small and almost invariably overwhelmed by diamagnetism. Gaseous N_2, for example, has an experimental susceptibility of $-16.7 \times 10^{-11}\ m^3\ mol^{-1}$ made up, according to *ab initio* calculations, of a diamagnetic contribution of $-54.8 \times 10^{-11}\ m^3\ mol^{-1}$ together with a TIP term of $+38.1 \times 10^{-11}\ m^3\ mol^{-1}$.

However, very strong TIP, sometimes sufficient to outweigh diamagnetism, is generated when the admixed excited states (represented by (b) in Fig. 3.4) are of relatively low energy[b] and $\delta \propto \Delta^{-1}$ correspondingly very large. The compounds in mind would ordinarily be coloured, a prominent instance being the permanganate anion $[MnO_4]^-$, which is paramagnetic (if very weakly); thus $KMnO_4$ has $\chi = +35 \times 10^{-11}\ m^3\ mol^{-1}$ (net of diamagnetism). The purple colour of $[MnO_4]^-$ is due to an optical (*charge-transfer*) transition to the first excited (singlet) state at *ca.* 15,000 cm^{-1}.

[b] – though not so low in energy as to be thermally populated, when the susceptibility would become a function of temperature. [See section 4(1).]

If, in ignorance, TIP were treated as Curie paramagnetism[c] then, recalling formula (3.4), it would be attributed to a magnetic moment

$$m_{eff} = (3kT\alpha)^{\frac{1}{2}} \equiv (T\chi_{TIP}/C')^{\frac{1}{2}} \qquad (3.20)$$

[c] It has been known, even in some prestigious journals.

The apparent m_{eff} associated with TIP vanishes as $T \to 0$.

(actually varying as $\sqrt{T}$) where C' is defined in (3.5b). The diamagnetic susceptibility of $KMnO_4$ has been estimated as -65×10^{-11} m^3 mol^{-1}, so that the $[MnO_4]^-$ anion has $\chi_{TIP} = +100 \times 10^{-11}$ m^3 mol^{-1}. According to (3.20), this is equivalent to a magnetic moment of *ca.* 0.44 μ_B at 300 K.

It should be recorded, in summary, that

A weak TIP is also observed for simple metals due to their itinerant (conduction) electrons [section 8(2a)].

Unpaired electrons are not a *requirement* for paramagnetism.

However, the vast majority of paramagnetic materials do, of course, possess unpaired electrons.

(b) Non-magnetic open-shell systems

The presence of unpaired electrons does not guarantee ordinary (*i.e.* first order) paramagnetism.

There are many *open*-shell systems incorporating strong spin-orbit coupling in which the ground state is a *singlet* showing TIP. In atoms such a ground state (J = 0) arises when the quantum numbers L and S happen to be equal and the outer sub-shell is less than half occupied [recall § 2.5(4)]. It is often described (if ambiguously) as "non-magnetic" since, with **L** and **S** directed exactly in opposition, leaving **J** = 0, the orbital and spin moments cancel and there is no *first* order paramagnetism.

However, there is invariably a considerable second order Zeeman effect involving mixing with the first excited multiplet (J = 1); when the system is sufficiently cold that only the singlet state is occupied, this is manifested by a substantial TIP. One dramatic example is the Eu^{3+} ($4f^6$) ion which has the ground singlet $^7F_0$, the first excited multiplet $^7F_1$ having the relative energy λ (the spin-orbit coupling constant), a mere 355 cm^{-1} in the free ion. As shown in § 4.3(4), the second order stabilisation of the ground singlet is

$$\delta = 8(\mu_B B)^2/\lambda \tag{3.21}$$

such that,[a] remarkably, $\chi_{TIP} = 7.4 \times 10^{-8}$ m^3 mol^{-1}, comparable with the susceptibility of a Curie paramagnet at ordinary temperatures.

a – using $\mu_0 N_A \mu_B^2 = 3.277 \times 10^{-6}$ cm^{-1} m^3 mol^{-1}.

The TIP estimated for Eu(III) compounds corresponds, *via* (3.20), to an apparent magnetic moment of about 3.76 μ_B at 300 K. This dominates the experimental moment at ambient temperatures, though the susceptibility varies with T (and in a complicated fashion) because of thermal population of the upper $^7F_1$ multiplet.[b]

b The details are pursued in § 4.3(4).

Open-shell compounds that show just TIP even at room temperature include, notably, octahedral complexes of Re(III) and Os(IV) – both $5d^4$ with S = 1. The osmium complexes have typically $\chi_{TIP} \gtrsim 10^{-8}$ m^3 mol^{-1}, suggestive of magnetic moments $m_{eff} \gtrsim 1.4\ \mu_B$ at *ca.* 300 K.[c]

c As explained in § 5.6(5), the ground singlet is here analogous to that of the atomic p^2 configuration.

3.4 Paramagnetism in general: open-shell systems

The bulk paramagnetism of open-shell compounds will always have a second order component unless both **S** and **L** are well-defined.[d] Two cases may usefully be distinguished (though the second is a generalisation of the first).

d – as in the case, most obviously, of spin-only paramagnetism [1(2)].

1. Compounds with thermally isolated ground states

In the paramagnetic domain ($T > T_N$ or T_C) the susceptibility here is straightforwardly of the form

$$\chi = \frac{C}{(T-\theta)} + D \tag{3.22}$$

where $D = \chi_{TIP}$ and C is again a constant (invariably >> D) – *a natural extension of the Curie-Weiss law.* Recalling (3.5a), the molar value of the C defines a *characteristic magnetic moment*

$$m_{eff} = (3kC/\mu_0 N_A)^{\frac{1}{2}} \quad (\text{in } \mu_B) \tag{3.23}$$

The dependency of χ^{-1} on temperature when incipient magnetic ordering is negligible (*i.e.* $\theta \approx 0$) is illustrated by Fig. 3.5 (which should be contrasted with Fig. 1.6). Note the curvature in the plot as the TIP term weighs in.

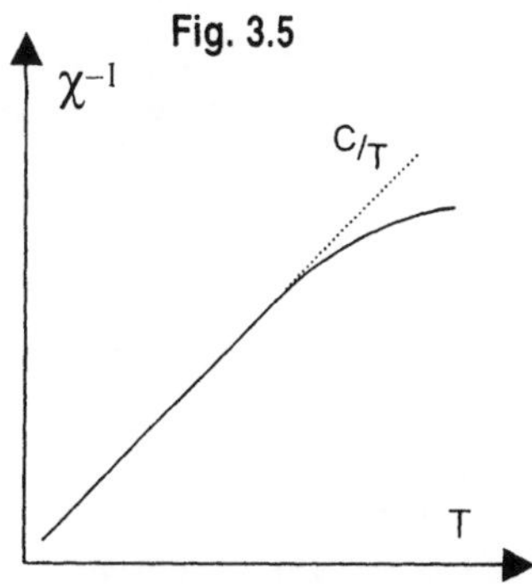

The inverse susceptibility *vs.* T when there is a significant TIP component (eqn. 3.21), but negligible θ.

$$\chi^{-1} = T/(C + DT)$$

(both C and D being constant)

Compounds showing such behaviour have ground states in which the orbital angular momentum **L** is quenched in first order. However, m_{eff} is *invariably different from the spin-only value* through the admixture of excited states by the spin-orbit coupling (acting in second order) which generates some orbital momentum in the ground state. The observed m_{eff} has then a minor *orbital* contribution. Also modified is the experimental g-factor from EPR spectroscopy, which deviates noticeably from the free spin value g_e.

Take, for example, the tetrahedral Co(II) complex anion, $[CoCl_4]^{2-}$. The $3d^7$ Co^{2+} ion has *three* unpaired electrons and therefore a total spin $S = \frac{3}{2}$. The susceptibility of the caesium salt $Cs_2[CoCl_4]$ in the range 300 – 80 K, corrected for diamagnetism, can be simulated quite accurately by (3.22) with $C = 31.2 \times 10^{-6}$ m^3 mol^{-1} K, $D = 0.85 \times 10^{-8}$ m^3 mol^{-1} and Weiss temperature $\theta \approx 0$.[a] (θ is certainly no less than –10 K.) According to (3.23), the characteristic moment $m_{eff} = 4.45\,\mu_B$, here considerably *greater* than the spin-only value of $g_e\sqrt{[S(S+1)]}\mu_B = 3.88\,\mu_B$.[b] Naïvely interpreted, the magnitude of the susceptibility at 300 K (11.24×10^{-8} m^3 mol^{-1}) suggests a larger *apparent* moment of 4.63 μ_B. At ordinary temperatures, TIP contributes some 8 % to the susceptibility.[c]

EPR measurements on $[CoCl_4]^{2-}$ diluted in diamagnetic $Cs_2[ZnCl_4]$ yield an average g-value of 2.29, much in excess of g_e. This points to a first order moment of $g\sqrt{[S(S+1)]}\mu_B = 4.43\,\mu_B$, in good agreement with the value determined by analysis of the temperature dependence of the susceptibility.

a However, the extended Curie-Weiss law doesn't apply at very low temperature because of fine splitting of the ground energy level.

b m_{eff} may however be *less* than the spin-only expectation (and $g < g_e$) when the *d* sub-shell is less than half-filled – *e.g.* octahedral complexes of V(II) and Cr(III), both $3d^3$. [See Table 5.3.]

c The TIP is naturally less important at low temperatures since the first order component of χ varies with the inverse of temperature. (Recall also eqn. 3.19).

2. Complex paramagnetism – the general case

The parameter C in (3.22) is by no means always constant. First, this is because there may well be thermal population of low-lying excited states with, in general, different magnetic moments. Invoking Botzmann statistics, a model two-level system (as per Fig. 3.4) will display a first order susceptibility χ_1 expressible in molar terms as

$$\chi_1 = \frac{\mu_0 N_A}{3kT} \frac{[\omega_a m_a^2 + \omega_b m_b^2 \exp(-\frac{\Delta}{kT})]}{[\omega_a + \omega_b \exp(-\frac{\Delta}{kT})]} \tag{3.24a}$$

where the ω denote the degeneracies of the states.[d] Secondly, the excited states admixed with the ground state in the second order Zeeman effect may themselves be thermally populated, when the resulting contribution to the susceptibility will also be dependent upon temperature. Pursuing the two-level system, the applied field induces a second order susceptibility

d The Curie limit is again assumed. [Recall (3.4) earlier.]

Note that $m_B = -\alpha B$ in the upper state.

$$\chi_2 = \frac{\mu_0 N_A \alpha[\omega_a - \omega_b \exp(-\frac{\Delta}{kT})]}{\omega_a + \omega_b \exp(-\frac{\Delta}{kT})} \qquad (3.24b)$$

The two components of χ_{mol} are not distinguishable empirically; in practice, they are merged in an observed susceptibility of notional form

$$\chi = \frac{C(T)}{(T-\theta)} + D \qquad (3.25)$$

In *complex paramagnetism* (a.k.a. *non-Curie paramagnetism*) m_{eff}, like C, is a function of temperature.

where D is the TIP term involving high-lying, *unpopulated* excited states [as in (3.21)]. This is here termed complex paramagnetism. The theoretical analysis of susceptibility data in terms of (3.24) is never straightforward, and often impossible without simplifying assumptions.

A major source of fine structure in the ground state that leads to C being a function of temperature (also m_{eff} naturally) is the spin-orbit interaction; this may cause a significant first order splitting both in atoms,[a] and also in molecules, if the orbital momentum **L** is incompletely quenched.

[a] – recall § 2.5

The classic instance is gaseous NO, which conforms to the two-level model (both levels being *doublets*).[b] The variation of m_{eff} with temperature is shown in Fig. 3.6. The ground level happens to be non-magnetic – here through cancellation of opposing spin and orbital *moments*[c] – and so m_{eff} is expected to decline in magnitude, ultimately to zero, as the temperature is continually lowered. Progress is interrupted by condensation of the gas but, the trend $m_{eff} \rightarrow 0$ has been confirmed in the NO/β-quinol clathrate "compound" (essentially a gas-solid solution).

[b] NO has a solitary unpaired electron occupying a π^* molecular orbital.

The ground state is split by the spin-orbit coupling into two doublets with a separation $\Delta = \zeta$, where ζ is the spin-orbit coupling constant for the π^* MO.

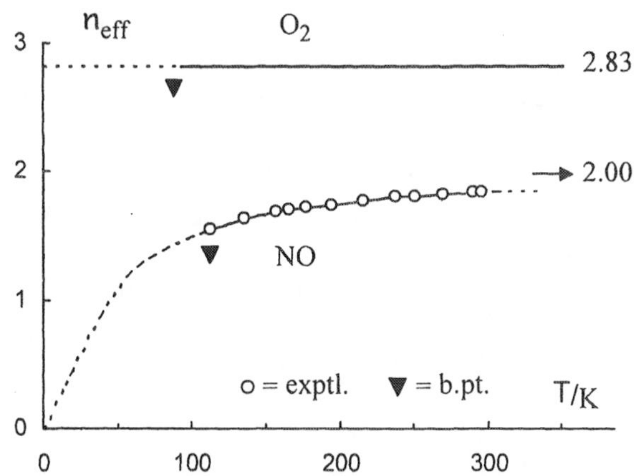

Fig. 3.6

Temperature variation of $n_{eff} = m_{eff}/\mu_B$ for gaseous NO ($S = \frac{1}{2}$) contrasted with the simple Curie behaviour of O_2 (S = 1).

The dotted lines are theoretical [after Van Vleck 1932)].

NO and O_2 liquefy at 121 K and 90 K, respectively. Liquid NO shows partial dimerisation and otherwise, as in liquid O_2, strong antiferromagnetic interactions.

In linear molecules the orbital angular momentum is tied to, and quantised with respect to the figure axis (z), the perpendicular components being quenched. An electron in a π orbital has $\ell_z = \pm\hbar$ and, in the ground multiplet of NO (denoted $^2\Pi_{\frac{1}{2}}$), the spin angular momentum is driven antiparallel to the orbital momentum and the only well-defined angular momentum is $j_z = \ell_z - s_z = \pm\frac{1}{2}\hbar$. The first order magnetic moment, confined to the molecular axis, is therefore $m_z = -(\ell_z + 2s_z)(\mu_B/\hbar) = -[(\pm 1) - 2(\pm\frac{1}{2})]\mu_B = 0$, the spin and orbital contributions cancelling exactly. A comparable non-magnetic

[c] This non-magnetic ground state is different in kind from that encountered in compounds of Eu(III) [section 3(2b) and § 4.3(4)], where it is the angular momenta that cancel.

ground multiplet state is encountered in octahedral complexes of d^1 cations such as Ti^{3+} [§ 5.6(5d)].

The successful interpretation of the susceptibility data for NO, due to Van Vleck (1932),[$] was one of the early triumphs of the quantum theory of molecular magnetism.

$ Van Vleck's theory gives

$$n_{eff} = 2\left\{\frac{1-e^{-y}+ye^{-y}}{y(1+e^{-y})}\right\}^{1/2}$$

where $x = \zeta/kT$.

Substantial splitting of an orbitally degenerate molecular ground state can occur for other reasons, notably because of distortion of the molecule to a lower symmetry. Very *fine* splittings due to second (or higher) order effects, existing in the absence of a magnetic field,[a] also have a marked affect on the susceptibility at low temperatures (and especially on EPR spectra).

a – hence commonly referred to as *zero-field splittings* [§ 5.6(1b) & (4c)].

3.5 The diversity of bulk magnetic properties – a roundup

Bulk magnetism takes a wide variety of forms and, as witnessed by Table 3.1 (overleaf), the particular element *iron* happens to display practically the whole panoply. It would be a useful exercise for the reader to compile comparable tables for other transition metals – *e.g.* V, Cr, Mn and Co. (Some research will be necessary!)

3.6 Further to cooperative magnetism

This section elaborates upon the elementary review of cooperative magnetism in § 1.3(3), with particular reference to the nature of dipolar coupling in solids. Attention is focussed on *non-metallic* substances – that is, insulators or semiconductors.[b]

b An adequate treatment of magnetic coupling in metallic materials is beyond the scope of the present work.

As pointed out in § 1.5(3), the direct, through-space coupling of magnetic dipoles is too weak to account for what is generally observed. Its effect is apparent only when no other mechanism is available and, in those cases, magnetic ordering occurs only at *extremely* low temperatures. A well-known instance is the cerium(III) salt, $[Ce(NO_3)_6]_2[Mg(OH_2)_6]_3.6H_2O$,[b] which fails to order spontaneously until 1.9 mK!

c – often known as "CMN" or "CeMN". The Ce^{3+} ions are spaced on average a distance of 857.5 pm apart; the field due to one ion at the site of a neighbour is then about 29 G per unit dipole moment.

It has long been clear that strong dipolar coupling is due principally to *exchange* interactions of some form.

1. Exchange coupling

The energy of interaction between two magnetic dipoles may be expressed as

$$U_{ij} = -\mathcal{J}_{ij}\mathbf{S}_i.\mathbf{S}_j/\hbar^2 \qquad (3.26)$$

where $\mathcal{J}_{ij}$ is the *exchange* coupling constant characteristic of the magnetic centres *i* and *j*. The $\mathcal{J}_{ij}$ are predominantly, or exclusively positive in a ferromagnetic lattice. The dipoles are here assumed to be due simply to the many-electron spins **S** of the various centres but, where appropriate, the interaction may instead be formulated in terms of their total angular momenta **J**.

In many accounts the exchange parameter is instead written as $2\mathcal{J}_{ij}$.

As originally conceived, by Heisenberg (1928), the coupling is due to quantum mechanical exchange depending on the *direct* overlap of relevant orbitals at the two sites; this may lead to ferromagnetic or antiferromagnetic (negative $\mathcal{J}$) ordering according to circumstances. However, it transpires that this simple mechanism is not usually powerful enough to account for the very substantial coupling existing in many substances. In the case of (insulating) compounds of the transition elements, it is frequently necessary to invoke

Table 3.1 The magnetic properties of iron and its compounds

Fe metal	ferromagnet	$T_C = 1043$ K ($m_{sat} = 2.22\ \mu_B$)
$LaFe_2Al_{10}$	Pauli paramagnet*	
Fe_3P	ferromagnet	$T_C = 716$ K ($m_{sat} = 1.84\ \mu_B$)
FeO	antiferromagnet	$T_N = 198$ K
FeS	antiferromagnet	$T_N = 613$ K
FeF_2	antiferromagnet ($T_N \approx 80$ K) / complex paramagnet$^{\alpha}$	
$Ca(Fe)F_2$ 2 %	complex paramagnet	
FeF_3	antiferromagnet	$T_N = 394$ K
$FeCl_2$	metamagnet ($T_N = 24$ K) / complex paramagnet$^{\alpha}$	
$FeCl_3$	simple paramagnet	$m_{eff} = 5.73\ \mu_B$, $\theta = -11.5$ K
$[Fe(OH_2)_6]^{2+}$	complex paramagnet	$m_{eff} = 5.2 - 5.4\ \mu_B$ at ~ 300 K
$[Fe(OH_2)_6]^{3+}$	simple paramagnet	$m_{eff} = 5.90\ \mu_B$ at ~ 300 K
$[Fe(CN)_6]^{4-}$	diamagnet	
$[Fe(CN)_6]^{3-}$	complex paramagnet	$m_{eff} = 2.25\ \mu_B$ at ~ 300 K
$[FeCl_4]^{2-}$	simple paramagnet	$m_{eff} = 5.4\ \mu_B$
$[FeCl_4]^{-}$	simple paramagnet	$m_{eff} = 6.00\ \mu_B$
Fe(1-*nor*)$_4$	diamagnet	
Fe(*mes*$^{\#}$)$_2$	simple paramagnet	$m_{eff} = 5.18\ \mu_B$
α-Fe_2O_3	antiferromagnet$^{\beta}$	$T_N = 950$ K
γ-Fe_2O_3	ferrimagnet	$T_{fN} = 856$ K $m_{sat} \approx 2.45\ \mu_B$ $^{\gamma}$
Fe_3O_4	ferrimagnet	$T_{fN} = 858$ K $m_{sat} \approx 4.1\ \mu_B$ $^{\gamma}$
$Y_3Fe_5O_{12}$	ferrimagnet	$T_{fN} = 553$ K $m_{sat} = 4.96\ \mu_B$ $^{\gamma}$
$LaFeO_3$	antiferromagnet	$T_N = 750$ K
$CaFeO_3$	paramagnetic metal above 115 K, $m_{eff} \approx 5.6\ \mu_B$ (?) #	
$K_2[FeO_4]$	simple paramagnet	$m_{eff} = 2.86\ \mu_B$
$K_3[FeO_2]$	simple paramagnet	$m_{eff} = 3.60\ \mu_B$
$Na_4[FeO_3]$	diamagnet	
$Ba_3[FeN_3]$	complex paramagnet	[$S = \frac{1}{2}$?]
$(C_5H_5)_2Fe$	diamagnet	
$Fe(CO)_5$	diamagnet	
$Fe_2(CO)_9$	diamagnet$^{\delta}$	
haemoglobin	complex paramagnet	$m_{eff} \approx 4.95\ \mu_B$
oxyhaemoglobin	diamagnet$^{\varepsilon}$	
haematin	complex paramagnet	$m_{eff} \approx 4.47\ \mu_B$

T_C = Curie point, T_N = Néel point, T_{fN} = ferrimagnetic Néel point, and θ = Weiss constant. * See section 8(2a).
1-*nor* = 1-norbornyl; *mes*$^{\#}$ = 2,4,6-*tris*(^{t}Bu)phenyl.

The table provides a suitable impression of the range of magnetic behaviour encountered in chemical compounds.

α FeF_2 and $FeCl_2$ have $m_{eff} \approx$ 5.6 μ_B and 5.4 μ_B, respectively, at ambient temperature.

β Also harbours a weak ferromagnetism between 265 K and 950 K.

γ The m_{sat} are here quoted per formula unit.

\# antiferromagnetic semiconductor below 115 K.

δ The diamagnetism of $Fe_2(CO)_9$ and $Fe_3(CO)_{12}$ was important early evidence for the existence of direct Fe–Fe bonds in these compounds.

ε Magnetic measurements had a major role in the early characterisation of haemoglobin and its various derivatives.

instead, or additionally, an *indirect* pathway for exchange coupling between metal ions that involves the intervening ligands and is mediated by covalency. In its simplest forms, this is known as *superexchange* [section (3)] and tends typically to promote antiferromagnetic coupling. But, whatever the details, the energy of the dipolar interaction may usually still formally be written as (3.26) for semi-empirical purposes.

The exchange interaction between a pair of open-shell ions declines rapidly with the distance between them, indeed more rapidly than the through-space interaction, though the latter is usually the smaller at short range.

2. Formulation of the Weiss field in terms of exchange coupling

The potential energy of any particular centre i is given by

$$U_i = -\mathbf{S}_i.\sum_{j\neq i} \mathcal{J}_{ij}\mathbf{S}_j/\hbar^2 \equiv -\left(\frac{\mathbf{m}_i}{g_e\mu_B}\right).\sum_{j\neq i} \mathcal{J}_{ij}\left(\frac{\mathbf{m}_j}{g_e\mu_B}\right)$$

If the system is ferromagnetic and its magnetic centres are equivalent, and only interactions between contiguous centres are important, then the potential energy of each site is simply

$$U = -\mathcal{J}\left(\frac{\mathbf{m}}{g_e\mu_B}\right).\left(\frac{z\langle\mathbf{m}\rangle}{g_e\mu_B}\right)$$

where $\mathcal{J}$ is the common coupling energy, z the number of nearest neighbours and $\langle\mathbf{m}\rangle$ their average magnetic moment.

Now $\langle\mathbf{m}\rangle$ is simply $\mathbf{M}/\mathrm{N}$, where N is the number of magnetic sites *per unit volume*. Therefore the energy may be formulated as

$$U = -\left[\frac{z\mathcal{J}}{N(g_e\mu_B)^2}\right]\mathbf{m}.\mathbf{M} \equiv -\mathbf{m}.\mathbf{B}_W \qquad (3.27)$$

whence the Weiss constant, defined by (3.9), is given by

$$\lambda = z\mathcal{J}/N(g_e\mu_B)^2 \qquad (3.28)$$

It is directly proportional to the exchange coupling constant $\mathcal{J}$. It follows, recalling (3.12), that

In a simple lattice, the Weiss constant λ is proportional to $\mathcal{J}$.

$$\theta = \lambda C/\mu_0 = N m_{eff}^2\lambda/3k = z\mathcal{J}S(S+1)/3k \qquad (3.29)$$

If the solid is instead antiferromagnetic (the $\mathcal{J}$ being largely negative) and, as assumed in section 2(3), involves simply two equivalent sub-lattices, then the corresponding expressions for λ and the Weiss temperature contain an additional factor of $-\frac{1}{2}$. [Recall (3.14).]

3. Superexchange#

\# Kramers (1934), Anderson (1950) and Van Vleck (1951).

[Anderson shared the Nobel prize for Physics with Van Vleck in 1977.]

The dominant source of dipolar coupling in non-metallic (*i.e.* most) transition metal compounds is superexchange, an indirect mechanism dependent on the simultaneous covalent bonding of the metal ions with their bridging ligands. It leads most commonly to antiferromagnetic ordering.

(a) Antiferromagnetic compounds

Consider a linear array such as $M^{n+}–O^{2-}–M^{n+}$ (as may occur, for instance, when MO_6 octahedra share vertices), assuming for simplicity that the indirectly coupled metal ions both possess just one electron.

The consequences of extended covalent bonding for magnetic coupling are most easily appreciated by reference initially to the ionic limit (Fig. 3.7, overleaf). The overlap between each metal d-orbital and the (filled) p-orbital of the intervening ligand leads to partial donation of electron density to the metal site. Suppose that, as in Fig. 3.7, the unpaired electron on the left-hand site (X) has its spin "up" (↑): then, by the Pauli exclusion principle, the electron density transferred from the ligand *p*-orbital must have the opposite spin (↓). To preserve spin neutrality at the ligand site (as ↑↓), the electron density donated simultaneously to the other metal site (Y) must be of up-spin (↑) character. But this is possible only if the extant electron of the second metal ion has its spin down (↓). One-centre exchange effects thus dictate, through covalency, an antiferromagnetic coupling between the spins of the bridged cations.#

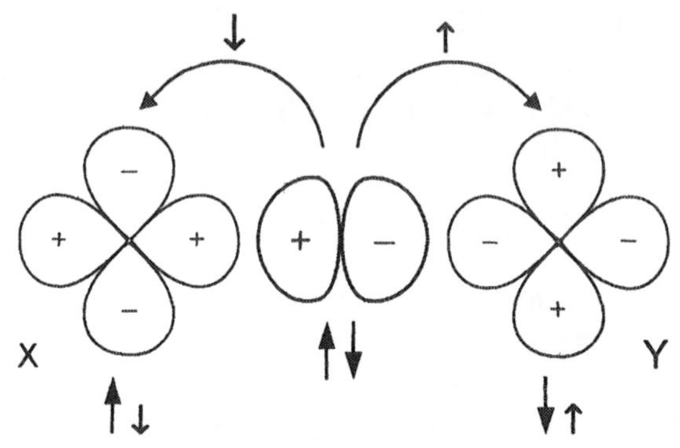

Fig. 3.7

A simple view of superexchange in a linear M–L–M system.

The exchange is here mediated by σ-bonding with *p*-orbitals on the bridging ligand.

The argument is couched in terms of σ-bonding but a comparable, if generally weaker interaction occurs *via* π overlap of orbitals (as below).

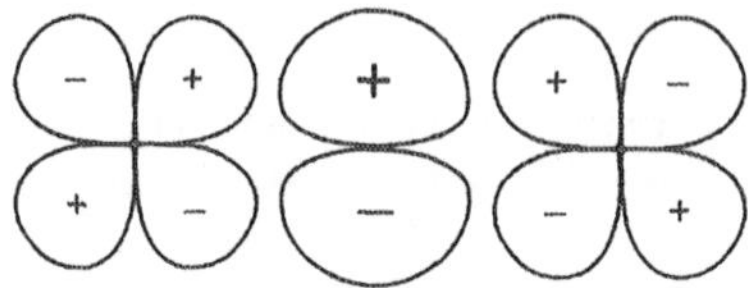

Extended π-bonding in a linear M–L–M system.

Table 3.2 Néel temperatures T_N (/K) for selected oxides and fluorides showing antiferromagnetic ordering.

VO	117					VF_3	<4
		CrF_2	53	$KCrF_3$	<77	CrF_3	80
MnO	122	MnF_2	67	$KMnF_3$	88	MnF_3	45
FeO	198	FeF_2	79	$KFeF_3$	113	FeF_3	394
CoO	291	CoF_2	40	$KCoF_3$	114	CoF_3	460
NiO	525	NiF_2	83	$KNiF_3$	275		
α-Cr_2O_3	307	$LaCrO_3$	320				
α-Fe_2O_3	953	$LaFeO_3$	750				

There is more than a hint of the influence of covalency in the T_N values assembled in Table 3.2. In particular, there is an overall trend of increasing T_N across the first transition series as the nuclear charge Z increases. Moreover, the T_N tend to be higher for M(III) compounds than those of M(II), and also higher for oxides than fluorides. However, other factors are involved, not least the details of the *d*-electron configuration, such that a correlation between ordering temperature and the expected (or, in some cases, known) degree of covalency is not always observed.

The strength of exchange coupling *via* a bridging ligand proves highly dependent on the M–L–M bond angle. The superexchange pictured in Fig. 3.7 is weakened through diminished orbital overlap when the bond angle is reduced towards 90°. Concomitantly, the opportunity for a significant *ferro*magnetic interaction arises [section (b) below].

In compounds that adopt the rocksalt (NaCl) structure, the upshot is that antiferromagnetic coupling between *nearest* neighbour cations – whether direct or through superexchange *via* a 90° bridge – is distinctly weaker than that between *next-nearest* neighbours connected by a linear M–L–M bridge. This has the consequence that the *magnetic* unit cell of the monoxides MO, the irreducible structural unit that conveys the nature of long range order, has twice the dimensions of the ordinary crystallographic unit cell (Fig. 3.8). In MnO the magnetic moments are orientated in layers perpendicular to the (111) direction, a body-diagonal of the cubic unit cell. As the figure shows, the magnetisation is reversed in successive layers.

Fig. 3.8

The magnetic unit cell of MnO.

This is twice the size of the "chemical" unit cell of the f.c.c. rocksalt lattice. [The intervening oxide ions are not shown.]

The antiferromagnetic superstructure in Fig. 3.8 is revealed by *neutron diffraction*, arising from the scattering both by atomic nuclei and also (the neutron having an intrinsic magnetic moment) by the electronic magnetic moments of the atoms.#

See, *e.g.* Cheetham (1987) and refs. therein.

(b) Ferromagnetic coupling through superexchange

There are circumstances in which superexchange leads to a ferro- rather than antiferromagnetic interaction between cations; one such, already alluded to, is the 90° M–L–M configuration (Fig. 3.9). In what follows, it is again assumed that the cations carry just one d electron each.

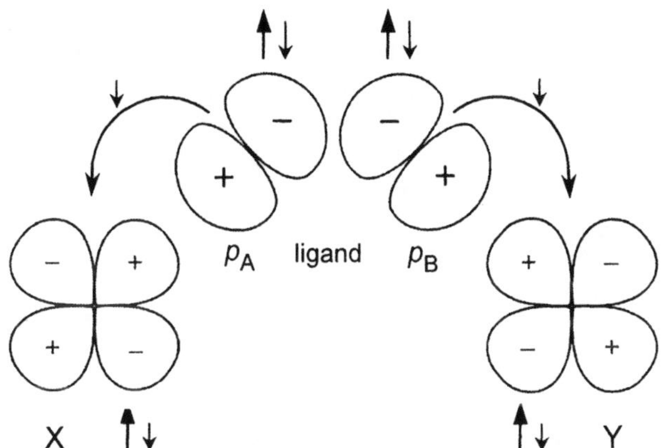

Fig. 3.9 Covalency driven ferromagnetic coupling in a 90° M–L–M array.

[The orthogonal ligand p_σ orbitals A and B are separated in the interests of clarity.]

The metal d-orbitals here overlap with *different* p-orbitals on the ligand. As in Fig. 3.7 σ-bonding between a half-occupied d-orbital at site X and a filled p-orbital (p_A) of the ligand results in the transfer of electron density with spin opposed (↓) to that of the metal electron; the p-orbital involved is accordingly deficient regarding spin of that sense. Spin correlation within the ligand, which always favours the parallel alignment of spins in different orbitals, then encourages an equivalent spin configuration in the second, orthogonal p-orbital (p_B) which is achieved by the donation to the d-orbital on the second metal ion (Y) of electron density having the same spin (↓) as that transferred to the first metal ion (X). The exclusion principle ensures in turn that the original d electron of the second cation (Y) assumes a spin anti-parallel to that of the incoming electron density, but parallel to the spin of the

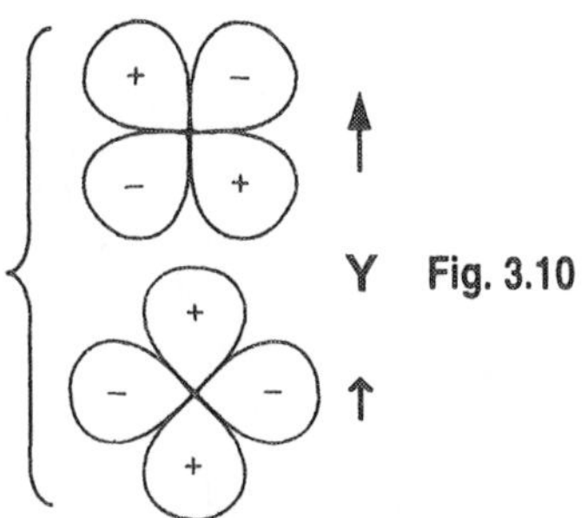

Fig. 3.10

Ferromagnetic coupling in a 180° M–L–M system (to be viewed in conjunction with Fig. 3.7).

The upper orbital, of π character with respect to the bridge, has zero overlap with the ligand p_σ orbital.

d electron of the first cation. The one-centre exchange effects thus operate here to promote ferromagnetic coupling.

Approximately 90° ligand bridges are commonplace in, for example, the *layer* lattices of chlorides, *etc.* As mentioned in § 1.3(3c), the metamagnetic $FeCl_2$ shows ferromagnetic ordering within the metal ion layers. The 90° configuration is also occasionally encountered in oxides – *e.g.* $LiNiO_2$, which (unusually) is a ferromagnetic insulator.

A ferromagnetic outcome is also to be anticipated in the case of a linear M–L–M array when the d_σ orbital at the second metal site (Y) happens to be empty but has an occupied *orthogonal* partner (Fig. 3.10). Spin correlation within the cation then drives its electron spin parallel to that of the transferred p_σ density and therefore parallel also to the spin of the *d* electron on the first site (X). Compounds of this general type include Rb_2CrCl_4 and $LaMnO_3$, both insulators (with T_C = 57 K and 100 K, respectively). Another, more recent example is $CsNi^{II}[Cr^{III}(CN)_6].2H_2O$ (T_C = 90 K), with 180° Ni–NC–Cr linkages. [Contrast the ferrimagnetic cyanides in section (4c).]

A quite different mechanism for ferromagnetic coupling, known usually as *double exchange*, is described in connection with Fe_3O_4 [section (4b)].[a]

[a] This is confined to certain metallic or near-metallic compounds of mixed oxidation state.

4. Ferrimagnetism revisited

(a) The transition to ferrimagnetism and the Néel model

The inverse susceptibilities of compounds inclined to order ferrimagnetically [§ 1.3(3d)] have typically a temperature variation of the form sketched in Fig. 3.11. The transition occurs at the ferrimagnetic Néel temperature T_{fN}.[b] The asymptotic paramagnetic behaviour is of Curie-Weiss form with *negative* θ, as with antiferromagnetism [Fig. 3.2(a)], but the progress towards magnetic order otherwise bears no resemblance to that of either ferromagnetic or antiferromagnetic ordering. Particularly striking is the rapid development of $1/\chi$ immediately above T_{fN}.

[b] – also often known as the *ferrimagnetic Curie temperature* and simply denoted T_C.

Fig. 3.11
The switch from paramagnetism to ferrimagnetism.

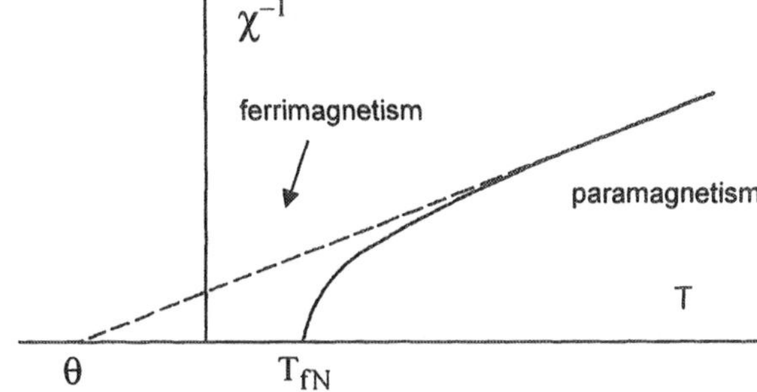

This distinctive behaviour was explained by Néel (1948)[c] who assumed that *the exchange interactions are predominantly, or exclusively antiferromagnetic*. Consider, as in section 2(3), a system having two sub-lattices (A and B) but instead allow them to be *inequivalent* in some essential respect – most obviously, perhaps, by harbouring different metal ions.[d] The internal fields experienced by the individual magnetic sites may be formulated as

$$B_A = B_0 - \lambda M_B - \lambda_A M_A \quad \text{and} \quad B_B = B_0 - \lambda M_A - \lambda_B M_B$$

[c] L Néel, Nobel laureate in Physics in 1970.

[d] – or having the same ion in crystallographically distinct locations that differ in number, or in another way. (Consider, for example, the case of YIG = $Y_3Fe_5O_{12}$ below.)

where the interactions between magnetic centres of the *same* kind are now included through the additional Weiss constants λ_A and λ_B.

It may then be shown, proceeding essentially as in sections 2(2,3) above, that the inverse (volume) susceptibility is of the form

$$\chi_m^{-1} = \frac{T}{C} + \rho - \frac{\sigma}{(T-\varphi)} \qquad (3.30)$$

where the parameters ρ, σ and φ are functions of the three Weiss constants involved and also the Curie constants, C_A and C_B, characteristic of the ions.[a] The latter are defined by

$$C_X = \mu_0 N_X (m_{eff}^2)_X / 3k \qquad (X = A \text{ or } B) \qquad (3.31)$$

(with the unit [C] = K) the N_X being the numbers of ions per unit volume at the different sites; the quantity C in eqn. 3.30 is equal to $C_A + C_B$. The equation is a *hyperbola*, though only the range of physical interest is drawn in Fig. 3.11. Its asymptote (as $T \to \infty$), represented by the dotted line in the figure, is

$$\chi_m^{-1} = \frac{T}{C} + \rho \qquad (3.32)$$

Furthermore, and crucially, Néel made the assumption that *the magnetic coupling between the different sub-lattices is much stronger than that within the sub-lattices, i.e.* that $\lambda > |\lambda_A|, |\lambda_B|$. It then follows that the χ^{-1} intercept ρ is ordinarily positive, even if λ_A or λ_B happens to be negative (because the coupling within the sub-lattice is ferromagnetic).[b] The asymptotic line intercepts the T-axis at $-\rho C \equiv \theta$ and (θ being negative) is equivalent to a Curie-Weiss law $\chi = C/(T+\theta)$.

Employed semi-empirically, the above theory enjoys widespread success in its applications to ferrites. The quality of the interpretation that is possible, without elaboration of the model, is illustrated by Fig. 3.12.

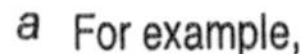

[a] For example,

$$\rho = \frac{1}{\mu_0 C^2}(C_A^2\lambda_A + C_B^2\lambda_B + 2C_AC_B\lambda)$$

See *e.g.* Morrish (1965) for the details.

[b] – as proves to be the case in one of the sub-lattices of Fe_3O_4 [see (b)(v)].

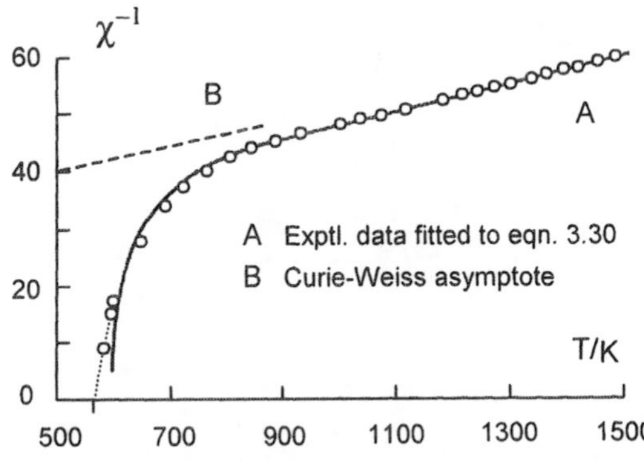

Fig. 3.12 Analysis of the inverse susceptibility of $Y_3Fe_5O_{12}$.
The parameters ρ and C may be determined from the asymptotic behaviour of χ^{-1}; these can then be used in the fit of the experimental data to (3.30).

The figure describes yttrium iron garnet (YIG), of composition $Y_3Fe_5O_{12}$ ($\equiv 3Y_2O_3.5Fe_2O_3$) and with $T_{fN} = 553$ K, which is important both to technology and also, given its comparatively simple structure, to the conceptual framework of the subject. The garnets have a cubic structure offering the Fe(III) ions alternative octahedral and tetrahedral sites (often distinguished as B and B', respectively),[c] the five ions of the formula unit actually being

[c] The structure may be expressed in the form $A_3B_2B'_3O_{12}$. The A-sites (8-coordinate) are occupied by the diamagnetic Y^{3+} ions.

distributed between these two sites in the ratio 2:3. The antiferromagnetic coupling proves to be dominated by that between the Fe(III) ions on *different* sites – as might have been anticipated since [recalling section (3a)] these are the most closely spaced while connected by a wide angle (127°) B–O–B' bridge.

The B and B' sub-lattices therefore have oppositely directed moments throughout the structure. The expected saturation moment m_{sat} is 5.01 μ_B (that of *one* Fe^{3+}), the net magnetisation being dominated by the more numerous Fe(III) with tetrahedral coordination. The experimental figure for $Y_3Fe_5O_{12}$ is 4.96 μ_B per formula unit.

a The nature of the magnetic order in YIG has been confirmed by neutron diffraction.

Further to Fig. 3.12, a full analysis of the magnetic data allows estimates of the Weiss constants λ and also, by further development of the theory in section (2), estimates of the exchange coupling constants J. The latter have been reported to be: B–B' –23.6, B'–B' –2.4, and B–B –0.2, all in cm^{-1}.[b]

b The exchange coupling constant is distinctly larger in ferromagnetic metals, typically ~ 100 cm^{-1}.

(b) Spinels incorporating transition ions

(i) Structure and composition

Essentially of stoichiometry AB_2O_4, spinels have a cubic structure based upon a c.c.p. lattice of oxide ions, the metal ions A and B occupying respectively 1/8 and 1/2 of the tetrahedral and octahedral interstices. The unit cell is large, containing eight formula units. In a so-called *normal* spinel, *e.g.* the mineral prototype *spinel* $MgAl_2O_4$, the A sites are occupied by divalent cations while the B sites accommodate trivalent ions (Fig. 3.13). An *inverse* spinel, often formulated $B(AB)O_4$, has instead the divalent ions sharing the octahedral holes, with half the trivalent ions occupying the tetrahedral holes. A prominent example is, of course, *magnetite*, $Fe^{III}(Fe^{II}Fe^{III})O_4 \equiv Fe_3O_4$.

The labels A and B are here used in a particular, crystallographically specific sense, as opposed to their general use in section (a).

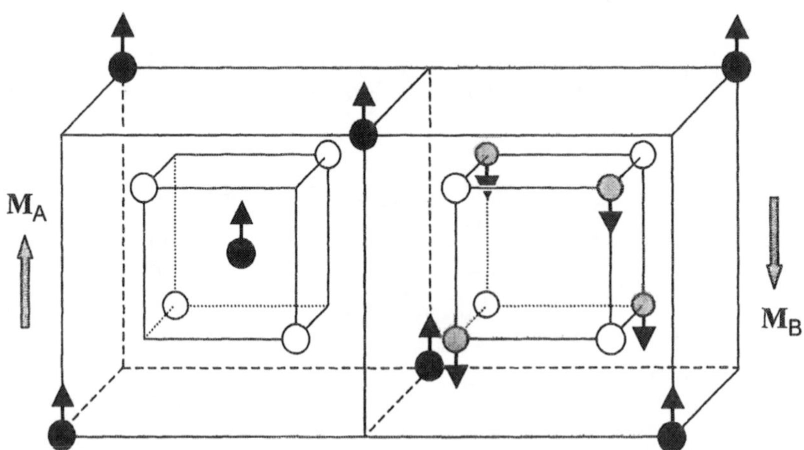

Fig. 3.13 Ferrimagnetic ordering in spinels AB_2O_4.
Two representative octants of the cubic unit cell are shown; these alternate throughout the cell.

The preferred distribution in any particular case is often markedly influenced by *ligand field effects*,[c] described in § 5.2(3). The structure adopted is also often dependent on the method of preparation and the thermal history of the sample; it may not be the thermodynamically most stable form.

c The ligand field splittings of the metal *d* orbitals affects the relative energies of cations at tetrahedral and octahedral sites. Fe^{2+} proves to be the more stable in an octahedral site, while Fe^{3+} has no pref-

Intermediate or hybrid structures also occur. If the two types of cation are randomly distributed the structure is of the form $A_{1/3}B_{2/3}(A_{2/3}B_{4/3})O_4$, which may be described as 2/3 or 66.7% inverted. A huge variety of substituted spinels is also known – for example, the system $Zn_\delta Fe^{III}{}_{1-\delta}(M^{II}{}_{1-\delta}Fe^{III}{}_{1+\delta}$ – many of which have significant applications.

(ii) Exchange interactions in the spinel lattice

Practically all spinels that have open-shell cations at *both* the A and B sites display ferrimagnetism.[a] There are three distinct, short-range exchange interactions in the lattice, namely those between: (1) neighbouring tetrahedral sites (A–A); (2) adjacent octahedral sites (B–B); and (3) the neighbouring sites of different kind (A–B). As pointed out above, Néel asserted that the exchange coupling is invariably antiferromagnetic, but that the A–B interaction is considerably stronger than either the A–A or B–B interactions.

[a] Co_3O_4 is perhaps an exception, the Co(III) ion being non-magnetic (a.k.a. *low-spin* – see § 5.3).

The relative magnitude of the different exchange terms can be rationalized in terms of the superexchange mechanism. The interaction is likely to be strong when the coupled ions are close and the M–O–M configuration linear [section (3a)]. Scrutiny of Fig. 3.13 suggests that the optimal pathway for superexchange is the A–O–B (126°) link; the B–O–B (90°) bridge is likely to be less efficient, and the closest A–O–A connection (79°) even less so because the ions are more widely separated.[b]

[b] The angle of the A–O–B bridge is almost identical with that of the B–O–B' linkage in the garnets [*cf.* section (a) above].

(iii) The spinel ferrites MFe_2O_4

The technologically important ferrites MFe_2O_4 (Table 3.3) have typically the inverse configuration $Fe^{III}(M^{II}Fe^{III})O_4$. If the A–B couplings are indeed dominant then, at saturation, the magnetisation of the B sublattice (M^{2+} & Fe^{3+}) is oppositely directed to that of the A sublattice (Fe^{3+}) and, using a self-evident notation, the net saturation moment per formula unit is

$$m_{sat} = m_M + (m_{Fe})_B - (m_{Fe})_A = m_M \tag{3.33}$$

the saturation moment of Fe^{3+} (5.01 μ_B) again being independent of location. The moments calculated on this basis, assuming spin-only values of the M^{2+} moments (m_M), are compared with the experimental moments in Table 3.3. The agreement is tolerable given that, from study of the paramagnetism of the octahedrally coordinated M^{2+} in other compounds, the m_M for Fe^{2+} – Cu^{2+} are very likely to be larger than the spin-only values. Moreover, the samples may not be perfectly inverse and in some cases – *e.g.* the Cu(II) ferrite – the structure may be prone to tetragonal distortion [§ 5.5(1b)].

Table 3.3
Spinel ferrites $M^{II}Fe^{III}{}_2O_4$

M(II)	S	m_{sat}/μ_B Néel#	expt.	T_{fN}/K
Mn	5/2	5.01	4.55	573
Fe	2	4.00	4.1	858
Co	3/2	3.00	3.94	793
Ni	1	2.00	2.3	858
Cu	1/2	1.00	1.3	726
Mg	0	0	1.1	713

After Craik (1995). # $m_{sat} = g_eS\mu_B$
The compounds with Fe – Cu have the inverse spinel structure.

The Néel interpretation is supported by the fact that $ZnFe_2O_4$, which has the *normal* spinel structure, does not display ferrimagnetism. It is instead a simple paramagnet above T_N = 9.5 K, the B-site Fe^{3+} ions ordering antiferromagnetically below this temperature. The low value of T_N reflects the weak B–B exchange interaction. The essential validity of the Néel picture, here and elsewhere, has otherwise been abundantly confirmed by, in particular, analysis of the magnetic scattering aspect of neutron diffraction.

If it were instead an inverse spinel, $ZnFe_2O_4$ would be ordered with $m_{sat} \approx 0$ (except, of course, at high temperatures > T_{fN}). The magnesium analogue $MgFe_2O_4$ in Table 3.3 is only about 90% inverse and so displays a non-zero saturation moment;[c] it has, moreover, a remarkably high T_{fN} = 713 K. The

[c] Zn(II) has a high relative affinity for tetrahedral coordination, much more so than Mg(II).

Mn(II) ferrite $MnFe_2O_4$ is also an intermediate spinel (actually about 80 % *normal*) though, since both Mn^{2+} and Fe^{3+} are d^5 ions (with S = 5/2), the theoretical estimate of m_{sat} is unaffected.

(iv) The special case of *magnetite* Fe_3O_4

Table 3.4

The electrical resistivities ρ of some spinel ferrites (/ Ω m)

$MnFe_2O_4$	$\sim 10^2$
Fe_3O_4	4×10^{-5}
$CoFe_2O_4$	$\sim 10^5$
$NiFe_2O_4$	$\sim 10 - 10^2$
$MgFe_2O_4$	$\sim 10^5$

By way of comparison, metallic iron has $\rho \sim 10^{-7}$ Ω m.
[After Craik (1995).]

Fe_3O_4 stands out among the spinel ferrites in showing an almost metallic conductivity of *ca.* 250 $(\Omega\ cm)^{-1}$ at ordinary temperatures (Table 3.4). This involves a hopping mechanism ($Fe^{2+} \rightarrow Fe^{3+}$) facilitated by a substantial *charge-transfer* interaction between the different cations *via* the bridging oxide ions, which stabilises the system. As evidenced by various physical techniques, the extra electron of Fe^{2+} is significantly delocalised throughout the B sub-lattice. The A-site ions (tetrahedral Fe^{3+}) have apparently no role in electrical conduction; the stoichiometric analogues, Mn_3O_4 and Co_3O_4, both having the *normal* spinel structure, are therefore essentially insulators.

(v) Double exchange

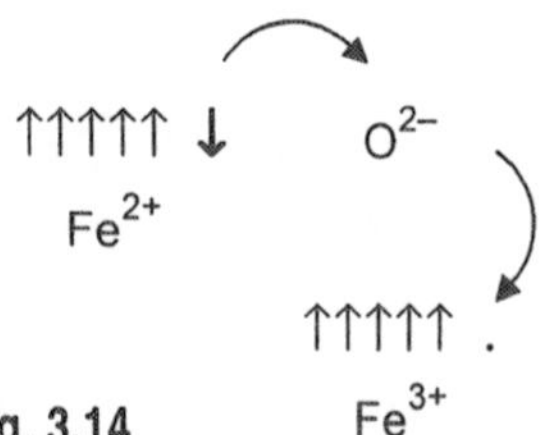

Fig. 3.14

Electron transfer between the octahedral (B) sites in Fe_3O_4 (see text).

The charge-transfer interaction particular to Fe_3O_4 promotes a *ferromagnetic* coupling within the B sub-lattice that naturally facilitates the overall ferrimagnetic ordering. The electron transfer of lowest energy from Fe^{2+} (d^6 S = 2) must be that of the minority spin electron (↓ in Fig. 3.14) since this preserves the stable parallel arrangement of the other five electrons (with S = 5/2). At the same time, to conform with the Pauli exclusion principle, the total spin of the neighbouring Fe^{3+} ion (d^5 S = 5/2) has to be orientated parallel to the majority spin aspect of the Fe^{2+}. In other words, the magnetic moments of the Fe^{2+} and Fe^{3+} cations are driven parallel by the charge-transfer interaction. This mechanism for ferromagnetic coupling is termed *double exchange* – because, presumably, it depends on the intra-atomic exchange effect (spin correlation) stabilising the S = 5/2 (d^5) parallel spin arrangement at *two* magnetic centres in a synchronised fashion.

Ferromagnetic ordering through double exchange is well authenticated in other, usually metallic oxides – *e.g.* the perovskite system $La_{1-\delta}Sr_\delta MnO_3$, containing Mn^{3+} and Mn^{4+} (having S = 2 and 3/2, respectively). The charge-transfer interaction is so strong that when $\delta \gtrsim 0.2$ the extra electron of Mn^{3+} is itinerant, and the compounds then therefore ferromagnetic metals with T_C in the range 300 – 400 K.

(c) Ferrimagnetism without Fe(III)

The presence of Fe(III) is not a requirement for ferrimagnetism, even among spinels. Thus both the normal spinel $Mn^{II}(Cr^{III}{}_2)O_4$ and the quasi-normal $Co^{II}(Co^{II}V^{IV})O_4$ are ferrimagnetic, if with relatively low ordering temperatures T_{fN} of 55 K and 158 K, respectively.[a]

Ferrimagnetism is moreover well established elsewhere – *e.g.* in complex cyanides of the type $A_nM[M'(CN)_6]_m.xH_2O$, where A is an alkali metal, typically with face-centred cubic structures incorporating linear M–NC–M' bridges.[b] A number of such compounds (so-called *cyanomanganates*) have M' = Mn, including $Mn^{II}[Mn^{IV}(CN)_6]\cdot 0.57H_2O$ with T_{fN} = 48.7 K and m_{sat} = 1.91 μ_B. Assuming spin-only moments, the Néel model has $m_{sat} = g_e(5/2 - 3/2)\mu_B = 2.0\ \mu_B$ for this compound.

[a] These Fe(III)-free ferrimagnets have m_{sat} equal to 1.20 μ_B and 1.70 μ_B, respectively, which (as the reader may care to confirm) are consistent with the Néel theory.

[b] They are immediate relatives of *Prussian blue*, with the constitution $Fe^{III}{}_4[Fe^{II}(CN)_6]_3\cdot 15H_2O$. (This orders *ferro*magnetically at 5.6 K.)

Distinctly higher transition temperatures are encountered in analogous *cyanochromates* such as $CsMn^{II}[Cr^{III}(CN)_6]$ (T_{fN} = 90 K) and $Cr_5(CN)_{12}$.-$10H_2O$ (T_{fN} = 240 K) with probably the make-up $Cr^{II}{}_3[Cr^{III}(CN)_6]_2.10H_2O$. The theory of superexchange *via* the cyanide bridge has been exploited in the design of ferrimagnets of this type.[a]

[a] See Kahn (1995) for a compilation of the T_C or T_{fN} for such compounds.

(d) Ferrites involving Fe(III) exclusively

A huge number of ferrimagnetic oxides, many with important applications, contain Fe(III) alone. Some are spinels, including $Li_{0.5}Fe_{2.5}O_4$ (T_{fN} = 943 K) and γ-Fe_2O_3 (T_{fN} = 856 K), which may be formulated as $Fe(Fe_{5/3}V_{1/3})O_4 \equiv Fe_{8/3}O_4$, V signifying a vacant B-site. The latter was discussed in § 1.3(3d), albeit from the classical viewpoint; but what was said there remains valid provided the magnetic moment m of Fe^{3+} is taken to be 5.01 μ_B. According to the Néel theory, the saturation moment for $Fe_{8/3}O_4$ should be just $m_{sat} = (5/3 - 1)m = \frac{2}{3}m$ and, therefore, m_{sat} per formula unit of γ-Fe_2O_3 should be $\frac{3}{4} \times \frac{2}{3}m = \frac{1}{2}m = 2.50\ \mu_B$. This agrees well with the experimental figure of 2.4 μ_B (Table 3.1).

$Li_{0.5}Fe_{2.5}O_4$ displays a saturation moment of 2.6 μ_B per Fe atom. (The reader might care to rationalise this figure.)

Also of considerable technical significance are Fe(III)-containing oxides of the *garnet* and *magnetoplumbite* families, respectively of general formulae $M_3Fe_5O_{12}$ (where M is a trivalent cation) and $MFe_{12}O_{19} \equiv MO.6Fe_2O_3$ (M being a divalent ion). A prominent garnet $Y_3Fe_5O_{12}$ was described in section (a). The second category[b] is more complex, the structure being hexagonal with three different sites available to the Fe(III) ions – octahedral, tetrahedral and trigonal bipyramidal (or "hexahedral"). The distribution is such that *eight* of the cations per formula unit have their magnetic moments aligned in one direction while the other *four* have moments orientated in the opposite direction. The expected saturation moment (assuming, of course, that the M^{2+} ion is diamagnetic) is then (8 – 4) x 5.01 μ_B = 20.04 μ_B, in essentially exact agreement with the observed value of 20.0 μ_B in $BaFe_{12}O_{19}$ (*barium ferrite*) a common magnet material.[c]

[b] – named after the lead-containing mineral *magnetoplumbite*, with the prototype hexagonal structure.

[c] See *e.g.* Morrish (1965) for more information about both the magneto-plumbites and the garnet ferrites.

5. Binuclear systems – local magnetic ordering

There is an enormous and still rapidly expanding literature concerning magnetic interactions within *polynuclear* metal complexes held together by bridging, commonly chelating, ligands. Their consideration is here limited to binuclear species, which bring out the major features.

The systems in mind here should be distinguished from so-called *cluster* compounds, a term usually implying significant direct metal-metal bonding.

The seminal case is hydrated Cu(II) ethanoate, having the stoichiometry $Cu(O_2CMe)_2.H_2O$ and involving a d^9 (S = $\frac{1}{2}$) cation. The compound originally excited interest because its susceptibility displays a maximum (at about 260 K) while declining very rapidly, apparently towards zero, at lower temperatures (< 100 K). Its EPR spectrum at ordinary temperatures is characteristic of a *triplet* state (S = 1), as in *e.g.* Ni(II) d^8 compounds, with an average g-value of 2.159. The spectrum is moreover temperature-dependent: the intensity is maximal at *ca.* 260 K and diminishes in intensity as the temperature is reduced, vanishing at ~ 77 K.

These properties are now understood to reflect a remarkable dimeric *molecular* structure with the formula $Cu_2(O_2CMe)_4.2H_2O$ (Fig. 3.15), in which coupling of the unpaired electron spins of the contiguous metal ions gives rise to singlet and triplet states significantly separated in energy. The

The binuclear structure shown in Fig. 3.15 is adopted at the "convenience" (so to say) of the carboxylate ligands, which cannot bind in bidentate mode with small cations. Hydrated Cr(II) ethanoate is analogous but happens to involve substantial metal-metal bonding.

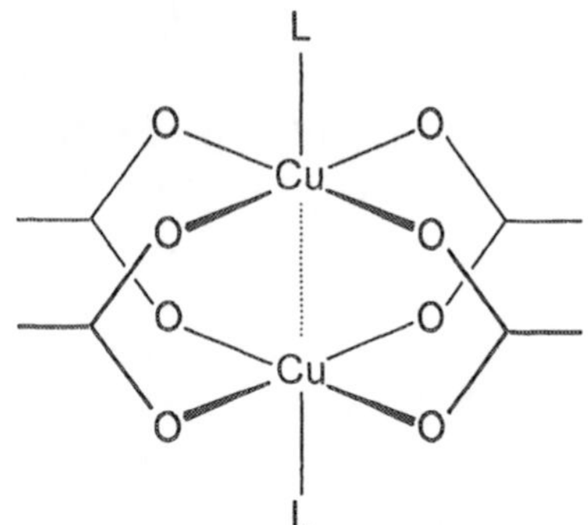

Fig. 3.15 The essential structure of binuclear Cu(II) carboxylates.

L = OH_2 in the case of hydrated Cu(II) ethanoate. It may otherwise be pyridine, for example.

interaction is apparently *antiferromagnetic* making the singlet (S = 0) the ground state (Fig. 3.16). The Cu–Cu distance in Fig. 3.15 is a mere 4% longer than in the metal.

The combination of angular momenta to form a resultant was introduced in § 2.5(2) in connection with the spin-orbit interaction. The coupling of the two spins $\mathbf{S}_1$ and $\mathbf{S}_2$ in a binuclear system is formally similar such that

$$\mathbf{S}_1.\mathbf{S}_2 = \tfrac{1}{2}[S(S+1) - S_1(S_1+1) - S_2(S_2+1)]\hbar^2 \qquad (3.34)$$

equivalent to (2.31). Cu(II) ethanoate represents the simplest case, $S_1 = S_2 = \frac{1}{2}$, with (recalling eqn. 3.26) the exchange energy

$$U(S) = -\tfrac{1}{2}\mathcal{J}[S(S+1) - \tfrac{3}{2}] \qquad (S = 1 \text{ or } 0) \qquad (3.35)$$

where the coupling constant $\mathcal{J}$ happens to be negative. Hence the energy level diagram in Fig. 3.16.

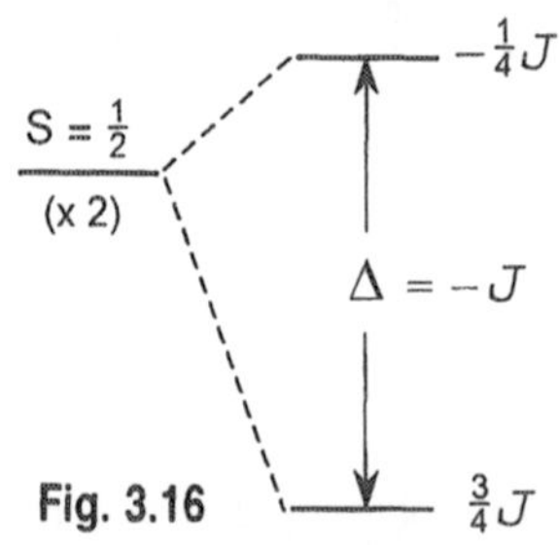

Fig. 3.16

Spin-spin coupling in a binuclear system having $S = \frac{1}{2}$ at each site.

Invoking (3.24), the molar susceptibility per binuclear unit is

$$\chi_{mol} = \frac{\mu_0 N_A \mu_B^2}{3kT}\frac{[0 + 3(2g^2)\exp(-\frac{\Delta}{kT})]}{[1 + 3\exp(-\frac{\Delta}{kT})]} + \mu_0 N_A \alpha \qquad (3.36a)$$

$$= \frac{\mu_0 N_A \mu_B^2}{3kT}(6g^2)[3 + \exp(\frac{\Delta}{kT})]^{-1} + \mu_0 N_A \alpha \qquad (3.36b)$$

where g is the spin g-factor for the triplet state. The second order contribution is the same in both states (unaffected by the spin configuration) and therefore independent of temperature. The coefficient α may be estimated through measurements on related, mononuclear Cu(II) compounds.

As shown in Fig. 3.17, the theory gives a good account of the susceptibility data for the range 100 – 400 K. Using the experimental value of the g-factor, the exchange coupling constant $\mathcal{J}$ is estimated to be $-295\ cm^{-1}$, in excellent agreement with the value $\mathcal{J} = -298\ (\pm 4)\ cm^{-1}$ determined by inelastic neutron scattering spectroscopy.#

\# – actually carried out on the deuterium-substituted compound, $Cu_2(O_2CCD_3)_4.2D_2O$.

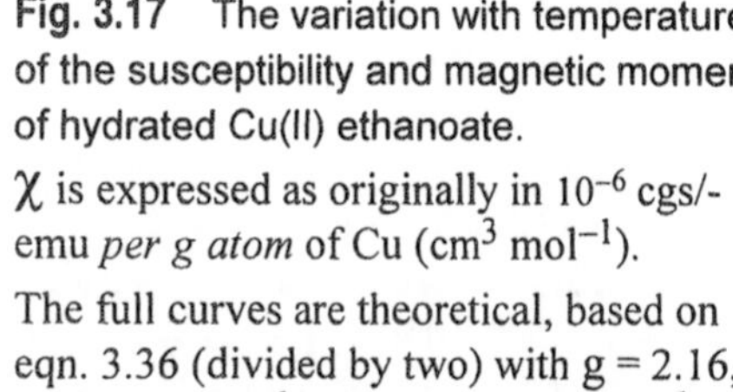

Fig. 3.17 The variation with temperature of the susceptibility and magnetic moment of hydrated Cu(II) ethanoate.

χ is expressed as originally in 10^{-6} cgs/-emu *per g atom* of Cu ($cm^3\ mol^{-1}$).

The full curves are theoretical, based on eqn. 3.36 (divided by two) with g = 2.16, $\mathcal{J} = -295\ cm^{-1}$, and $\chi_{TIP} = 60 \times 10^{-6}$.

[The χ data may be converted into SI units through multiplication by $4\pi \times 10^{-6}$.]

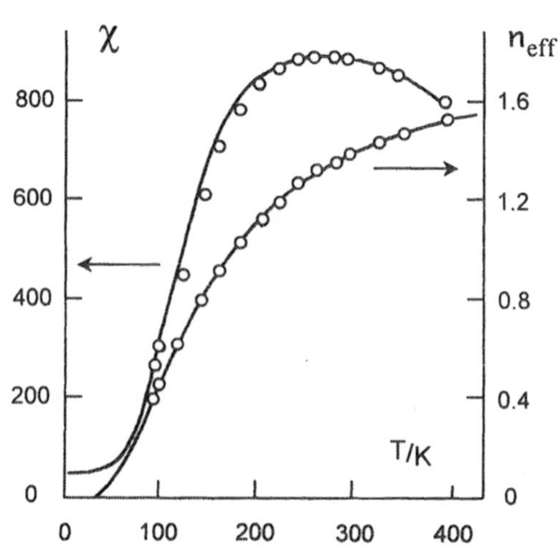

The first order magnetic moment of the binuclear unit is given by

$$m_{eff} = \{6g^2[3 + \exp(\frac{\Delta}{kT})]^{-1}\}^{1/2}\mu_B \tag{3.37}$$

which has the limiting value $\sqrt{6}g/2\ \mu_B = 2.45\ \mu_B$, or $1.87\ \mu_B$ per Cu(II) ion, as $T \rightarrow \infty$.# There is no maximum in the variation of m_{eff} with temperature as is observed in $\chi(T)$.

\# But the plot of m_{eff} in Fig.3.17, based on the susceptibility data, also contains a TIP contribution (as per eqn. 3.20).

Despite the short Cu–Cu distance, it appears that the coupling is due primarily to superexchange *via* the bridging carboxylate ligands, rather than direct (Heisenberg) exchange or incipient chemical bonding. Innumerable studies in which both the donor ligand L in Fig. 3.15 and the alkyl group are varied indicate that the coupling constant J is less sensitive to the concomitant variation (albeit minor) in the Cu–Cu separation than to changes in the electronic properties of the ligand system.

That superexchange is generally the dominant mechanism, at least in binuclear Cu(II) complexes, is strongly supported by studies of systems with bridging η^2–azide ions having the planar network (A). One such compound, in which the Cu–Cu distance (514.5 pm) is almost twice that in the ethanoate (261.6 pm), shows only a weak TIP even at ordinary temperatures. Evidently, only the singlet state is populated, this implying in turn that $-J \gtrsim 10^3$ cm^{-1}. The interaction must here be due exclusively to superexchange.

(A) Cu(N–N–N)(N–N–N)Cu

(B) Cu(N–N–N)(N–N–N)Cu

The interaction is almost invariably antiferromagnetic. One rare instance of a binuclear Cu(II) compound displaying instead ferromagnetic coupling is the η^1–azide bridged species $[Cu_2(\eta^1\text{–}N_3)_2(5\text{-}^t\text{Bu-}py)_4](ClO_4)_2$ (B), with a Cu–Cu distance of 304.5 pm. Both susceptibility measurements and EPR reveal a triplet ground state, the coupling constant J being $\simeq 110$ cm^{-1}.

There is no maximum in $\chi(T)$ in the ferromagnetically coupled case; χ simply declines progressively with temperature. The variation in the susceptibility is characterised by a product χT which *increases* as the temperature is lowered, terminating in a plateau as $T \rightarrow 0$. On the other hand, in binuclear systems with antiferromagnetic coupling (the vast majority), χT *decreases* with falling temperature to zero at $T = 0$. There is a common high temperature limit $\chi T = C$, where C is the Curie constant (Fig. 3.18).

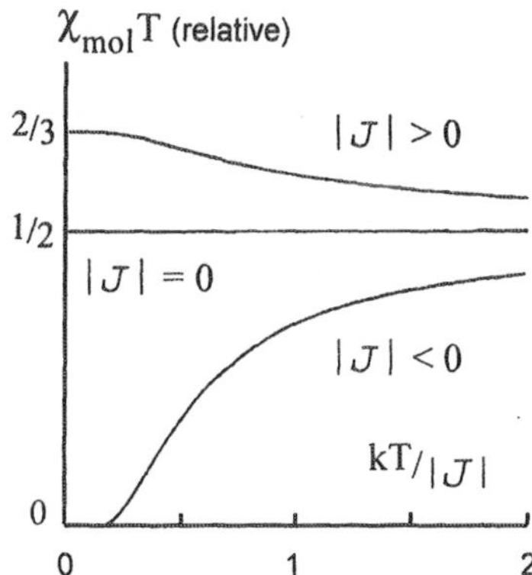

Fig. 3.18 The product $\chi_{mol}T$ per binuclear unit expressed in terms of $\mu_0 N_A \mu_B^2 g^2/k$.
Such plots of the experimental χ data indicate the sign of J immediately.

6. Postcript to magnetic ordering

The forgoing account has only scratched the surface of the subject of cooperative magnetism. Any compound that is paramagnetic at ordinary temperatures apparently orders magnetically at a sufficiently low temperature. But there exists in an enormous variety of solid materials, both organic and inorganic, which show cooperative magnetism (of actual or potential importance to technology) at temperatures remote from the absolute qzero. These include many compounds with chain structures exhibiting *low dimensional* magnetic behaviour. For further information consult, in the first instance, the texts by Carlin (1986) and Kahn (1993).

Restrictions on space have precluded enquiry into the effects of magnetisation on the heat capacity (*a.k.a.* specific heat) and therefore of practically significant phenomena such as *adiabatic demagnetisation*. (It was in part the latter effect, once extensively exploited to attain temperatures below 1 K, which first triggered the interest of physicists in paramagnetism.) The subject is raised to allow the comment that the onset of magnetic order is typically marked by an anomaly in the variation of the heat capacity with temperature in the form of a λ-shaped discontinuity.[a] Carlin (1986) provides an excellent introduction to this important and informative aspect of magnetism.

a The temperature of the magnetic phase transition is accordingly often referred to as a λ-point.

3.7 Diamagnetism

As pointed out in § 1.3(1), diamagnetism involves a negative magnetisation and is a universal property of materials. It is however overshadowed by paramagnetism *etc.*, in compounds having open valence electron shells.

1. The Larmor precession of electron orbits

Further insight into the origin of diamagnetism may be achieved through consideration of the reaction of electron orbits to an applied magnetic field.

The Larmor precession is of some general importance since it occurs for any form of angular momentum **G** with an associated magnetic moment proportional to and parallel with itself, *i.e.* with $\mathbf{m} = \gamma\mathbf{G}$, where γ is the *gyromagnetic ratio* characteristic of **G**. One example is *nuclear spin* angular momentum **I**, on which the enormously important technique of nuclear magnetic resonance (NMR) depends.

The angular velocity of precession of **G** about **B** is $\omega_L = \gamma B$. [In the case of *orbital* motion, γ is simply $-e/2m_e$.]

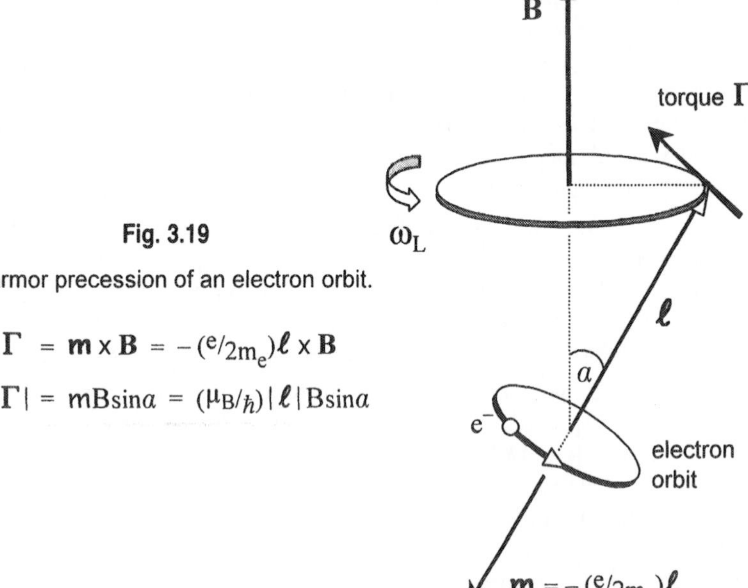

Fig. 3.19

Larmor precession of an electron orbit.

$$\boldsymbol{\Gamma} = \mathbf{m} \times \mathbf{B} = -(e/2m_e)\boldsymbol{\ell} \times \mathbf{B}$$

$$|\boldsymbol{\Gamma}| = mB\sin a = (\mu_B/\hbar)|\boldsymbol{\ell}|B\sin a$$

The vector interaction between the magnetic dipole moment **m** and the external field **B** (recall eqn. 1.7) causes the angular momentum vector $\boldsymbol{\ell}$ to precess about the field direction (Fig. 3.19). This motion is known as the *Larmor precession*. The tip of $\boldsymbol{\ell}$ moves in a circular path about **B** with an angular velocity ω_L and frequency $\nu_L = \omega_L/2\pi$, the *Larmor frequency*. The argument set out below shows that[b]

$$\omega_L = \left(\frac{e}{2m_e}\right)B \tag{3.38}$$

b – while $\nu_L = \mu_B B/h$.

The orbital angular momentum $\boldsymbol{\ell}$ must change at a rate equal to the torque or couple acting – that is,

$$\frac{d\boldsymbol{\ell}}{dt} = -(\frac{e}{2m_e})\boldsymbol{\ell} \times \mathbf{B} \qquad (3.39)$$

and $d\boldsymbol{\ell}/dt$ is perpendicular to both **B** and $\boldsymbol{\ell}$ itself with a constant magnitude of $(e/2m_e)|\boldsymbol{\ell}|B\sin\alpha$. Therefore $\boldsymbol{\ell}$ must move with a constant angular velocity, $\omega_L = d\omega/dt$, about the field direction and at a constant inclination α determined by the zero-field conditions (*i.e.* the tip of $\boldsymbol{\ell}$ describes a circle). An incremental movement ds of the tip of $\boldsymbol{\ell}$ along its circular path (Fig. 3.20) is related to the change in angle $d\omega$ by $ds = |\boldsymbol{\ell}|\sin\alpha.d\omega$.

$$\therefore \qquad \frac{d|\boldsymbol{\ell}|}{dt} = |\boldsymbol{\ell}|\sin\alpha\frac{d\omega}{dt} = |\boldsymbol{\ell}|\omega_L\sin\alpha$$

The formula for ω_L then follows from (3.39). The magnetic moment vector **m** naturally precesses in phase with $\boldsymbol{\ell}$.

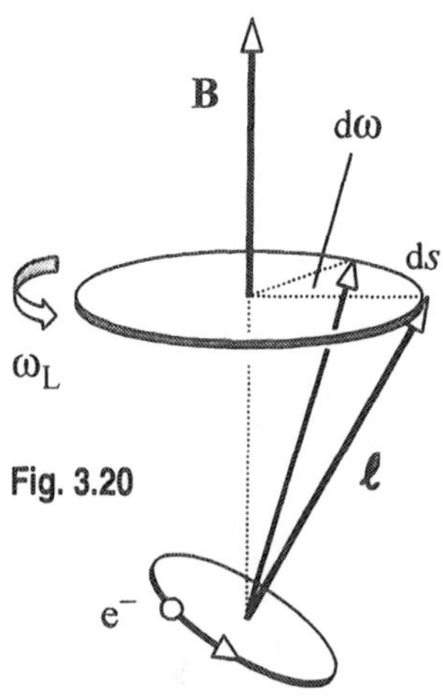

Fig. 3.20
Evaluation of ω_L (see text).

2. The semi-classical theory of diamagnetism

(a) Formulation of the susceptibility for atoms and molecules

The precessional motion of electron orbits results in an electric current about the direction of the applied magnetic field (Fig. 3.21) – the *diamagnetic current*. That due to a single electron orbit is given by

$$I = -e(\omega_L/2\pi) = -e^2B/4\pi m_e$$

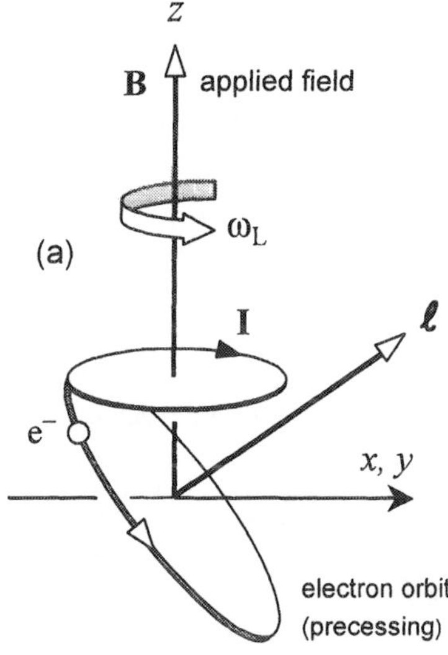

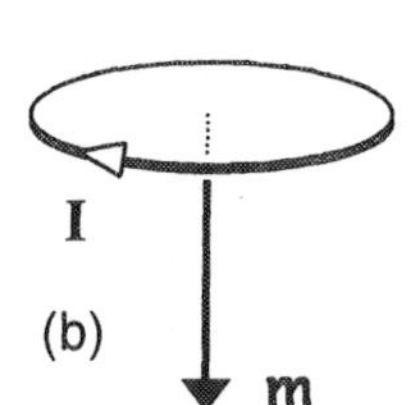

Fig. 3.21
The origin of diamagnetism:
(a) the diamagnetic current;
(b) the diamagnetic moment.

The diamagnetic current contrasts starkly with the paramagnetic current responsible for TIP in molecules [*cf.* section 3(2a)]. However, they are related in that the TIP element of the magnetisation may be viewed as a *reduction in the diamagnetism* due to hindrance of the precession of orbits by the presence of more than one nucleus. In general, a precessing electron will be scattered by off-axis nuclei.

The Larmor precession is unimpeded only in the case of linear molecules orientated *parallel* to the external magnetic field; the TIP is confined to the perpendicular direction.

The area of the current loop is $A = \pi\langle x^2 + y^2\rangle$, where x and y define the position of the electron in a plane perpendicular to the field direction z [Fig. 3.21(a)]. Assuming a randomly orientated orbit, when $\langle x^2\rangle = \langle y^2\rangle = \langle z^2\rangle = \frac{1}{3}\langle r^2\rangle$, the area $A = \frac{2}{3}\pi\langle r^2\rangle$, the quantity $\langle r^2\rangle$ being the mean square radius of the orbit (to be determined *ab initio* by quantum mechanical calculation). It follows that the magnetic moment generated by the precession is

$$\mathbf{m} = -\frac{e^2}{6m_e}\langle r^2\rangle\mathbf{B} \qquad (3.40)$$

which is inherently diamagnetic (independently of the sign of the electronic charge). The *molar* diamagnetic susceptibility ($\mu_0 N_A m/B$) is given by

$$\chi^{d}_{mol} = -\frac{\mu_0 N_A e^2}{6m_e}\sum_k n_k\langle r_k^2\rangle \qquad (3.41)$$

which involves a summation over all the electrons, n_k being the occupancy of the kth orbit.[a] This result, which applies equally to atomic and molecular substances, is known as the *Langevin-Pauli equation* [P Langevin (1905), W Pauli (1920)].[b] Of course, in the case of *closed-shell* systems all $n_k = 2$.

[a] The (negative) magnetisability α_d defined in (1.18) is therefore

$$\alpha_d = \frac{e^2}{6m_e}\sum_k n_k\langle r_k^2\rangle$$

[b] Pauli is credited merely because he corrected a minor slip by Langevin, who took the average value of $x^2 + y^2$ to be $\frac{1}{3}r^2$ rather than $\frac{2}{3}r^2$.

The characterisation of the diamagnetic current, and also of its paramagnetic counterpart [section 3(2a)], is of fundamental importance in the theory of chemical shifts in the NMR spectra of molecules.

(b) The susceptibilities of atoms

As illustrated by the data in Table 3.5, experimental *atomic* susceptibilities are found to be roughly 10^{-11} Z m^3 mol^{-1} (where Z is the atomic number). In the case of complex atoms having many different electron orbits occupied it is naturally those electrons in "outer" orbits (having the largest $\langle r^2\rangle$) that contribute most to the diamagnetic susceptibility. In the case of xenon, *ab initio* calculations show that the outer $(5p)^6$ and $(5s)^2$ subshells make up some 58.5% and 15.0%, respectively, of the observed susceptibility; the highly contracted $(1s)^2$ shell contributes a mere 0.001%.

Table 3.5

Magnetic susceptibilities of the noble gases.

	χ_{mol}	Z
He	−2.54	2
Ne	−8.75	10
Ar	−24.3	18
Kr	−36.4	36
Xe	−57.2	54

Unit = 10^{-11} m^3 mol^{-1}.

These diamagnetic susceptibilities may be compared with a typical *paramagnetic* susceptibility of *ca.* 10^{-8} m^3 mol^{-1} or greater for an open-shell atom or molecule.

The theory can be extended to atomic ions in crystals, most readily to those isoelectronic with the noble gases. The susceptibilities of K^+ and Cl^-, for example, have been estimated to be −16.8 and −31.8 x 10^{-11} m^3 mol^{-1}, respectively. Their sum reproduces accurately the experimental susceptibility of KCl(*s*), equal to −49.0 x 10^{-11} m^3 mol^{-1}.

(c) The correlation between χ^d and the polarisability α

The polarisability of an atom also increases with atomic number (see Table 3.6, for example). Indeed, the data for the noble gases and many high symmetry molecules show an approximate proportionality between $-\chi^d$ and α. A more accurate semi-empirical relationship, so it is claimed, is $-\chi^d \propto \sqrt{(n\alpha)}$, where n is the total number of electrons in the system. The latter has been exploited to estimate χ^d from polarisability data and *vice versa*.

Table 3.6

Polarisabilities of the noble gases.

	α	Z
He	0.23	2
Ne	0.45	10
Ar	1.85	18
Kr	2.97	36
Xe	4.55	54

Unit = 10^{-40} J^{-1} C^2 m^2.

(d) Delocalised electrons and hyper-anisotropy

The magnetic properties of molecules are in general anisotropic and diamagnetic susceptibilities markedly so when there are electrons delocalised in a particular molecular plane.

Delocalised electrons commonly have substantial effective radii and can therefore make especially large contributions to the diamagnetism of molecules. Thus in benzene the so-called "ring current" associated with the π-electrons allows a magnetisation perpendicular to the molecular plane which is almost three times that observed when the magnetic field is applied within the plane. The *anisotropy* $\Delta\chi = \chi_\perp - \chi_\parallel$ is −75.0 x 10^{-11} m^3 mol^{-1} which, in absolute terms, exceeds an average susceptibility $\chi_{av} = \frac{1}{3}(\chi_\perp + 2\chi_\parallel)$ of −68.9 x 10^{-11} m^3 mol^{-1}. [a]

A not unrelated, and more dramatic instance is graphite. This has a layer lattice consisting of hexagonal, benzenoid sheets or nets bound together by van der Waals (or dispersion) forces. The extensive delocalisation of the π electrons within the sheets leads to an electrical conductivity that is enormous relative to that of diamond (though weak in comparison to most metals); it also results in the largest magnetic anisotropy known in any material. When the magnetic field is applied perpendicular to the hexagonal layers the susceptibility $\chi_{\perp} = -27.3 \times 10^{-11}$ m^3 (g atom)$^{-1}$; but the susceptibility measured with the field directed parallel to the sheets is a trivial $\chi_{\parallel} = -0.4 \times 10^{-11}$ m^3 (g atom)$^{-1}$.[b]

Unusually, $\chi_{\perp}$ is found to vary significantly with temperature, like the electrical conductivity within the layers. The explanation of such behaviour requires a description of the π electrons in terms of 2D *energy bands* that cannot be ventured into here.

[a] The perpendicular and parallel components of the susceptibility are -43.9 and -119×10^{-11} m^3 mol^{-1}, respectively.

[b] The remarkably small negative susceptibility observed within the graphitic layers is presumably due mainly to a large TIP.

(e) Decomposition of molecular χ^d (Pascal constants)

The diamagnetic susceptibilities of inorganic compounds may be broken down semi-empirically into contributions χ_a from individual atoms or groups (*e.g.* ligands) such that the susceptibility may be simulated as

$$\chi^d = \sum_a n_a \chi_a \qquad (3.42)$$

where n_a is the number of atoms or groups a in the formula unit. The quantities χ_a, commonly referred to as *Pascal constants* [after Pascal (1910)],[c] are employed routinely to correct the susceptibility of a paramagnetic compound for the underlying diamagnetism.

In the case of *organic* compounds, the susceptibilities are instead usually simulated as

$$\chi^d = \sum_a n_a \chi_a + \sum_b n_b \lambda_b \qquad (3.43)$$

where the λ_b are structural increments or *constitutive corrections* recognising the influence of particular bonding features, *e.g.* the presence of a C=C or some other multiple bond.

[c] An extensive compilation of these parameters is provided by Mulay and Boudreax (1976). Much further information concerning the different aspects of diamagnetic behaviour may be found in this text.

3. Superconductors: the Meissner effect

Many solid materials become *superconducting* at low temperatures; in other words, they exhibit zero electrical resistance. These include many transition metal oxides, among them some remarkable copper-containing compounds[d] which display superconductivity at temperatures above the boiling point of liquid nitrogen (77 K) In the superconducting state the materials are moreover *perfectly diamagnetic* in that an applied magnetic field is excluded by a specimen, a unique magnetic response known as the *Meissner effect*. The relative permeability μ_r (defined by eqn. 1.17) is then zero and the volume susceptibility $\chi_m = -1$, the maximum conceivable negative value. One well known manifestation of the phenomenon, popular in the teaching laboratory, is the levitation of a small magnet placed near the surface of a specimen.

[d] – *e.g.* $YBa_2Cu_3O_{7-\delta}$, the so-called "1:2:3 compound", which shows superconductivity below 95 K. See Cox (1992) for an introductory account of this and related materials.

[W Meissner and R Ochsenfeld (1933)]

For further basic information on superconductivity and the Meissner effect see *e.g.* Bleaney and Bleaney (1976).

Invoking Lenz's law, the perfectly diamagnetic response might be attributed simply to the perfect electrical conductivity. However, the reasons for the Meissner effect are much more complex. It transpires that the external magnetic field cannot penetrate a superconducting material to a depth of more than about 1 μm, so the diamagnetic current that screens the field is confined to a region very close to the surface of the specimen.

It is appropriate at this stage to introduce some outstanding aspects of paramagnetism with, first, a brief simple account of the magnetic properties of the *metallic* elements.[a] The concepts broached apply also to *compounds* that happen to be metallic.

[a] These were of course included in the survey of Fig. 1.12.

3.8 Elemental metals

1. Energy bands

In any monatomic solid chemical bonding generates one or more sets of quasi-continuous electronic energy levels, corresponding to delocalised orbitals or *lattice orbitals* (of energy ϵ, say). Any such set is termed a *(valence) band*, and is characterised by its density of states (DoS) $n(\epsilon)$, the number of *one-electron* energy levels available in unit volume per unit energy range[b]. In the case of metallic materials the valence band, known more usually as the *conduction band*, is only partly occupied (Fig. 3.22), the level to which it is filled being called the *Fermi energy* and denoted ϵ_F. The high electrical conductivity typical of a metal is associated with the inherently *itinerant* nature of the electrons concerned, which therefore respond readily to an applied electric field gradient.

[b] $n(\epsilon).d\epsilon$ = the number of energy levels between ϵ and $\epsilon + d\epsilon$.

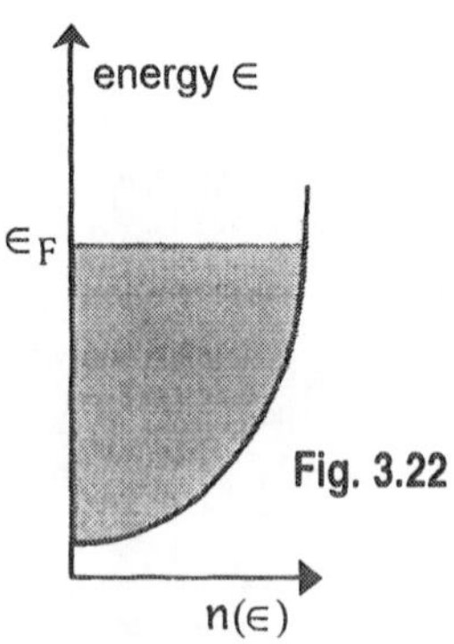

Fig. 3.22

The conduction band of a simple metal.

As regards their electronic structures the elemental metals fall into *two* broad categories.

(1) "Simple" metals, such as those of the pre-transition groups (especially groups 1 and 2) in which *all* the valence electrons are itinerant, the ion cores being of course diamagnetic.

(2) The more complex metals – notably transition, lanthanide and actinide elements – in some of which the valence electrons are partially localised about the ion cores.

Attention is here confined to the simple metals. An account of complex metals would require development of the band model beyond the scope of this text.[c] Band theory can explain, for example, a trend observed in the first transition series in which, at ordinary temperatures, the paramagnetism of the early metals (Ti and V) gives way to antiferromagnetism (at Cr and Mn), to be followed by ferromagnetism in Fe, Co and Ni. The absence of ferromagnetism among the heavier transition elements can also be rationalised.

[c] See *e.g.* Cox (1987) for an introductory account of the properties of complex metals.

2. Simple metals: the free-electron model

Many *simple* metals show weak TIP, virtually by definition. Major insights regarding this behaviour may be gained through the *free-electron model*, in which the itinerant electrons are treated in effect as a gas, the electrons moving in an almost constant potential. However, whereas a *molecular* gas conforms to Boltzmann statistics, the denser *electron gas* is subject to the Pauli exclusion principle and must be described by Fermi-Dirac statistics.

The application of the latter (quantum) statistics leads to analytical expressions for the DoS $n(\epsilon)$ and therefore the Fermi energy ϵ_F.[a]

[a] – namely $\epsilon_F = (3N/\pi)^{2/3}h^2/8m_e$ while $n(\epsilon_F) = 3N/2\epsilon_F$.

In the absence of a magnetic field, the electron spins are perfectly paired in the conduction band (as assumed in Fig. 3.22) while, because of the irregularity of their translational motions, the orbital momentum of the conduction electrons also averages to zero.

(a) Pauli (spin) paramagnetism

Application of a field induces a marginal spin polarisation of the electrons manifested by a small positive magnetisation, termed *Pauli paramagnetism* [W Pauli (1927)]. The associated volume susceptibility proves to be simply

$$\chi_m(P) = \mu_0\mu_B^2 n(\epsilon_F) \qquad (3.44a)$$

It is expressible alternatively as

$$\chi_m(P) = 12(\pi/3)^{2/3}\mu_0 m_e N^{1/3}\mu_B^2/h^2 \qquad (3.44b)$$

where N is the concentration of conduction electrons (per unit volume).

The effect is illustrated by Fig. 3.23, which shows separate conduction bands for electrons of alternative spin orientation with respect to the applied magnetic field. The spin polarisation depends simply on the greater stability (by the amount $2\mu_B B$) of electrons with spins directed against the applied field (LHS) than those with parallel spins (RHS); by the Pauli exclusion principle, the former must outnumber the latter at equilibrium. The effect, which is universal in metals, is due to the excess of electrons in the *down*-spin band.

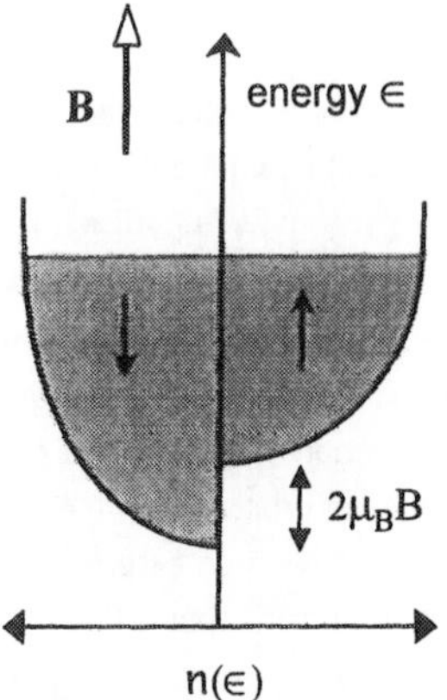

Fig. 3.23 The origin of Pauli (spin) paramagnetism.

The alternative spin states are assigned separate bands. (The scale of the spin polarisation is much exaggerated.)

In contrast to Curie paramagnetism, where the unpaired electrons are localised at atomic or molecular centres, Pauli paramagnetism is essentially independent of temperature.

(b) Landau diamagnetism

At the same time, as a result of a lateral magnetic force,[b] the itinerant electrons are driven in helical trajectories about the field direction and, as with the Larmor precession of electron orbits, this gives rise to a negative magnetisation; the effect is known as *Landau diamagnetism* [Landau (1930)]. Remarkably it transpires (according to the theory) that in absolute magnitude the diamagnetic susceptibility of the electron gas is exactly one-third that of the spin susceptibility, that is

[b] – the *Lorentz force*, $\mathbf{F} = -e(\mathbf{v} \times \mathbf{B})$, directed ⊥ to both the field **B** and the velocity **v**.

$$\chi_m(L) = -\tfrac{1}{3}\chi_m(P) \qquad (3.45)$$

The expected outcome is thus a net *paramagnetic* susceptibility of $\tfrac{2}{3}\chi_m(P)$. However, the deficiencies of this simple theory are such that commonly the diamagnetic term is numerically less significant.

A combination of experiment and more sophisticated theory show that the susceptibility of metallic lithium is made up as follows:

$$\chi(P)\ 34.0 \quad \chi(L)\ -2.3 \quad \chi(\text{He core})\ -0.8 \qquad \text{exptl. } \chi_{mol}\ 30.9$$

(in 10^{-11} m^3 mol^{-1}).[c]

[c] The paramagnetic component of the susceptibility can be distinguished experimentally through what is called the *Knight effect* in NMR.

For an explanation of this effect, and more information concerning the magnetic properties of metals see, in the first instance, Morrish (1965).

3.9 Paramagnetism due to nuclear spin

1. Nuclear spin

Many atomic nuclei (or *nuclides*), in particular those having odd isotopic mass numbers *A*, possess intrinsic angular momentum. In its essential properties this *nuclear spin*, usually denoted **I**, resembles electron spin [§ 2.2(1) & (2)]. Thus, as *per* eqn. 2.11,

$$|\mathbf{I}| = [I(I+1)]^{\frac{1}{2}}\hbar \qquad (3.46)$$

(while $I_z = M_I\hbar$),# where I is the nuclear spin quantum number.

M_I is the associated magnetic quantum number, which ranges from + I to – I in integral steps

The quantum number I can be integral or half-integral, depending on the number of protons and neutrons (the *nucleons*) present in the nucleus and their configuration in the nuclear ground state. The atomic number Z gives the number of protons while the number of neutrons equals *A* – Z. Both nuclear particles have the spin quantum number $\frac{1}{2}$, like the electron.

Some insight into the nuclear spin values can be acquired through a simple *shell* model not dissimilar to that in the theory of atomic structure. So far as possible, the spins of the nucleons are paired up in the ground state. An odd nucleon with a spin $\frac{1}{2}$ moves in an orbit within the nucleus with an angular momentum $l\hbar$. The observed nuclear spin is then $l+\frac{1}{2}$ or $l-\frac{1}{2}$.

Nuclides with odd mass numbers *A* are therefore found to have a net *half-integral* spin. These include abundant nuclei such as ^1H (= the proton), ^{19}F and ^{31}P, each with I = $\frac{1}{2}$, and minority isotopes such as ^{13}C and ^{14}N which also display I = $\frac{1}{2}$. Examples of nuclei having higher spin values are ^{55}Mn, with I = $^5/_2$, and ^{133}Cs, which has I = $^7/_2$. Nuclei with even mass numbers show, on the other hand, *integral* values of I. However, those which have also *even* charges (*i.e.* even atomic number Z), such as ^{16}O and ^{12}C, have I = 0, corresponding to the complete pairing of nucleon spins, and do not therefore show nuclear spin. Nuclei of *odd* atomic number and even mass number do possess spin but are relatively uncommon; they include ^2H (the deuteron) and ^{14}N, both of which have I = 1, and ^{10}B with I =3.

Since atomic nuclei are charged, their spins give rise to nuclear magnetic moments $\mathbf{m}^N$ defined by

$$\mathbf{m}^N = g_N(\mu_N/\hbar)\mathbf{I} \qquad (3.47)$$

Alternatively, $\mathbf{m}^N = \gamma_N\mathbf{I}$, where the magnetogyric ratio $\gamma_N = g_N e/2m_p$.

where μ_N is the *nuclear magneton,*

$$\mu_N = \frac{e\hbar}{2m_p} = 5.0508 \times 10^{-27}\ \mathrm{J\,T^{-1}} \qquad (3.48)$$

m_p being the mass of the proton. The proton mass being 1837 times that of the electron, $\mu_N \sim 0.5 \times 10^{-3}\ \mu_B$. g_N is the *nuclear spin g-factor*, a dimensionless quantity (like the electronic g_e) which is characteristic of the nuclide in question.

The proton and neutron have the g-factors g_p = 5.586 and g_n = – 3.826, respectively.

2. The nuclear spin susceptibility

The collective paramagnetic behaviour of an assembly of non-interacting nuclear spins should be wholly analogous to that of an assembly of electron spins. Therefore, translating eqn. 3.4,

$$\chi^N_{mol} = \frac{\mu_0 N_A}{3kT} g_N^2 I(I+1)\mu_N^2 \qquad (3.49)$$

The limited data available relate to molecular systems where it is appropriate to define a susceptibility per mole of compound, summing over the various magnetic nuclei

which implies an effective magnetic moment $m_{eff} = g_N[I(I+1)]^{\frac{1}{2}}\mu_N$.

Nuclear spin paramagnetism actually proves to be negligible in open-shell systems: since $\mu_N/\mu_B \sim 10^{-3}$, nuclear spin susceptibilities are some 10^{-6} times first order electronic paramagnetic susceptibilities. The paramagnetism due to nuclear spin is usually eclipsed even by electronic diamagnetism, which is typically 10^3 times larger at ordinary temperatures. In the case of water, for example, the protons make a contribution of merely *ca.* $+4 \times 10^{-15}$ $m^3\ mol^{-1}$ to a total molar susceptibility of -9×10^{-12} $m^3\ mol^{-1}$. However, since $\chi^N \propto 1/T$, the nuclear paramagnetism can assume some importance at very low temperatures. Thus the diamagnetic susceptibility of solid hydrogen decreases markedly on cooling below 4 K and is neutralised by the (positive) nuclear spin susceptibility at about 0.5 K (when $\chi^N \simeq 2 \times 10^{-11}$ $m^3\ mol^{-1}$).

The original experiment led to one of the first estimates of g_P, as *ca.* 5.6.

3.10 Appendix: the general (Brillouin) theory of electronic spin paramagnetism

1. The average magnetic moment $\langle m_B \rangle$

Semi-classical *Boltzmann statistics* are invoked. The number of atoms or molecules populating the M_S Zeeman level at thermal equilibrium is

$$n(M_S) = N_A \frac{\exp[-E(M_S)/kT]}{\sum_{M_S} \exp[-E(M_S)/kT]} \qquad (3.50)$$

(assuming 1 mole of material), where the summation involved extends from $M_S = -S$ to $M_S = S$. The energies $E(M_S)$ are defined by (3.2).

The *molar* magnetization $M = N_A \langle m_B \rangle$, where $\langle m_B \rangle$ is the mean molecular or atomic magnetic moment induced in the field direction. The contribution to M from molecules occupying the M_S Zeeman level is $-g_e M_S \mu_B \times n(M_S)$. Summing over all Zeeman levels,

The molar magnetization is the magnetic moment induced per unit applied field in 1 mole of substance [recall § 1.2(4a)].

$$\langle m_B \rangle = -\frac{g_e \mu_B}{N_A} \sum_{M_S} M_S n(M_S) \qquad (3.51)$$

If the variable $x = g_e \mu_B B/kT$ is now introduced, then the average magnetic moment in the field direction is

$$\langle m_B \rangle = -g_e \mu_B \frac{\sum_{M_S} M_S \exp[-M_S x]}{\sum_{M_S} \exp[-M_S x]} \qquad (3.52)$$

The variation of $\langle m_B \rangle$ with B/T, the physical quantity which determines the parameter x, is illustrated by Fig. 3.24. This is based on seminal experiments by W Henry (1952) on various crystalline salts containing open-shell cations (*e.g.* Fe^{3+} d^5 with $S = 5 \times \frac{1}{2} = 5/2$) conducted at, what were for the time, very high field strengths of up to 5 tesla and very low temperatures,[a] down to 4 K and below. The data points shown relate to 1.3 K, the magnetic field strength being varied at constant temperature. The full curves drawn are the general solutions to eqn. 3.52 (due to Brillouin) for the different values of S arising.

[a] Low temperatures are essential. For $\mu_B B$ to even equal kT at ordinary temperatures, B would have to be about 10^2 tesla.

It is not difficult to show,[b] returning to eqn. 3.52, that the average atomic or molecular magnetic moment in the field direction is of a the general form

[b] See, for example, Carlin (1986) or Crangle (1977, 1991),

$$\langle m_B \rangle = g_e \mu_B S B_S(y) \qquad (3.53a)$$

where $B_S(y)$ is the *Brillouin function*,

$$B_S(y) = \frac{2S+1}{2S} \coth\left[\frac{(2S+1)}{2S} y\right] - \frac{1}{2S} \coth\left[\frac{y}{2S}\right] \qquad (3.53b)$$

$$\coth z = \frac{e^z + e^{-z}}{e^z - e^{-z}} = \frac{1}{z} + \frac{z}{3} - \frac{z^3}{45} + \ldots..$$

with $y = Sx = g_e\mu_B SB/kT$. Note that $B_S(y) \to 1$ when $y \to \infty$ (because $\coth y \to 1$), which confirms the saturation limit (eqn. 3.7).

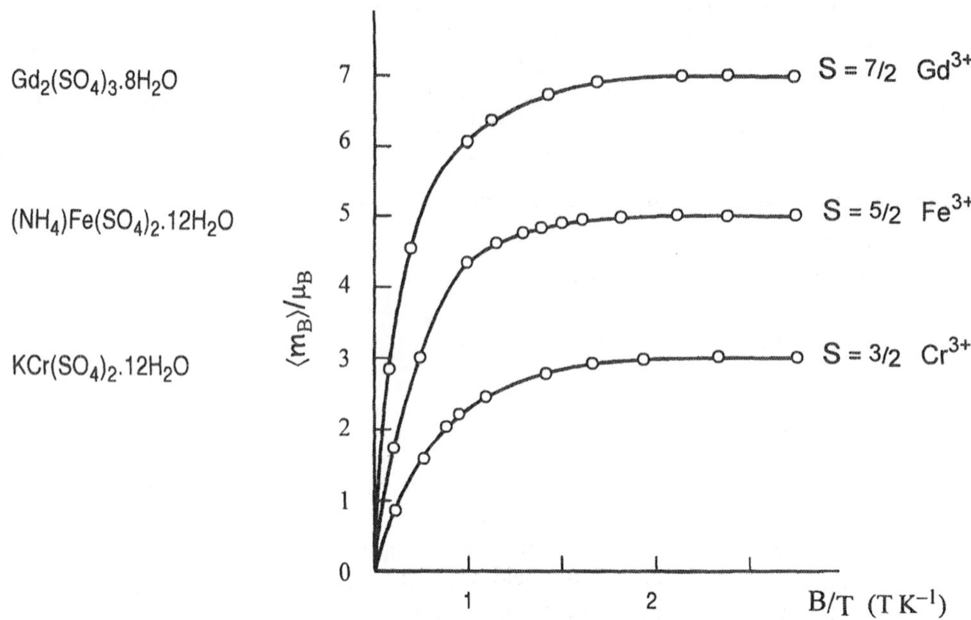

Fig. 3.24 The average magnetic moment $\langle m_B\rangle$ as a function of B/T

2. The Curie limit revisited

As for the alternative limit, $y \to 0$, the reader may care to verify that since, for very small z, $\coth z \to 1/z + z/3$,

$$B_S(y) \to (S+1)y/3S = g_e(S+1)\mu_B B/kT$$

(The $1/z$ elements cancel.) The average magnetic moment is hence

$$\langle m_B\rangle = \frac{B}{3kT}\, g_e^2 S(S+1)\mu_B^2 \tag{3.54}$$

$\chi_{mol} = \mu_0 N_A \langle m_B\rangle/B$

and the formula for the susceptibility# in the Curie limit (eqn. 3.4) is thereby established on a rigorous quantum mechanical basis.

The magnetic moment in the Curie limit may alternatively be deduced directly from (3.52). Since $e^x = 1 + x + \frac{1}{2}x^2$ .....,

$$\lim_{x\to 0} \frac{\Sigma_{M_S} M_S \exp[-M_S x]}{\Sigma_{M_S} \exp[-M_S x]} = \frac{\Sigma_{M_S} M_S \exp[1 - M_S x]}{\Sigma_{M_S} \exp[1 - M_S x]}$$

But $\Sigma_{M_S} M_S = 0$ and $\Sigma_{M_S} 1 = 2S + 1$ and, furthermore,

$$\sum_{-S}^{+S} M_S^2 = \tfrac{1}{3}S(S+1)(2S+1) \qquad \text{[recall section 1(2b)]}$$

In the limit of extremely small Zeeman intervals compared with kT, the mean magnetic moment in the field direction (eqn. 3.52) is therefore

$$\langle m_B\rangle = g_e\mu_B[-\tfrac{1}{3}S(S+1)]x$$

which, on substitution of $x = g_e\mu_B B/kT$, leads to (3.54).

Also naturally of interest is the connection between the Brillouin analysis and the classical Langevin-Debye theory [§ 1.5]. The latter may be obtained by allowing S in eqn. 3.53b to become infinite, so that $(2S+1)/2S \to 1$, while as appropriate letting the Zeeman interval $\Delta_Z \to 0$. Then

$$B_S(y) \to \coth y - \frac{1}{y} = L(y)$$

This provides an example of the *correspondence principle* of Bohr. [See Cox (1996), Friedman and Atkins (1996), or Green (1997) – cited in § 2.]

The semi-classical and quantum theories are compared in the figure beneath.

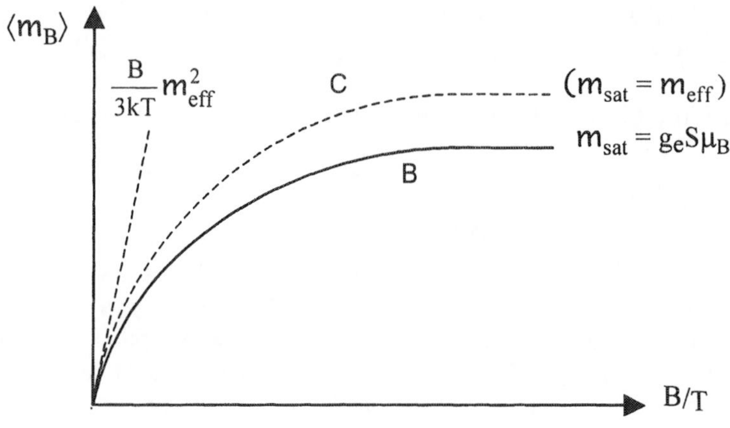

Fig. 3.25
Sketch comparing the Brillouin theory (B) with the Langevin-Debye theory (C) of paramagnetism.
Note the common Curie limit as $B/T \to 0$.
The plot C is *semi*-classical in as much as the magnetic moment is taken to be the constant $m_{eff} = 2[S(S+1)]^{\frac{1}{2}}\mu_B$.

3.11 Bibliography

B I Bleaney and B Bleaney, *Electricity and Magnetism*, Oxford 1976, 1989 #

E A Boudreaux and L N Mulay, *Theory and Applications of Molecular Paramagnetism*, Wiley-Interscience 1976

R L Carlin, *Magnetochemistry*, Springer-Verlag 1986

A Carrington and A D McLachlan, *Introduction to Magnetic Resonance*, Harper and Row 1967; Chapman and Hall 1979

A K Cheetham, ch. 2 of *Solid State Chemistry: Techniques*, ed. A K Cheetham & P Day, Oxford 1987

P A Cox, *The Electronic Structure and Chemistry of Solids*, Oxford 1987 #

P A Cox, *Transition Metal Oxides*, Oxford 1992

D Craik, *Magnetism: Principles and Applications*, Wiley, 1995

J Crangle, *The Magnetic Properties of Solids*, Edward Arnold 1977; *Solid State Magnetism*, Van Nostrand Reinhold (New York) 1991 #

J B Goodenough, *Magnetism and the Chemical Bond*, Interscience, 1963

J E Greedan, *Magnetic Oxides*, p 2017, vol. 4 of *The Encyclopaedia of Inorganic Chemistry*, ed. R B King, Wiley 1994

O Kahn, *Molecular Magnetism*, Wiley-VCH 1993

A H Morrish, *The Physical Principles of Magnetism*, Wiley 1965; re-issued 2001 by the Inst. of Electrical & Electronics Engineers, New York

L N Mulay and E A Boudreaux, *Theory and Applications of Molecular Diamagnetism*, Wiley-Interscience 1976

introductory texts

P W Selwood, *Magnetochemistry*, Interscience, New York, 1943, 1956 #

L Smart and E Moore, *Solid State Chemistry*, Chapman & Hall 1992, 1995 #

J H Van Vleck, *Physica*, 1973, **69**, 177

A R West, *Solid State Chemistry and its Applications*, Wiley 1984 #

Further information on magnetically ordered materials, including so-called molecular magnets:

O Kahn, *Advances in Inorganic Chemistry* (Academic Press, ed. A G Sykes), 1995, **43**, 180 [concerning heterobimetallic compounds]

J S Miller and A J Epstein, *Angewandte Chemie Intnl. Edn.*, 1994, **33**, 385 [on *Organic and Organometallic Molecular Magnetic Materials – Designer Magnets*]

J S Miller and M Drillon. *Magnetism: Molecules to Materials*, vols. I and II, Wiley-VCH 2001

3.12 Exercises

$\frac{\mu_0 N_A \mu_B^2}{3k} = 1.57 \times 10^{-6}\ m^3\ mol^{-1}\ K$

1. Interpret the following observations.

(a) The susceptibility of oxygen gas varies with temperature according to the Curie law. It amounts to $4.22 \times 10^{-8}\ m^3\ mol^{-1}$ at 25°C.

[$m_{eff} = 2.83\ \mu_B$; ∴ S = 1]

(b) A neutral mononuclear complex of manganese has a susceptibility which conforms closely with the Curie law between 300 K and 70 K. At the former temperature it has the value $3.68 \times 10^{-7}\ m^3\ kg^{-1}$; its relative molecular mass (atomic wt.) is 352.

[$m_{eff} = 4.97\ \mu_B$; ∴ Mn(III) d^4]

(c) $(cp)_3Lu$ and $(cp)_2Yb$ are diamagnetic, but $(cp)_2Eu$ is paramagnetic and obeys the Curie law with C = $93.6 \times 10^{-6}\ m^3\ mol^{-1}$ K.

$cp = \eta^5\text{-}C_5H_5$
[$m_{eff} = 7.72\ \mu_B$; ∴ Eu(II) f^7]

(d) $(cot)_2Pu$ is diamagnetic with a bulk susceptibility of $-0.24 \times 10^{-8}\ m^3\ mol^{-1}$. $(cot)_2Np$ is paramagnetic having a susceptibility at 293 K of $1.44 \times 10^{-8}\ m^3\ mol^{-1}$.

$cot = \eta^8\text{-}C_8H_8$ [$m_{eff} = 1.77\ \mu_B$]

(e) The susceptibility of the molecular compound NpF_6 is given by $\chi = (44.2/T + 0.207) \times 10^{-8}\ m^3\ mol^{-1}$ over a considerable temperature range. Its EPR spectrum yields $|g| = 0.605$.

[$m_{eff} = 0.52\ \mu_B$]

(f) The complex cyanides $K_2CuFe(CN)_6$ and $KFe_2(CN)_6$* have effective magnetic moments of 1.81 μ_B and 5.91 μ_B, respectively. *Prussian Blue*, of stoichiometry $Fe_7(CN)_{18}$ has an m_{eff} of 11.8 μ_B. These values, expressed per formula unit in each case, are essentially independent of temperature.

* – known as *soluble* Prussian Blue. (The compounds are normally hydrated.)

2. Sketch the variation of the product χT with temperature T for a paramagnetic substance, and comment on its utility.

3. Comment on the following observations.

The Curie-Weiss law [§ 3.2] is

$$\chi = C/(T - \theta)$$

(a) The paramagnetic susceptibility of solutions of oxygen in liquid nitrogen shows a Curie-Weiss dependence on temperature. The Weiss temperature θ is negative, |θ| increasing with oxygen concentration.$ At high pressures oxygen gas shows comparable behaviour, |θ| increasing with the pressure.

However, in its β-quinol clathrate "compound", oxygen obeys the unadulterated Curie law (θ = 0) down to at least 10 K.

$ Actually |θ| ≃ 40 x its mole fraction [*cf.* Fig. 3.26, facing].

(b) CoO orders antiferromagnetically at about 292 K. However, the susceptibility of Co(II) diluted in ZnO obeys the Curie-Weiss law. When its molar concentration is 20 %, $\theta = -68$ K, and $|\theta|$ falls continuously with increasing dilution to ~ 0 K at 5 % Co(II).

(c) The susceptibilities of $Gd_2(SO_4)_3.8H_2O$ and $(cp)_3Gd$ conform with the simple Curie law, but analogous compounds of other lanthanide metals have susceptibilities that instead largely obey the Curie-Weiss law (at least for 300 > T > 100 K) with values of $-\theta$ in the range 15 – 70 K.

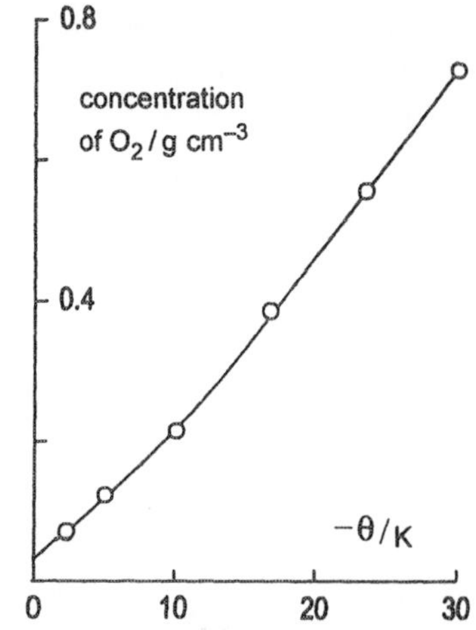

Fig. 3.26 Early data on the interaction between O_2 molecules in liquid N_2. [After Selwood (1943, 1956).]

4. Explain why the purple anion $[MnO_4]^-$ is paramagnetic while the congeneric, colourless anion $[ReO_4]^-$ is diamagnetic.

5. Comment on the following susceptibility data (10^{-11} m^3 mol^{-1}):

$Li_2[CrO_4]$	+26.0	$K_2[CrO_4]$	−4.5	$K[MnO_4]$	+34.9
$Na_2[CrO_4]$	+19.4	$Cs_2[CrO_4]$	−65.0	$Cs[MnO_4]$	+4.4

6. Comment on the following values of the Néel temperature T_N (/K).

CrF_3 80 & MoF_3 185; FeO 198 & FeS 613

7. Recall the two sub-lattice model that was used in § 3.6(3a) to explain the origin of ferrimagnetism. Show that, if the only significant coupling is that *between* the sub-lattices (this being antiferromagnetic), the inverse susceptibility is of the form

$$\chi_m^{-1} = \frac{T}{C} + \rho - \frac{\sigma}{(T - \rho C)}$$

where $\rho = \dfrac{2C_A C_B \lambda}{\mu_0 C^2}$ and $\sigma = \dfrac{\rho\lambda}{2\mu_0 C}(C_A - C_B)^2$.

8. γ-Fe_2O_3 has a structure intimately related to that of magnetite Fe_3O_4. Show that, if the Fe^{3+} ions were randomly distributed between the tetrahedral (A) and octahedral (B) sites occupied in the spinel lattice, then $m_{sat} = \frac{2}{3} \times 5\ \mu_B = 3.33\ \mu_B$ (per formula unit). Explain the discrepancy between this estimate and the figure of $\frac{1}{2} \times 5\ \mu_B = 2.50\ \mu_B$ implied in § 1.3(3d).

The experimental m_{sat} for γ-Fe_2O_3 is 2.4 – 2.5 μ_B at 0 K, but it seems to be dependent on the history of the sample (or purity?). Figures as high as 3.2 μ_B have been reported.

9. Explain the following observations.

(a) Incorporation of Ni(II) by Fe_3O_4 to form $Ni_\delta Fe_{3-\delta}O_4$ reduces its electrical conductivity progressively, there being a dramatic fall when $\delta \approx 0.2$.

(b) Doping of $MnFe_2O_4$ with Fe(II) rapidly increases its conductivity.

10. Interpret the diagram (Fig. 3.27) concerning the saturation moments per formula unit of the ferrites (a) $Ni_xMn_{1-x}Fe_2O_4$, and (b) $Zn_xCo_{1-x}Fe_2O_4$.

Fig. 3.27

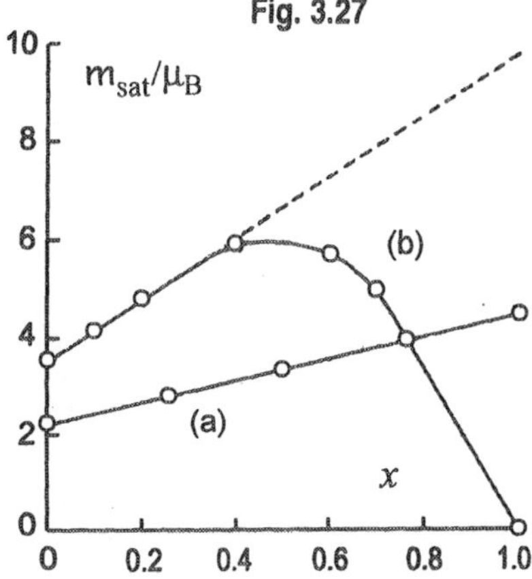

11. Comment on the observations that

(a) the lanthanide garnets $Ln_3Fe_5O_{12}$ have very similar ferrimagnetic Néel temperatures $T_{fN} \sim 565 \pm 15$ K; and

(b) the gadolinium compound has a saturation moment of 15.2 μ_B at 0 K.

12. The compound $CsMn_2(CN)_6.\frac{1}{2}H_2O$ orders ferrimagnetically below 31 K with $m_{sat} \lesssim 3.7\ \mu_B$. At higher temperatures it is paramagnetic with $m_{eff} = 7.04\ \mu_B$. Discuss.

13. An iron-sulphur protein (*ferredoxin*) with a core $[Fe_2S_2(SR)_4]^{2-}$ (where SR denotes a cysteine residue) is only feebly paramagnetic and fails to give an EPR spectrum. On reduction, however, the protein displays an EPR spectrum characteristic of an $S = \frac{1}{2}$ system. Interpret this information.

14. Comment on the following observations.

(a) Hydrated Cr(II) ethanoate $Cr(O_2CMe)_2.H_2O$ is isostructural with its Cu(II) analogue [section 6(5)], but is reported to show only a weak paramagnetism.*

* The compound is often described as diamagnetic, but that is almost certainly untrue. Moreover, NMR studies indicate that the residual paramagnetism varies with temperature.

(b) Chromium(III) forms a binuclear complex $[Cr_2Cl_9]^{3-}$ having a confacial bis-octahedral structure with a Cr–Cr distance of 312 pm. It is a simple paramagnet, at least above *ca.* 90 K. The salt $(Et_4N)_3[Cr_2Cl_9]$, for example, has $m_{eff} = 3.91\ \mu_B$ at 300 K.

Tungsten(III) forms an analogous complex with a W–W distance of 246 pm but apparently this invariably displays only TIP. The salt $K_3[W_2Cl_9]$ has $m_{eff} = 0.47\ \mu_B$ at 300 K.

(The interatomic distances in the *metals* are: Cr 250 pm and W 274 pm.)

15. The magnetic susceptibilities χ of two copper compounds, A and B, are given below (in $m^3\ mol^{-1} \times 10^9$) as a function of temperature. Interpret the data in terms of the electronic and molecular structures of the compounds.

T/K	299	233	203	82	
χ_A	19	24	28	69	A = $Cu(O_2CCH_2OEt)_2.2H_2O$
χ_B	11	10.9	10.2	2.4	B = $Cu(O_2CEt)_2.H_2O$

16. The reaction of Ti(III) chloride with sodium cyclopentadienide in tetrahydrofuran yields, after work-up, a compound of stoichiometry $(C_5H_5)_2TiCl$. Fig. 3.28 shows its susceptibility as a function of temperature.

Account for the essential form of the plot. Also, comment on the observation that the maximum in $\chi(T)$ occurs at lower and higher temperatures, respectively, in the analogous fluoride and bromide complexes.

Fig. 3.28

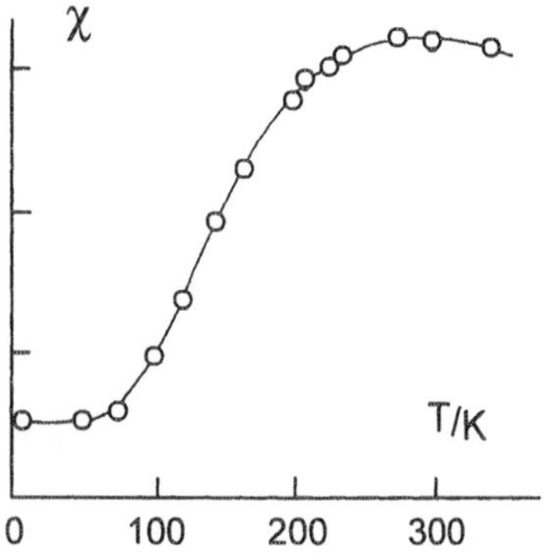

[χ(T) in arbitrary units.]

17. The susceptibility of helium gas is given in Table 3.5. Estimate the mean square radius of the helium atom in the atomic unit a_0^2 (where a_0 is the Bohr radius, 5.291×10^{-11} m).$ Compare your answer with the values for hydrogenic atoms determined by quantum mechanics, namely $\langle r^2\rangle = 3a_0^2/Z^2$.

$ $\dfrac{\mu_0 N_A e^2}{6m_e} = 3.55 \times 10^9\ m^3\ mol^{-1}$.

$[\langle r^2\rangle/a_0^2 = 1.28]$

18. (a) Discuss the following atomic susceptibilities (in $10^{-11}\ m^3\ mol^{-1}$):

K^+	−16.8	Ar	−24.3	Cl^-	−31.8
Cs^+	−46.0	Xe	−57.2	I^-	−69.6

(b) Comment on the fact that potassium metal is paramagnetic while copper is diamagnetic.

19. Linear molecules show anisotropic susceptibilities with *negative* values of $\Delta\chi = \chi_{\parallel} - \chi_{\perp}$ (where $\parallel$ refers to the figure axis). Experimental values of $\Delta\chi$ for diatomic species include: N_2 −10.8, CO −10.3 and CS -30.2×10^{-11} $m^3\ mol^{-1}$. Discuss.

20. Confirm that eqns. 3.41 and 3.44(a,b) are dimensionally correct.

4 Paramagnetism – part I

Atomic and quasi-atomic systems

The net paramagnetism observed in open-shell compounds is, for the most part, dominated by that due to electron spin, described in the previous chapter (§ 3.1). However, free open-shell atoms or ions have also well-defined orbital momentum **L** (§ 2.4) and this contributes on an equal footing. The nature of the individual spin and orbital components of the susceptibility will naturally depend on the details of the coupling of **S** and **L** in the ground state. The necessary theory is addressed below (in sections 1 and 2).

The rest of this book is concerned almost exclusively with paramagnetism in inorganic compounds and, while reference will often be made to pertinent g-data from EPR, it is mainly about bulk paramagnetism in the Curie limit.

The applicability of the theory to ions of the *lanthanide* elements in their compounds is investigated in section 3, and certain discordant data scrutinised further. It is convenient,# subsequently, to compare – or actually, to a large extent, contrast – the magnetic properties of compounds of the *actinide* elements. The chapter concludes with a brief description (section 5) of the paramagnetism observed in the lanthanide and actinide *metals* and also in prominent metallic compounds of the lanthanide elements.

It will transpire, however, that the atomic approximation is of only limited value in the actinide context.

4.1 The combined effect of spin and orbital angular momentum

Since the spin-orbit interaction can be a significant influence, it is advisable to proceed first in terms of the total angular momentum **J** [§ 2.5(4)]. The initial concern is therefore to determine the effect of an applied magnetic field on the multiplet states distinguished by the quantum number J.

1. The total electronic magnetic moment

The combined magnetic moment $\mathbf{m} = \mathbf{m}^o + \mathbf{m}^s$ will depend on the relative orientation of the angular momenta **L** and **S** as determined by the J quantum number. A particular combination is pictured in Fig. 4.1, where it is evident that, because of the spin g-factor $g_e \simeq 2$, *the total magnetic moment* **m** *lies in a different direction from* **J**.

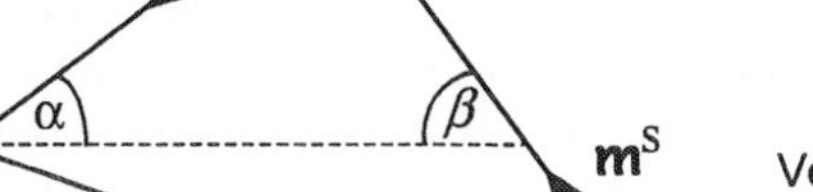
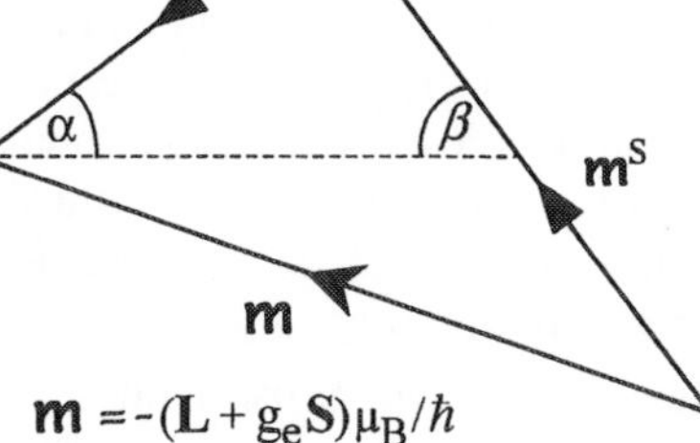

Fig. 4.1

Vector diagrams for $\mathbf{J} = \mathbf{L} + \mathbf{S}$ and the total magnetic moment $\mathbf{m} = \mathbf{m}^o + \mathbf{m}^s$ in a particular multiplet state of an atom.

In the absence of an external magnetic field **J** remains constant in magnitude and direction. The spin-orbit interaction energy $E_{so} = \lambda\mathbf{S.L}/\hbar^2$ (eqn. 3.26) may be interpreted as the potential energy of the spin magnetic moment $\mathbf{m}^s$ in a field $\mathbf{B}^o = -\lambda\mathbf{L}/g_e\mu_B\hbar$ due to the net orbital motion or,

[Recall that the scalar interaction of a magnetic dipole with a magnetic field is $E = -\mathbf{m.B}$ (eqn. 1.6).]

equivalently, as that of the orbital moment $\mathbf{m}^o$ in a field $\mathbf{B}^s = -\lambda\mathbf{S}/\mu_B\hbar$ due to the total electron spin. The vector interaction of $\mathbf{m}^s$ with $\mathbf{B}^o$, and of $\mathbf{m}^o$ with $\mathbf{B}^s$, lead to a gyration (Larmor precession)# of the magnetic moments about the respective fields. But the couples acting on $\mathbf{m}^s$ and of $\mathbf{m}^o$ must be equal and opposite since there is no external couple acting on the atom. It follows that both $\mathbf{m}^s$ and $\mathbf{m}^o$ must precess around the direction of **J**, the constant of motion (Fig. 4.2); so also, naturally, do the angular momentum vectors **L** and **S**. The internal dynamics of the system are such that $\mathbf{m}^s$ and $\mathbf{m}^o$ are effectively both subject to an *internal field* $B_{int} \sim \lambda/\mu_B$ parallel to the resultant angular momentum **J**.

Recall § 3.7(1). [The precessional frequencies are $\nu_L = g\mu_B B/\hbar$, where g is the appropriate g-factor.]

The precession about **J** occurs with an angular velocity of $\lambda\mathbf{J}/\hbar^2$. For a detailed analysis of the motion see, *e.g.*, Bleaney & Bleaney (1989) cited earlier in § 3.

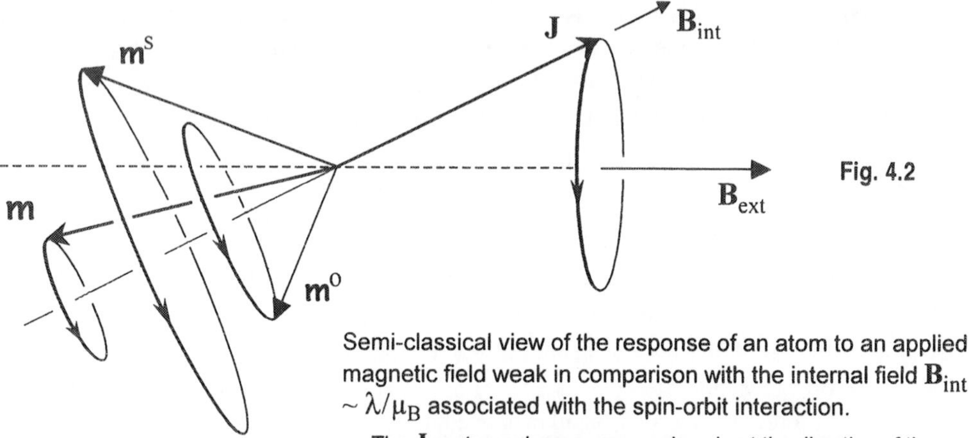

Fig. 4.2

Semi-classical view of the response of an atom to an applied magnetic field weak in comparison with the internal field $\mathbf{B}_{int} \sim \lambda/\mu_B$ associated with the spin-orbit interaction.

The **J** vector undergoes precession about the direction of the applied field $\mathbf{B}_{ext}$ but the magnetic moment vectors (together with **L** and **S**) continue to process around $\mathbf{B}_{int}$, co-directional with **J**.

The total magnetic moment **m** has a steady component parallel to **J**, denoted m_J, and a perpendicular precessing component $m_\perp$ (Fig. 4.3). In the cases of practical interest, involving $\lambda \approx 300$ cm^{-1} or more, B_{int} is very large – at least 600 T (6 x 10^6 G). The period of the precession of $m_\perp$ – *ca.* h/λ, roughly 10^{-14} s when $\lambda \approx 300$ cm^{-1} – is then extremely short in relation to a typical experimental time-scale, and accordingly averages to zero.

The two components of the magnetic moment, m_J and $m_\perp$, are known respectively as the *low frequency* and *high frequency* components (Van Vleck 1932). They are pictured in focus in Fig. 4.3 (facing).

The magnetic fields applicable in the laboratory are of modest proportions compared to B_{int} above; in particular, the field strengths utilised for the routine measurement of bulk susceptibilities are usually no larger than 1 T (10^4 G), at least in the every-day chemical context. Hence, while an external field B_{ext} [denoted B_0 in § 1.2(4b)] will induce the total angular momentum **J** to precess about its direction, the other vectors will continue to gyrate, and much more rapidly, about **J**.

$\lambda = 300$ cm^{-1} is a "ball park" figure for $3d^N$ cations of the first transition series. But λ is commonly larger in $4f^N$ ions of the Lanthanide elements [section 3(3)], and an order of magnitude greater in the $5d^N$ ions of the third transition series.

When $\lambda \approx 300$ cm^{-1}, the frequency of precession of $m_\perp$ about **J** is $\nu_L \sim \lambda/h = ca.$ 10^8 MHz; the frequency with which **J** precesses about an applied field of 1 T ($\nu_L \sim \mu_B B_{ext}/h$) is however a mere 10^4 MHz. In these circumstances, only the component m_J of the resultant moment **m**, rotating *slowly* with **J**, can be observed. Therefore, by analogy with the treatment of spin paramagnetism [§ 3.1(2)] m_J *will be* m_{eff} *for an assembly of atoms characterised by a common J value.*

The projection of **m** onto **J** may be expressed as

$$m_J = \frac{\mathbf{m.J}}{|\mathbf{J}|}$$

[See exercise 1, section 7.]

When instead $B_{ext} >> B_{int}$, a rare situation encountered only for light atoms, where the spin-orbit interaction is weak, the angular momenta **L** and S are uncoupled and each precesses independently about the external field. This complete breakdown of the (magnetic) **L**−**S** coupling, usually requiring a strong magnetic field, is known as the *Paschen-Back effect.* If $B_{ext} \sim B_{int}$ no simple description of the system is possible.

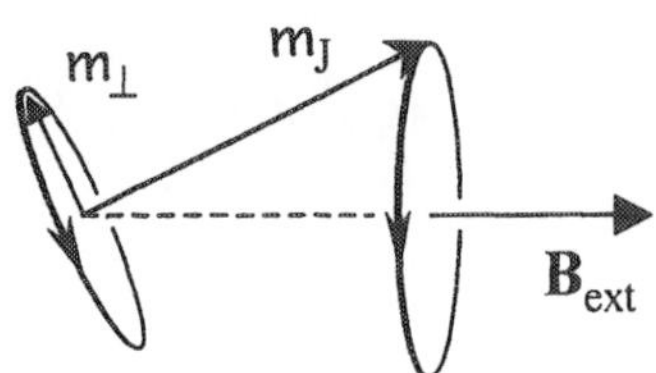

Fig. 4.3 The components of **m** precessing in a magnetic field of modest strength.

2. The Landé g-factor

It can be seen in Fig. 4.1 that the projection of **m** in the direction of **J** is just

$$\mathbf{m}_J = \mathbf{m}^o\cos\alpha + \mathbf{m}^s\cos\beta \qquad (4.1)$$
$$= -(\mu_B/\hbar)(\mathbf{L}\cos\alpha + g_e\mathbf{S}\cos\beta)$$

which, since $\mathbf{J} = \mathbf{L}\cos\alpha + \mathbf{S}\cos\beta$, may instead be expressed as

$$\mathbf{m}_J = -(\mu_B/\hbar)\{\mathbf{J} + (g_e - 1)\mathbf{S}\cos\beta\}$$

But, by the *cosine rule*, $\cos\beta = \dfrac{\mathbf{J}^2 + \mathbf{S}^2 - \mathbf{L}^2}{2\mathbf{J.S}}$, whence

Cosine rule:

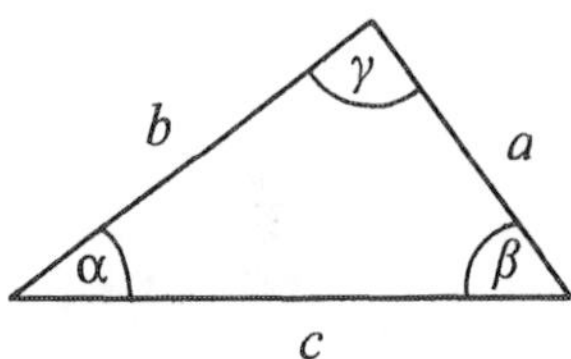

$$c^2 = a^2 + b^2 - 2ab\cos\gamma$$

$$\mathbf{m}_J = -(\mu_B/\hbar)\{1 + (g_e - 1)\frac{\mathbf{J}^2 + \mathbf{S}^2 - \mathbf{L}^2}{2\mathbf{J}^2}\}\mathbf{J}$$

Therefore $\mathbf{m}_J$ is of the form,

$$\mathbf{m}_J = -g_J(\mu_B/\hbar)\mathbf{J} \qquad (4.2)$$

with magnitude $g_J[J(J+1)]^{\frac{1}{2}}\mu_B$, where

$$g_J = 1 + (g_e - 1)\frac{J(J+1) + S(S+1) - L(L+1)}{2J(J+1)} \qquad (4.3)$$

The magnetogyric ratio is thus

$$\gamma_J = -g_J(\mu_B/\hbar)$$

the quantized values $\mathbf{J}^2 = J(J+1)\hbar^2$, *etc.*, having been made explicit. The parameter g_J is known as the **Landé g-factor,** and is of fundamental importance regarding both the Zeeman effect in atomic spectra and also bulk paramagnetism. Eqn. 4.3 is not, of course, applicable when J = 0.

Note also that g_J correctly reduces to g_e when L = 0, and to 1 when S = 0.

3. The Zeeman effect in multiplet states

(a) The first order splitting

The energy of interaction of an atom with an applied field of ordinary strength is given by

$$E = -\mathbf{m}_J.\mathbf{B}_{ext} = +g_J(\mu_B/\hbar)\mathbf{J}.\mathbf{B}_{ext}$$
$$= g_J(\mu_B/\hbar)J_zB_{ext} = g_JM_J\mu_BB_{ext} \qquad (4.4)$$

$g_J \sim 1$. For example, in Lanthanide ions, with configurations $4f^N$, g_J for the ground multiplets ranges from 2/7 to $g_e \simeq 2$. See Table 4.3 later.

Hence the Zeeman splitting pictured in Fig. 4.4 (overleaf). As with electron spin [§ 3.1(1)], the lowest Zeeman level is that having $J = -M_J$.

(b) The second order Zeeman effect

Also shown in Fig. 4.4 are the displacements δ of the Zeeman levels of the ground multiplet due to the mixing effects of the applied field – the *second order Zeeman effect* (recall § 3.3). It turns out that:

The external magnetic field can mix together Zeeman states with a common M_J value if they belong to *adjacent* multiplets.

The mixing implies a small degree of *uncoupling* of **L** and **S**.

This necessarily stabilises the *ground* multiplet.

In the context of paramagnetism, the second order energy shift (*negative* in the case of the ground multiplet) is often written as

$$\delta = \tfrac{1}{2}\alpha B^2 \equiv \tfrac{1}{2}\alpha'(\mu_B B)^2 \qquad (4.5)$$

From here on, for B read B_{ext} (or B_0), unless otherwise indicated.

[*cf.* § 3.3(2a)] where α, like δ, is a function of both J and M_J. The analysis by perturbation theory shows that, for any particular multiplet J, the *average* value of α' is

For an isolated multiplet, the only one populated, α is identifiable with the *magnetizability* [§ 3.3(2a)].

$$\alpha'_J = \frac{2}{3\lambda}(g_J - 1)(g_J - 2) = \alpha_J/\mu_B^2 \qquad (4.6)$$

– a remarkably simple and elegant relation in the circumstances. g_J is to be

The theoretical basis of this figure, and the derivation of eqn. 4.6, are provided by, *e.g.*, Griffith (1961). See also Van Vleck (1932), which remains a very good read.

The scale is unrealistic in that, first of all, the multiplet interval $(J+1)\lambda$ is invariably much greater than the Zeeman splitting Δ_Z (often by two orders of magnitude or more).

Secondly, the magnitude of δ, the displacement of the ground levels through the second order Zeeman effect, has been exaggerated in relation to the first order splitting Δ_Z.

The second order shifts of the M_J levels of the higher multiplet are not shown.

The important particular case of Eu^{3+} is pictured in Fig. 4.9.

[The mixing effects illustrated here should be carefully distinguished from the mixing of muliplet states J that occurs between terms under the spin-orbit interaction [§ 2.5(5)]. The latter takes place in the absence of a magnetic field.]

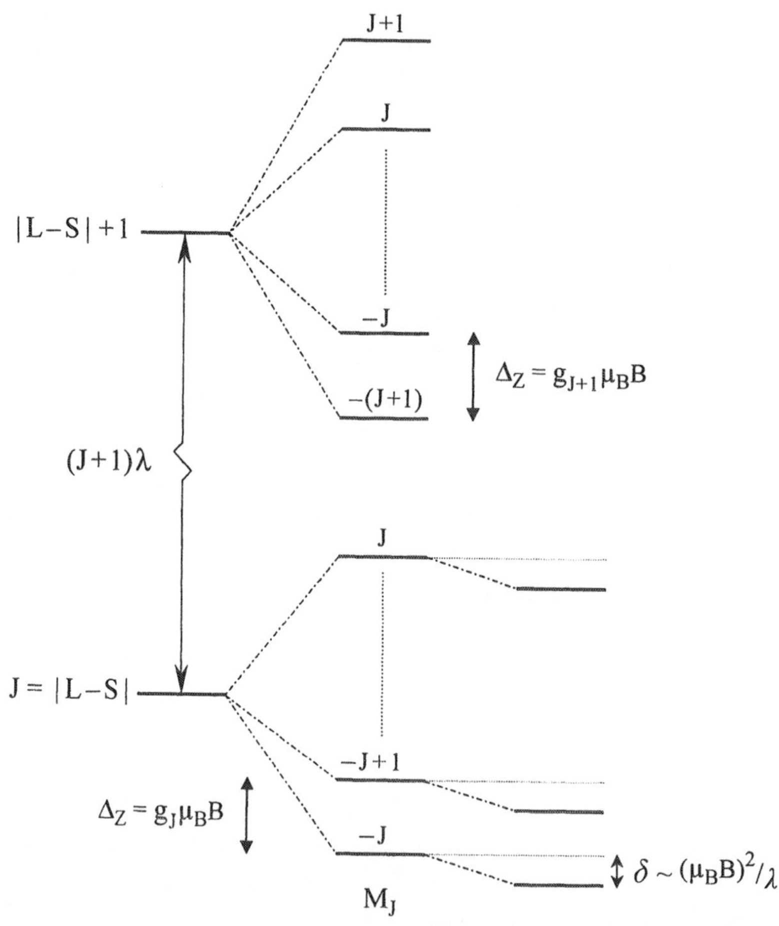

Fig. 4.4 Energy level scheme illustrating the effect of an external magnetic field B on multiplet states. The diagram is constructed for the case of *regular* (normal) multiplets, though only the first of the excited mutiplets is featured. Moreover, only representatlve Zeeman levels are shown.

obtained from eqn. 4.3 (with $g_e \simeq 2$) except that, for the purposes of the formula (4.6), $g_J = L + 2$ when $J = 0$.

In atoms the second order Zeeman effect depends upon the existence of a multiplet splitting of the (ground) Russell-Saunders term. *It will not occur, therefore, if either* L *or* S *is zero* (or indeed when $\lambda \to 0$). Accordingly, returning to eqn. 4.6, it should be noted that $\alpha' = 0$ when $L = 0$ and $g_J = g_e \simeq 2$. The second order Zeeman effect vanishes in the case of "spin-only" atoms, those having S ground states because a degenerate subshell is half-filled (*e.g.* Gd^{3+} $4f^7$ ^8S). That the Zeeman effect is *purely* first order in such systems was, of course, implicit in § 3.1(1).

Similarly, the quantity α' vanishes when $S = 0$ and $g_J = g_L = 1$. The second order effect is absent also in *closed-shell* atoms, with the (singular) ^1S ground state, as is the first order Zeeman effect. There is no paramagnetic response whatsoever to an applied magnetic field. As emphasised earlier [in § 3.3(2a)], closed-shell atoms are *purely diamagnetic.*#

Note also, in this connection, that $^1(L)$, with $L \geqslant 1$ (*e.g.* ^1D of p^2 or d^2), never occurs as a *ground* term.

4.2 The paramagnetic susceptibility of an assembly of atoms

1. Some prevalent issues

The exposition builds on the treatment of spin paramagnetism in § 3.1(2), and the susceptibility will be formulated by analogy with eqn. 3.4. As before, it is assumed that the ensemble is magnetically dilute. *Attention is confined, however, to the Curie limit* ($kT >> \mu_B B$) – the circumstances in which most of the available data of chemical interest were obtained.

Only the *ground* Russell-Saunders term is normally relevant. Term separations are commonly of the order of 10,000 cm^{-1} as compared with $kT \sim 200$ cm^{-1} at ordinary temperatures. The Boltzmann factor $\Delta E/kT$ determining the population of an excited term is thus of the order of $10 - 10^2$ so that, as indicated by the sample figures in Table 4.1 below, there can be no significant occupancy of excited terms at thermal equilibrium.

[$kT = 0.6952$ T cm^{-1}]

Table 4.1 The relative populations of two energy levels separated by an energy $\Delta E = E_2 - E_1$ as given by the Boltzmann distribution law.

$\Delta E/kT$	n_1/n_2	$\Delta E/kT$	n_1/n_2
10^2	2.688×10^{43}	10^{-1}	1.105
10	2.203×10^4	10^{-2}	1.010
1	2.718	10^{-3}	1.001

$E_2 > E_1$; $n_1/n_2 = \exp[\Delta E/kT]$.

Since, in general, the ground electronic states are subject to multiplet splitting, much depends on the relative magnitude of the spin-orbit coupling constant λ and the ambient Boltzmann energy kT of the ensemble. It is appropriate, also convenient, to follow Van Vleck (1932) by distinguishing carefully *three* cases. In order of increasing complexity, they are:

(1) multiplet intervals small compared to kT $\lambda << kT$

(2) multiplet intervals large in comparison with kT $\lambda >> kT$

(3) multiplet intervals comparable to kT $\lambda \sim kT$

The different situations are now explored in turn, commencing with Case 1 where the spin-orbit coupling turns out to be irrelevant.

2. Case 1 (narrow multiplets)

This is the simplest situation, its extreme form being a vanishingly small multiplet splitting of the ground atomic term $^{2S+1}(L)$ – *pure*, unadulterated Russell-Saunders coupling, in other words.[a] If the spin-orbit interaction can be neglected then the spin and orbital angular momenta, **S** and **L**, may be taken to respond independently to the external magnetic field and therefore to contribute separately to the magnetic susceptibility. For the electronic *orbital* motion,

[a] It is as if, for each atom, there were instead two atoms, one carrying only spin and the other only orbital angular momentum.

$$\chi^{o}_{mol} = \frac{\mu_0 N_A \mu_B^2}{3kT} L(L+1)$$

deduced by analogy with the Hund expression (3.4) for the spin susceptibility. The total molar paramagnetic susceptibility of the magnetically dilute, atomic "gas" is therefore

$$\chi_{mol} = \frac{\mu_0 N_A \mu_B^2}{3kT} \{g_e^2 S(S+1) + L(L+1)\} \qquad (4.7)$$

with an effective Bohr magneton number, $n_{eff} = m_{eff}/\mu_B$, given by

$$n_{eff} = \sqrt{g_e^2 S(S+1) + L(L+1)} \qquad (4.8)$$

(where $g_e^2 \simeq 4$). The paramagnetism is expected to be of a *purely first order* kind, as in the spin-only case.[b]

[b] – described by Van Vleck (1932) as *low frequency* paramagnetism, a reference to the relatively slow (and independent) precessions of **L** and **S** about the external field.

In practice the spin-orbit coupling is rarely negligible. But eqn. 4.7 still applies if the temperature of the ensemble is sufficiently high that there is an almost even Boltzmann distribution among the different multiplet levels. As can be seen by reference again to Table 4.1, the latter situation is realised when $kT \gtrsim 10^3 \lambda$. When the multiplet states are almost equally populated (with their appropriate 2J + 1 weightings) the *ensemble* should behave as though the spin-orbit interaction were absent.

3. Case 2 (wide multiplets)

This is the inverse of Case 1.[c] The multiplet intervals are now assumed to be so large that the ensemble is effectively "frozen" into the lowest multiplet state. To all intents and purposes, this happens when $\lambda \gtrsim 10kT$ (*cf.* Table 4.1). Each atom is accordingly characterised by a definite value of J, in addition to its parent quantum numbers, L and S, defining the ground term. This is a very special, knife-edge situation,[d] at least so far as measurements at ordinary temperatures are concerned. The spin-orbit interaction must be very strong, yet *not* large enough to subvert the Russell-Saunders coupling to the extent of invalidating the quantum numbers, L and S.

[c] – but, like Case 1, is formally of the category envisaged in § 3.4(1).

[d] – encountered only among ions of the lanthanide elements (section 3 following).

(a) The first order susceptibility

It was established in section 1(1) that, for external fields of modest strength, only the parallel component m_J of the magnetic moment (eqn. 4.2) can be observed, this being identifiable with m_{eff} in the Curie limit. (The perpendicular component $m_\perp$ is tied in precession about **J**.) Thus,

$$n_{eff} = g_J[J(J+1)]^{\frac{1}{2}} \tag{4.9}$$

where g_J is defined by eqn. 4.3. Since $g_e \simeq 2$, for all practical purposes

$$g_J = \frac{3}{2} + \frac{S(S+1) - L(L+1)}{2J(J+1)} \tag{4.10}$$

Eqn. 4.9 is sometimes known as the *Hund-Landé formula.*[a] (It is only applicable, it should be emphasised, when $\lambda >> kT >> \Delta_Z$.)

[a] It was first used by F Hund (1925) to interpret magnetic moment data for compounds of the lanthanide ions Ln^{3+}.

(b) The second-order aspect of the susceptibility

According to eqn. 4.5, the ground multiplet experiences a stabilisation $\delta = \frac{1}{2}\alpha'(\mu_B B)^2$ through the field-induced mixing with excited J states. It then follows, recalling § 3.3(2a), that the molar susceptibility of the system has an additional contribution from *temperature independent paramagnetism* (TIP),

$$\chi_{TIP}(J) = \mu_0 N_A \alpha_J \tag{4.11}$$

where $\alpha_J = \alpha_J' \mu_B^2$.

The complete expression for the susceptibility in Case 2 is therefore

$$\chi_{mol} = \frac{\mu_0 N_A \mu_B^2}{3kT}\{g_J^2 J(J+1)\} + \mu_0 N_A \alpha_J$$

$$= \frac{\mu_0 N_A \mu_B^2}{3kT}\{g_J^2 J(J+1) + 3kT\alpha_J'\} \tag{4.12}$$

with an *apparent* magnetic moment corresponding to

$$n_{eff}(J) = \sqrt{g_J^2 J(J+1) + 3kT\alpha_J'} \tag{4.13}$$

It should be noted that the second order contribution to m_{eff} varies with temperature and may be a source of confusion if not carefully distinguished from the first order component, $g_J[J(J+1)]^{\frac{1}{2}}\mu_B$. The second order term can be important, especially at higher temperatures and, of course, when the first order term is small.

Viewed in terms of the vector model, the TIP term arises from a limited *direct* response of **L** and **S** to the external magnetic field, despite the strong coupling between them, and their consequently rapid precession about the resultant **J**. In other words, the applied field partially uncouples **L** and **S**. The outcome, in effect, is that the component $m_\perp$ also makes a contribution to the observed magnetic moment. This is likely to be the greater the larger $m_\perp$ is in relation to m_J, *i.e.* when J is small compared with L and S, a situation which arises for *normal* multiplets when L and S have similar values (*e.g.* Eu^{3+} and Sm^{3+}, discussed in section 3(4) below).

Recognising that TIP derives from rapidly precessing angular momenta, Van Vleck described the effect as *high frequency* paramagnetism, distinguishing it from the simple low frequency (Curie) paramagnetism associated with angular momenta gyrating comparatively slowly about the external field.

4. Case 3 (multiplet widths ~ kT)

This is the general case[b] in which the excited multiplets of the ground term have appreciable thermal population (though the circumstances may be remote from the Case 1 limit, $\lambda/kT \to 0$). Each individual multiplet state

[b] – and belongs to the category of § 3.4(2). The Curie law is not obeyed when $\Delta_{so} \sim kT$.

contributes to the susceptibility a term of the form (4.12), including both first and second order elements, and so, for the ensemble,

$$\chi_{mol} = \frac{\mu_0 N_A \mu_B^2}{3kT} \frac{\sum_J (2J+1)\, n_{eff}^2(J) \exp[-\frac{J(J+1)\lambda}{2kT}]}{\sum_J (2J+1) \exp[-\frac{J(J+1)\lambda}{2kT}]} \qquad (4.14)$$

\# – namely, $-\frac{1}{2}[L(L+1)+S(S+1)]$

where the $n_{eff}(J)$ are given by eqn. 4.13. The energies of the multiplet states $E_{so}(J)$ are expressed relative to the constant term in eqn. 2.42.#

Clearly, the susceptibility will have a more complex dependency on the temperature than the simple Curie ideal of $\chi \propto T^{-1}$; this exemplifies the *non-Curie* or *complex paramagnetism* anticipated in § 3.4(2). It is noteworthy, in particular, that the second order terms α_J for both ground and *excited* multiplets make a contribution that varies with temperature.

5. Scope of the theory

The beam studies fully confirm the theoretical expectations regarding the nature of the electronic ground states.

There are no experimental data for the susceptibilities of atomic *vapours* that adequately test the foregoing theory. Early experimental effort in this area, fraught with difficulties, was quickly abandoned with the advent of atomic *beam* techniques [section (6)], which provide a more accurate and more widely applicable means of investigating the magnetic properties of atoms and also small molecules.

Susceptibility data are however available for certain paramagnetic gases, notably O_2 and NO. The approach of the previous section provides a full explanation of the observations when adapted for the limited quantization of *orbital* angular momentum (about the figure axis) in linear molecules. The case of NO proves particularly instructive [§ 3.4(2)].

The theory embraces transition metal cations with a partly filled $t_2(d)$ subshell, having usually a ground state ^{2S+1}T.

Arguably, the theory finds its most important application in connection with the prominent category of transition metal compound containing cations with electronic ground states analogous to the ^{2S+1}P states of the p^N configuration in free atoms. It is well known that, in *octahedral* and *tetrahedral* transition metal complexes, the *d* orbitals are split by the ligand field into two well-separated sets of orbitals denoted e and t_2, respectively. The latter set, comprising *three* of the *d* orbitals, displays angular momentum equivalent (save for a change in phase) to that of the set of atomic *p* orbitals. Cations with a partly filled $t_2(d)$ subshell may therefore be described by a straightforward modification of the above theory. This aspect of the subject is taken up in § 5.6(5).

– originally the work of Hund (1925), but extended and improved upon by Van Vleck and others.

However, the most celebrated application of the theory concerns the paramagnetism of compounds of the *lanthanide* (or *rare earth*) elements at ordinary temperatures. As detailed in the next section, the magnetic moments of a wide variety of lanthanide compounds can be approximated by eqn. 4.9 (of Case 2) and reliably enough for magnetic measurements to prove an essential guide to electronic structure in this area of solid-state chemistry.

6. Postscript: a digression on other sources of information about electronic magnetic moments or related quantities

The present text is focussed on the interpretation of bulk susceptibilities, the principal source of magnetic moment data for materials. There are, however, other means by which the magnetic moments of open-shell atoms and molecules (or the constituent parameters) can be characterised.

As regards the *gas phase*, the methods include:

(1) the Zeeman effect in electronic spectroscopy;

(2) electron paramagnetic resonance (EPR) of gases or beams;

(3) deflection experiments with beams.

EPR is also applicable to liquid samples and to *solids*, with enormous free-standing significance, and providing a powerful complementary technique to bulk susceptibility measurements. In the case of solids, particular mention should be made of:

(4) neutron diffraction;

(5) Mössbauer spectroscopy;

(6) Magnetic circular dichroism (MCD).

The Zeeman effect in atomic spectra allows the determination of g_J (and of course J) together with, at high resolution and using strong external fields, the second order energy shifts $\delta(J,M_J)$ [*cf.* eqn. 4.5]. Atomic emission spectroscopy embraces atomic *cations* as well as the neutral atoms.

The Zeeman effect may also be exploited in the high resolution spectroscopy of *molecules*.

The same parameters can be obtained for neutral species from the EPR spectra of gases and beams. In one experimental *tour de force* (1963), EPR spectra of oxygen atoms were found to display the *four* one-quantum transitions expected of a $^3P_2$ (p^4) ground multiplet (and consistent with $g = \frac{3}{2}$) plus, though with much reduced intensity, a further six explicable multi-quantum transitions. The EPR spectrum of thallium atoms, whether investigated as the vapour or in the form of an atomic beam, shows just *one* transition corresponding to $g = \frac{2}{3}$, confirming the $^2P_{\frac{1}{2}}$ (p^1) ground multiplet.

The atoms were investigated in the effluent from an electric discharge in O_2 gas.

If there were no 2nd order Zeeman effect, only a single one-quantum resonance would be observed.

In *deflection* experiments, now largely superseded by spectroscopic measurements, the atomic (or molecular) beam is subjected to a magnetic field with a substantial *gradient* directed at right angles to its path. The atoms will be space quantized by the field, the permitted values of M_J being equally probable. At the same time, each atom will experience a translational force in the perpendicular direction (z, say) of magnitude

$$F_z = m_z \frac{dB}{dz} = -M_J g_J \mu_B \frac{dB}{dz} \qquad \text{[recall (1.8)]}$$

which will deflect the atom from its path to a degree reflecting the value of M_J. The incident beam will therefore splay laterally into $2J+1$ separate beams (including the unaffected beam of atoms having $M_J = 0$). Careful detection of the effect allows the determination of both g_J and J. The overall spread of the beam is proportional to the quantity $g_J J$. A beam of thallium atoms thus splits into two, with a divergence yielding $g_J J = \frac{2}{3} \cdot \frac{1}{2} = \frac{1}{3}$.

In pioneering experiments by Stern and Gerlach (1921 – 24), the splitting of a beam of silver atoms ($5s^1$) into *two* separate beams demonstrated directly the existence of electron spin and its space quantization, with the quantum number $m_s = \pm\frac{1}{2}$.

The exploitation of neutron diffraction in the determination of magnetic moments is restricted to magnetically *ordered* materials. It depends upon the component of the scattering due to the interaction of the neutron magnetic moment with the moments of the open-shell ions present.

Consult Ebsworth *et al.* (1991) and Gibb (1976) for the details.

Mössbauer spectroscopy (*nuclear γ resonance spectroscopy*) is based on transitions between *nuclear* spin states. Remarkably, the transitions (belonging to the γ region of the electromagnetic spectrum) often prove sufficiently sharp that a slight dependency of their energies on the chemical environment of the nucleus can be characterised in some detail. The Mössbauer resonances may then provide information about the electronic structure of the atom concerned through *hyperfine* structure resulting from the interaction of the nuclear and electronic magnetic moments.[a] A magnetic field is sometimes applied to clarify features of the spectrum by means of the Zeeman effect. In the case of *ordered* solids, the Zeeman effect due to the internal Weiss field B_W [§ 3.2] alone allows the measurement of that field.

[a] The energy of the interaction is of the form $E_{en} = a\,\mathbf{J.I}/\hbar^2$, where a is the hyperfine coupling constant. The fine structure thus created is a prime feature of EPR spectra.

Circular dichroism (CD) is the differential absorption of left- and right-circularly polarized radiation, a property of optically active molecules. The effect may be induced in any system by a magnetic field and, for a particular optical transition, the difference in absorption strength for the alternative polarizations reflects the details of the Zeeman splittings of the electronic states involved. MCD spectra, usually confined to the visible and near UV region (but applicable to *solutions*), can thereby provide similar information to conventional Zeeman spectroscopy. In favourable circumstances, the magnetic moments of both ground and excited states may be determined.

See Denning (1987) and references cited there.

4.3 Paramagnetism in compounds of the Lanthanide elements

1. General considerations

Table 4.2 n_{eff} values (300 K) for a selection of compounds of Ce(III) $4f^1$ and Yb(III) $4f^{13}$.

CeF_3	2.47	YbF_3	4.54
$CeCl_3$	2.44	$YbCl_3$	4.47
Ce_2O_3	2.35	Yb_2O_3	4.30
Ce_2S_3	2.44	Yb_2S_3	4.52
$(cp)_3Ce$	2.46	$(cp)_3Yb$	4.00

[$cp = C_5H_5$]

The paramagnetism observed almost ubiquitously among compounds of the Lanthanide elements – *cerium* through to *lutetium* (general symbol Ln) – is due predominantly to the electrons of the incomplete $4f$ subshell. The magnetic moments reported for the compounds of any one of these elements do not vary greatly with the partner ligand, usually an anion, at least at ordinary temperatures. Illustrating this point is a selection of data concerning cerium(III) and ytterbium(III) compounds in Table 4.2. The data are fairly representative: in general, the magnetic moments measured at around 300 K for compounds of a particular Lanthanide metal show a variation of no more than 15 %, frequently less. Some of this can be blamed upon ambiguities associated with the processing of raw susceptibility data, *e.g.* whether in terms of the Curie-Weiss law or by assuming the simple Curie law. (Occasionally, moreover, there may be doubts concerning the purity of the materials in early measurements, *e.g.* Ce_2O_3 and $(cp)_3Yb$.) By way of contrast, the m_{eff} values at ordinary temperatures reported for compounds of the first transition series may vary by a factor of two or more[b].

[b] The particularly large variations are found for the $d^4 - d^7$ cations which have alternative *high-spin* and *low-spin* forms [§ 5.3].

It is clear from the data in Table 4.2 that the spin-only formula has little scope here. (Any compound containing Ce^{3+} or Yb^{3+} would be expected to have $n_{eff} = \sqrt{3} = 1.73$.) However, as demonstrated in section (3) below, the atomic theory apparently applies universally to susceptibility data gathered at ordinary temperatures for both "ionic" and molecular compounds of the lanthanide metals. This remarkable outcome, that a theory developed for atomic gases can be employed to interpret the paramagnetic behaviour of molecules and even solid materials, indicates that the outer $4f$ electrons of the Lanthanide cations (commonly Ln^{3+})[a] must have very contracted radial distribution functions; only then can one understand their magnetic properties being comparatively insensitive to the chemical environment of the ions.

a – but not exclusively. Ln^{2+} species are well known for Sm, Eu, Tm & Yb, and Ln^{4+} for Ce, Pr & Tb.

Compounds containing Nd^{4+} & Dy^{4+}, and (apparently) Pr^{2+}, Nd^{2+} & Dy^{2+}, can also be made.

The contracted, core-like nature of the $4f$ orbitals in the Lanthanide elements has been abundantly confirmed by *ab initio* quantum mechanical calculations using self-consistent field methods. For example, Fig. 4.8 shows the result of a seminal calculation (1962) for the neutral gadolinium atom, Gd (Xe) $6s^2 5d 4f^7$. (The vertical line labelled # is drawn at the experimental metallic radius for Gd.)

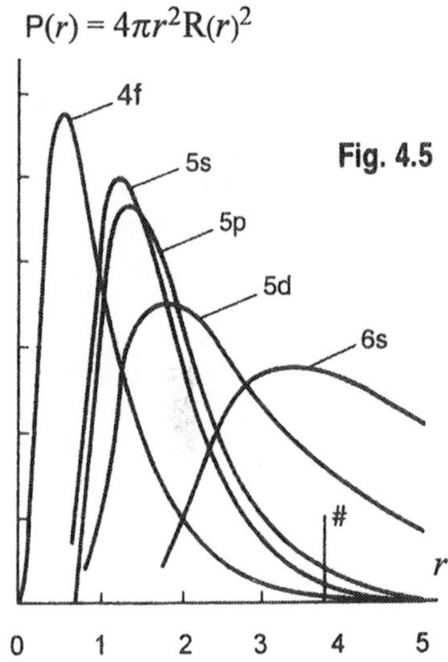

Fig. 4.5 The outer maxima of the radial distribution functions $P(r)$ (arbitrary scale) for the outer electrons of the gadolinium atom.
The radial coordinate is expressed in the atomic unit a_0.

Notice that the $4f$ distribution peaks well inside the maximum in the outer $5p^6$ "peel" of the xenon core, and has very little density protruding from the core. This proves true also of the Lanthanide cations, with the configuration (Xe) $4f^N$. Covalency in compounds of the Lanthanide elements must be based mainly on the $5d$, $6s$ and (perhaps) $6p$ orbitals, rather than $4f$. The low-lying $5d$ orbitals are a particularly significant feature of the electronic structure of Lanthanide atoms and ions (and are also occupied in the ground states of neutral La, Ce and Lu.)

The viability of the atomic theory in unadulterated form will also depend on the absence of significant magnetic interaction between neighbouring Lanthanide ions, the compounds being invariably solids. But the Lanthanides pose few problems in this respect: the reclusive nature of the $4f$ electrons ensures that most compounds show cooperative magnetic behaviour only at very low temperatures (commonly less than 10 K).[b]

b However, *metallic* and strongly semi-conducting compounds of these elements often show relatively high ordering temper-atures. This is particularly true of the Ln^{2+} compounds: *e.g.* EuO undergoes a ferro-magnetic transition at 69 K.

2. Ground states of $4f^N$ configurations

Application of the theory of section 4(2) requires prior characterisation of the ground multiplets of the $4f^N$ cations. This is a straightforward task, given the principles established in § 2.4(3,4) and § 2.5(4). (Recall especially the three rules attributed to Hund).[c] To begin with, by the hole theorem, the ground terms of the f^1 and f^{13} configurations (*cf.* Table 4.2) are both simply ^2F, and the multiplets of lowest energy are $^2F_{5/2}$ and $^2F_{7/2}$, respectively; and, as deduced earlier, the ground multiplet of f^7 (the half-filled subshell) is the unique $^8S_0$. Filling in some further detail, the ground multiplets of f^2 (& f^{12}) *etc.* may be determined through diagrams such as those in Fig. 4.6 which convey the salient features of the ground configurations.

c NB (D) of § 2.2(4), NB (G) of § 2.4(3) and NB (J) of § 2.5(4).

m_l	+3	+2	+1	0	−1	−2	−3		
	↑	↑	—	—	—	—	—	f^2	^3H
	↑	↑	↑	—	—	—	—	f^3	^4I
	↑	↑	↑	↑	—	—	—	f^4	^5I

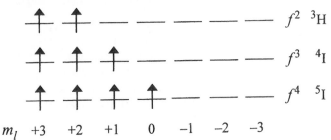

Fig. 4.6 The ground terms of the configurations $f^2 - f^4$, pictured as the microstates with the maximum values of $M_S = S$ and $M_L = L$.

In the case of f^3 (Nd^{3+}), for example, the ground term must have S = $\frac{3}{2}$, the maximum possible S (Hund's 1st rule). The maximal value of M_S is therefore $\frac{3}{2}$ also, when the maximum possible M_L = 6, which can only be derived from L = 6. Hence a ^{4}I ground term with J ranging in value from 9/2 to 15/2. The ground multiplet for f^3 is therefore $^4I_{9/2}$, and for f^{11} (Er^{3+}) (by Hund's 3rd rule) it is $^4I_{15/2}$. The hole theorem may again be invoked to identify the ground multiplets for the cations with $N > 7$. The complete specification of ground states is given in Table 4.3 opposite.

3. Applicability of the Hund-Landé equation

Spin-orbit coupling is substantial in the Lanthanide 4f subshell and increases in strength some five-fold across the series (Fig. 4.10). The large ζ_{4f} that are observed (640 – 2940 cm^{-1} in the Ln^{3+} ions) reflect both high nuclear charges and the large values of $\langle r^{-3} \rangle$ associated with relatively contracted "outer" orbitals (recall eqn. 2.34). However, the spin-orbit coupling is not quite large enough to invalidate the Russell-Saunders model. #

But the Russell-Saunders scheme is on the verge of breakdown because the spin-orbit interaction is sufficiently large to cause a noticeable mixing of multiplet states having the same value of J. The ground terms are modified by an admixture of ≲ 2 % of the excited terms.

Fig. 4.7

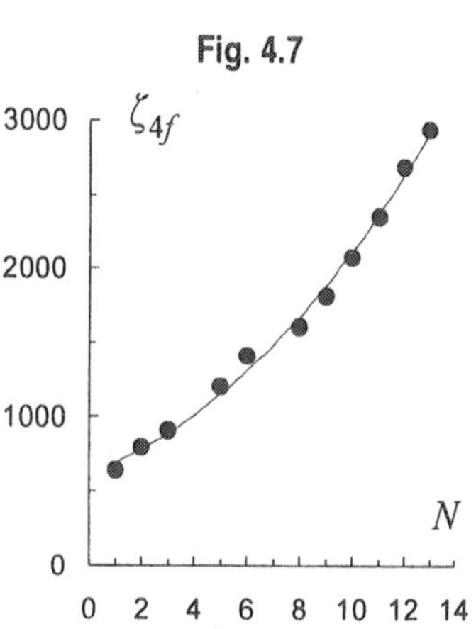

Variation of ζ_{4f} in the ground terms of the Ln^{3+} cations.

The somewhat erratic trend across the series is due to the 2nd order effects of the spin-orbit interaction which are difficult to unravel when estimating ζ_{4f} from atomic energy level data.

[The ζ_{4f} values plotted above are those which give the best *overall* fit to the multiplet structure of the ground terms.]

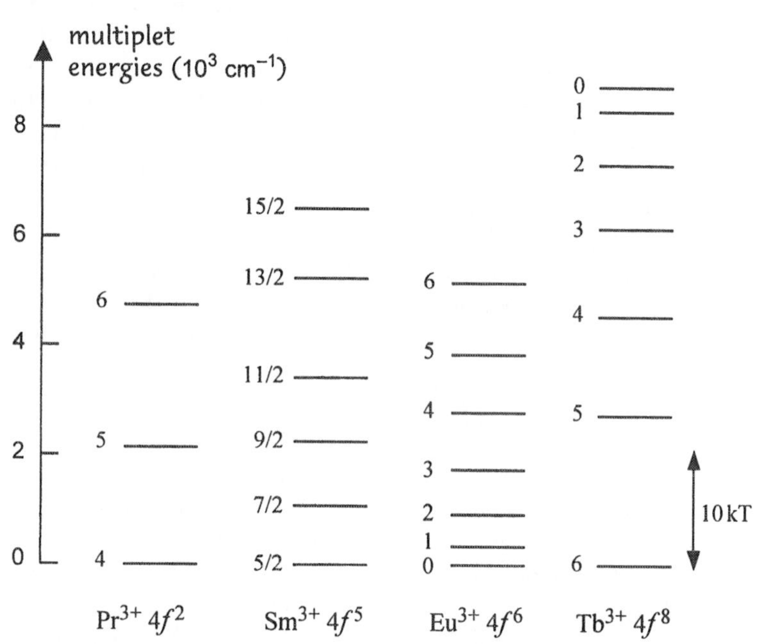

Fig. 4.8 Comparison of the energies of ground state multiplets for selected Ln^{3+} ions [kT = 208.6 cm^{-1} at 300 K]

The greater *width* (energy spread) of the Tb^{3+} multiplets as compared with Eu^{3+} is due essentially to an increase in ζ_{4f}.

As a result of the strong spin-orbit interactions, the multiplet widths for the ground terms lie in the range 5000 – 10,000 cm^{-1} (as illustrated by Fig. 4.8) and, with the obvious exceptions of Eu^{3+} and Sm^{3+}, the separation of the ground and first excited multiplets invariably exceeds 2000 cm^{-1}, roughly 10kT at ordinary temperatures. Thus the simple theory of **Case 2** [section 2(3)] should be applicable in most instances.

The theoretical expectations for the magnetic moments $m_{eff} = n_{eff}\mu_B$ are detailed in Table 4.3. Estimates by both the Hund-Landé equation (4.25) and the more sophisticated formula (4.12)[a], which incorporates second order effects (TIP), are included. However, discounting Sm^{3+} and Eu^{3+}, the alternative estimates differ but slightly, and only to a significant extent for Ln^{3+} ions with a less than half-filled subshell ($N < 7$) where, by the Landé interval rule (eqn. 2.42), the initial multiplet intervals are relatively small.

[a] The theoretical predictions (B) in Table 4.3 should be evaluated by reference to n_{eff} data obtained by assuming a simple T^{-1} dependency of the bulk susceptibility.

Table 4.3

Ground states of the ions Ln^{3+} and the values of n_{eff} at 300 K predicted by the atomic theory*.

ion	no. of $4f$ electrons	ground state L	S	J	g_J	theoretical n_{eff} (A)	(B)
Ce^{3+}	1	3	1/2	5/2	6/7	2.54	2.56
Pr^{3+}	2	5	1	4	4/5	3.58	3.62
Nd^{3+}	3	6	3/2	9/2	8/11	3.62	3.68
Pm^{3+}	4	6	2	4	3/5	2.68	2.86
Sm^{3+}	5	5	5/2	5/2	2/7	0.85	[~ 1.60] #
Eu^{3+}	6	3	3	0	–	0	[~ 3.45] #
Gd^{3+}	7	0	7/2	7/2	2	7.94	7.94
Tb^{3+}	8	3	3	6	3/2	9.72	9.72
Dy^{3+}	9	5	5/2	15/2	4/3	10.63	10.63
Ho^{3+}	10	6	2	8	5/4	10.60	10.60
Er^{3+}	11	6	3/2	15/2	6/5	9.57	9.57
Tm^{3+}	12	5	1	6	7/6	7.63	7.63
Yb^{3+}	13	3	1/2	7/2	8/7	4.50	4.50
Lu^{3+}	14	0	0	0	–	0	0

* Case 2 (eqn. 4.12): (A) without TIP; (B) with TIP.

\# Case 3 (eqn. 4.14), as originally estimated by Van Vleck (1932).

Note that $g_J < 1$ for $N < 7$ while $1 < g_J < 2$ when $N > 7$.

The effect of the spin-orbit interaction is such that $m_{eff} = g_J[J(J+1)]^{\frac{1}{2}}\mu_B$ is less than the spin-only value before the half-full subshell, since **L** and **S** are coupled in opposition in the ground state, but greater thereafter, when **L** and **S** have a concerted alignment. The Eu^{3+} ($4f^6$) case is special in that **L** and **S** cancel *exactly* in the $^7F_0$ ground multiplet so that the Hund-Landé equation predicts (incorrectly) a zero paramagnetic moment.

The scope of the atomic theory may be judged by Figure 4.9 where the predicted m_{eff} values (Table 4.3) are compared with the range of experimental data typical of the Ln^{3+} ions. Also plotted in the figure are the naïve expectations of the spin-only formula and eqn. 4.8 (Case 1).

Some specific data for particular compound types, sweeping across the lanthanide series, are provided in Table 4.4. Note that no magnetic data are available for Pm^{3+} compounds.[b]

[b] All the known isotopes of the element *promethium* (Pm) are highly radioactive (having very unstable nuclear structures); none of its compounds is well characterised. Pm was first identified as a fission product of ^{235}U.

It can be seen that, for the most part (Sm^{3+} and Eu^{3+} again excepted), the Hund-Landé formula works remarkably well. The measure of agreement obtained for molecular complexes such as, for example, the *tris*-cyclopentadienides $(C_5H_5)_3Ln$, or for the hydrated ions in solution, is every bit as good as it is in the case of comparatively ionic materials such as the fluorides LnF_3 or the sequioxides Ln_2O_3.

○ $n_{eff} = g_J[J(J+1)]^{\frac{1}{2}}$

■ $n_{eff} = [4S(S+1) + L(L+1)]^{\frac{1}{2}}$

● $n_{eff} = 2[S(S+1)]^{\frac{1}{2}}$

I – the error bar defines the range of experimental values of n_{eff} at about 300 K. [Much of the variation is due to *crystal field effects*, as explained in section (5).]

The asymmetry of the ○ plot reflects *multiplet inversion.*
[See NB (J) of § 2.5(4).]

The symmetry about $N = 7$ of the other two plots is consequent upon the *hole theorem.*
[See NB (H) of § 2.4(3).]

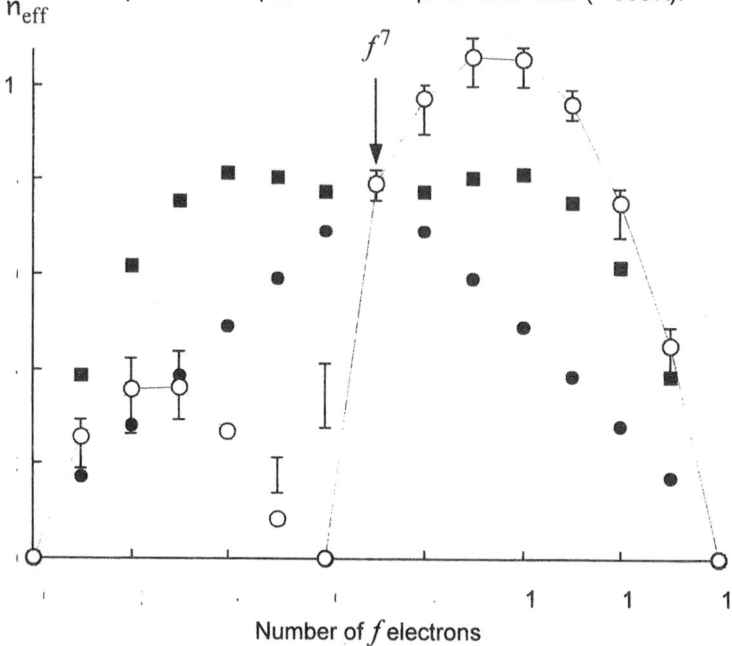

Fig. 4.9 Simple estimates of n_{eff} for Ln^{3+} ions $4f^N$ compared with representative experimental data (~ 300 K).

Table 4.4 n_{eff} data (300 K) for selected Ln(III) compounds

ion	no. of $4f$ electrons	$g_J[J(J+1)]^{\frac{1}{2}}$	LnF_3	$(cp)_3Ln$	A	B	C
Ce^{3+}	1	2.54	2.47	2.46	–	2.27	2.06
Pr^{3+}	2	3.58	3.59	3.61	3.33	3.38	3.34
Nd^{3+}	3	3.62	3.60	3.63	3.34	3.61	3.52
Pm^{3+}	4	2.68	–	–	–	–	–
Sm^{3+}	5	0.84	–	1.60	1.59	1.59	–
Eu^{3+}	6	0	3.40	–	3.34	3.35	3.42
Gd^{3+}	7	7.94	7.79	7.98	7.92	8.13	7.93
Tb^{3+}	8	9.72	9.47	9.00	9.64	9.89	9.67
Dy^{3+}	9	10.63	10.42	10.00	10.69	10.82	10.46
Ho^{3+}	10	10.60	10.47	10.36	10.37	10.65	–
Er^{3+}	11	9.57	9.34	9.45	8.93	9.69	9.54
Tm^{3+}	12	7.63	7.35	7.24	7.01	–	7.47
Yb^{3+}	13	4.50	4.54	4.00	4.06	4.59	4.28

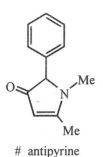

$A = Ln(dpm)_3(4\text{-}pic)_2$ where *dpm*H = *dipivaloylmethane* (2,2,6,6-tetra-methyl-heptane-3,5-dione) and 4-*pic* = 4-picoline (4-methyl-pyridine); $B = [Ln(Me_2SeO)_8](ClO_4)_3$; $C = [Ln(antipyrine)_6]I_3$ where *antipyrine* (#) = 2,3- dimethyl-phenyl-Δ^3-pyrazolin-5-one; and $cp = C_5H_5$

O N Me Me

antipyrine

Though the chemistry of the lanthanide elements is indeed dominated by the +3 oxidation state, certain of these elements display well characterised +2 and +4 oxidation states. Some of the data available for compounds of the "divalent" ions Eu^{2+} and Sm^{2+}, and for the "tetravalent" ions Pr^{4+} and Tb^{4+}, are listed in Table 4.5. The simple first order theory applies equally well to these compounds, save for the case of Sm^{2+}, isoelectronic with Eu^{3+}. Well-characterised cations representing other than the +3 oxidation state include Ce^{4+} ($4f^0$) and Yb^{2+} ($4f^{14}$), but both have a $^1S_0$ ground state and prove to be diamagnetic, like La^{3+} and Lu^{3+}.

The effective magnetic moments at ordinary temperatures are sufficiently characteristic of a given $4f^N$ configuration that, where a lanthanide element may exhibit alternative oxidation states, bulk susceptibility measurements can usually be employed on a purely empirical basis to diagnose the oxidation state. Thus Ce(IV) compounds (diamagnetic) are easily distinguished from those containing Ce^{3+} (paramagnetic), and it is equally straightforward to distinguish between the europium ions, Eu^{2+} and Eu^{3+}, even though the simple atomic theory fails for the "trivalent" ion. However, magnetic measurements are unhelpful in identifying Nd^{4+}, a rare species isoelectronic with Pr^{3+}, because its m_{eff} value at ordinary temperatures is not significantly different from that of Nd^{3+}. (As can be inferred from Table 4.3, they are both about 3.6 μ_B.)

Table 4.5 n_{eff} data (300 K) for selected Ln(II) and Ln(IV) compounds

Ion	Compound	n_{eff}
(a) "Divalent" ions Ln^{2+}		
Eu^{2+} $4f^7$	EuF_2	7.46
	$EuCl_2$	7.92
	$EuBr_2$	7.93
	EuI_2	7.85
	EuO	7.94
	EuS	7.49
	$(cp)_2Eu$	7.63
Sm^{2+} $4f^6$	$SmBr_2$	3.6
	SmS	3.4
	$(cp)_2Sm$	3.60
(b) "Tetravalent" ions Ln^{4+}		
Pr^{4+} $4f^1$	PrO_2	2.47
	$BaPrO_3$	2.32
Tb^{4+} $4f^7$	TbO_2	7.90
	$BaTbO_3$	7.75

It remains to explain the apparently discordant magnetic properties of the $4f^6$ and $4f^5$ cations – in particular, Eu^{3+} and Sm^{3+} – for which the experimental magnetic moments are very much greater than predicted by the Hund-Landé formula. According to the latter, Eu^{3+} should be diamagnetic.

The reason for these discrepancies is not hard to divine. As shown in Fig. 4.8, the Eu^{3+} and Sm^{3+} cations have unusually small energy separations between the lower multiplet states. This is especially true of Eu^{3+} (illustrated also in Fig. 4.10), where the $^7F_1$ excited multiplet lies a mere 355 cm^{-1} above the ground multiplet $^7F_0$.[a] At ordinary temperatures, where kT ~ 200 cm^{-1}, there must accordingly be a significant population of the excited multiplet states, and this immediately invalidates the Hund-Landé equation. At the same time the unusually small multiplet intervals suggest the possibility of important second order contributions to m_{eff}. In short, the room temperature data for the Eu^{3+} and Sm^{3+} ions (and also Sm^{2+}) require the more elaborate, general theory (Case 3) of section 2(4).

a In the Sm^{3+} ($4f^5$) case, the first multiplet interval is somewhat larger (almost 1000 cm^{-1}) though still quite small in comparison with the other Ln^{3+} ions.

The close spacing of the lower multiplet levels of the $4f^6$ and $4f^5$ ions was to be expected from the Landé interval rule (2.43), according to which a multiplet interval is proportional to the larger J quantum number involved. If the quantum numbers L and S happen to be comparable in value so that $J = |L-S|$ is small while, at the same time, both quantum numbers are relatively large, the separation of the lowest multiplet levels may well be small compared with kT despite a multiplet width, $W_{so} = E(L+S) - E(L-S)$, that is very much larger than kT. The Eu^{3+} ($4f^6$) cation, with its J = 0 ground multiplet, is an extreme example of this: the multiplet width is 21λ,[b] but the spacing of the ground and first excited multiplets is just λ (known to be 355 cm^{-1}). There is no difficulty, however, with the ions Tb^{3+} ($4f^8$) and Dy^{3+}

b Recall the definition of *multiplet width* W_{so} in eqn. 2.44.
From first principles,

$$W_{so} = \sum_{J=1}^{6} J\lambda = \sum_{J=0}^{5} (J+1)\lambda = 21\lambda$$

($4f^9$), the electronic configurations of which are the hole counterparts of Eu^{3+} and Sm^{3+}, since the multiplet patterns are there inverted.

As anticipated in Table 4.3, and demonstrated below, the "room" temperature data for compounds containing Eu^{3+} and Sm^3 can indeed be satisfactorily explained by means of the general theory (Case 3). They are, so to say, exceptions that "prove" the rule. It may be noted also, referring back to Fig. 4.9, that the m_{eff} values observed for Eu^{3+} and Sm^{3+} are still well short of what would be expected in the high temperature limit (Case 1). As was emphasised earlier, the multiplet widths remain large in these ions.

4. A closer look at Eu(III) and Sm(III) compounds

The magnetic properties at ordinary temperatures of compounds of Eu^{3+} ($4f^6$), ground term ^7F, may be explained adequately by allowing for the thermal population of just the first excited multiplet (J = 1). The relevant energy levels are those pictured in Fig. 4.10.

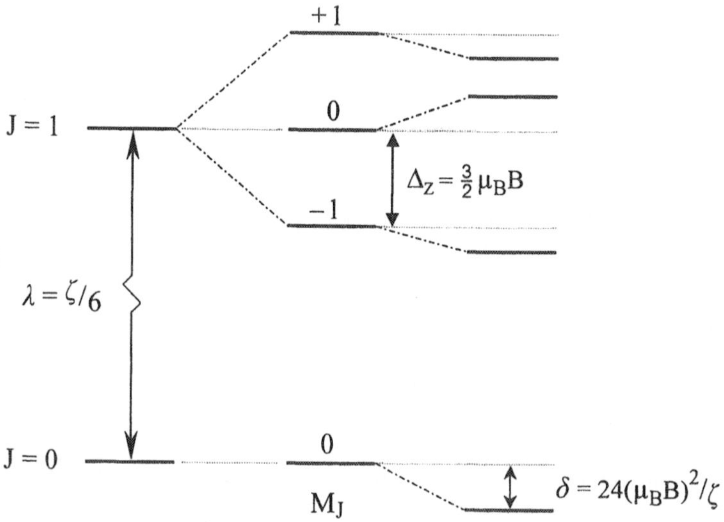

Fig. 4.10 The effect of an applied magnetic field on the ground and first excited multiplet states (J = 0 & 1) of Eu^{3+} ($4f^6$), with ground term ^7F. ($g_J = \frac{3}{2}$ for the excited multiplet.)

This is a special, almost unique instance of the general energy level scheme given in Fig. 4.4 earlier, the ground multiplet being non-degenerate. Again, the scale of the diagram is potentially misleading: the multiplet spacing of λ (experimentally 355 cm^{-1}) is some two orders of magnitude greater than the Zeeman interval Δ_Z – which, in turn, is typically two orders of magnitude larger than the second order displacements δ.

[For a field strength of 1 T (= 10^4 G), $\mu_B B \sim 1$ cm^{-1} and, for the case in point, $\delta \gtrsim 0.02$ cm^{-1}.]

The upward second order shift of the higher $M_J = 0$ Zeeman level is due to stronger mixing with the ground level than with the $M_J = 0$ level of the J = 2 multiplet (not shown), reflecting the Landé interval rule. The *average* second order displacement of the triplet of Zeeman levels is $(\mu_B B)^2/2\zeta$, and upward.

The second order terms α', which depend on the g-factor g_J, are described by eqn. 4.6 earlier. In the special case J = 0, the quantum mechanical theory has $g_J = L + 2 = 5$#, so that $\alpha' = 8/\lambda$. The effective magnetic moment in the ground multiplet, formulated generally by eqn. 4.13, is then defined by the purely second order expression

This result, of necessity invoked *ex cathedra*, is purely formal. The ground multiplet (J = 0) is not subject to the *first order* Zeeman effect.

$$n^2_{eff}(0) = 0 + \frac{24kT}{\lambda}$$

where $\lambda = \zeta_{4f}/2S = \zeta_{4f}/6$. For the excited multiplet, with (in the normal way) $g_J = \frac{3}{2}$, the second order coefficient α' is instead $-1/6\lambda$, and therefore

$$n^2_{eff}(1) = \frac{9}{2} - \frac{kT}{2\lambda}$$

The latter having a degeneracy 2J + 1 = 3, it follows from eqn. 4.14 that

$$\chi_{mol} = \frac{\mu_0 N_A \mu_B^2}{3kT} \frac{\left[\frac{24kT}{\lambda} + 3\left(\frac{9}{2} - \frac{kT}{2\lambda}\right)\exp[-\lambda/kT]\right]}{1 + 3\exp[-\lambda/kT]}$$

and, substituting $x = \lambda/kT$,

$$\chi_{mol} = \frac{\mu_0 N_A \mu_B^2}{3kT} \frac{[24 + (13.5x - 1.5)e^{-x}]}{x(1 + 3e^{-x})} \qquad (4.15)$$

The effective magnetic moment for the two-level system is hence given by

$$n^2_{eff} = \frac{24 + (13.5x - 1.5)e^{-x}}{x(1 + 3e^{-x})} \qquad (4.16)$$

and may be strongly dependent on temperature.

Estimates of the spin-orbit coupling constant λ for Eu^{3+} vary somewhat but it is here taken to be simply 355 cm^{-1}, the first multiplet interval as determined *via* atomic spectroscopy. Then, for T = 300 K, n_{eff} is estimated to be 3.26, in reasonable agreement with experiment. The complete analysis, in which all excited multiplets are included, leads to a slight increase in n_{eff} to nearly 3.27. The "room" temperature moments of Sm(II) compounds are expected to be a little larger since ζ_{4f} is smaller for the Sm^{2+} ion. The multiplet interval is actually 293 cm^{-1} so that the predicted $n_{eff} = 3.41$.

It should be made clear that the deviation from the Hund-Landé expectation ($n_{eff} = 0$) is due principally to the second order Zeeman effect in the *ground* multiplet, this contributing some 86 % or so of the bulk susceptibility at ordinary temperatures.[a] If there were no thermal population of the excited multiplet, the Eu^{3+} cation would simply display TIP with, *apparently*, the magnetic moment $m_{eff} = \sqrt{(24kT/\lambda)}\mu_B = \sqrt{(24/x)}\mu_B$.[b] At 300 K this amounts to 3.76 μ_B, so large is the second order Zeeman effect in this case. Thus, at ordinary temperatures, thermal population of excited multiplets causes a *reduction* of the apparent moment of Eu^{3+}. However, as the temperature is reduced the apparent magnetic moment declines continuously according to $\sqrt{T}$ and eventually vanishes.

The enhanced magnetic moment of Sm^{3+} ($4f^5$) is also due mainly to the second order Zeeman effect in the ground multiplet, now $^6H_{5/2}$ (J = 5/2 = S, L = 5). The complete theory, with $\lambda = 284$ cm^{-1}, predicts $n_{eff} = 1.62$ at 300 K. Thermal excitation to the excited multiplet is here more limited because of a greater multiplet interval (993 cm^{-1}) and has a much diminished influence on m_{eff}. However, in spite of the large multiplet interval, the difference between the magnetic moments observed for Sm(III) compounds (typically 1.5 – 1.65 μ_B) and the Hund-Landé prediction (0.85 μ_B) is due mainly to the second

[a] It is stated in many textbooks (or at least clearly implied) that the magnetic moment of Eu^{3+} is due simply to the thermal population of higher multiplets. It therefore decreases progressively towards zero as the temperature is lowered because the excited multiplets are depopulated.
Clearly, this is highly misleading.

[b] This actually happens in analogous compounds of americium(III), $5f^6$ – *e.g.* Am_2O_3 – where the increased λ precludes occupancy of the excited multiplet at ordinary temperatures. [See section 4 below.]

order Zeeman effect in the ground multiplet. The latter contributes almost 70 % to the total bulk susceptibility at room temperature.

5. Crystal field effects

The exposition thus far has ignored *crystal field effects* – by which is meant the splitting of multiplet energy levels by the electrostatic field of the surrounding ligands (usually anions).[a] It has been assumed implicitly that these splittings are negligible in relation to the Boltzmann energy kT, at least at ordinary temperatures. This is remote from the *actualité*, however, as may be seen in Figs. 4.11 and 4.12, which are representative of the evidence from high resolution optical spectroscopy. The crystal field splittings are often ⩾ 500 cm^{-1}. It must be admitted[b] that *the good agreement between the atomic theory and experiment described earlier is largely fortuitous.*

a Bethe (1929), Van Vleck (1932)

b – with due apology to the reader.

[Quantum mechanics is however well equipped to cope with the complex reality.]

The reader should not be distracted by the labels used in Figs. 11 and 12; they are included mainly for completeness. It may be noted, however, that A, E, T and U simply indicate degeneracies of 1, 2, 3 & 4, respectively – here *overall* spin-orbital degeneracies. They may be regarded as states with J-values of 0, $\frac{1}{2}$, 1 and $\frac{3}{2}$, respectively, in the crystal field.

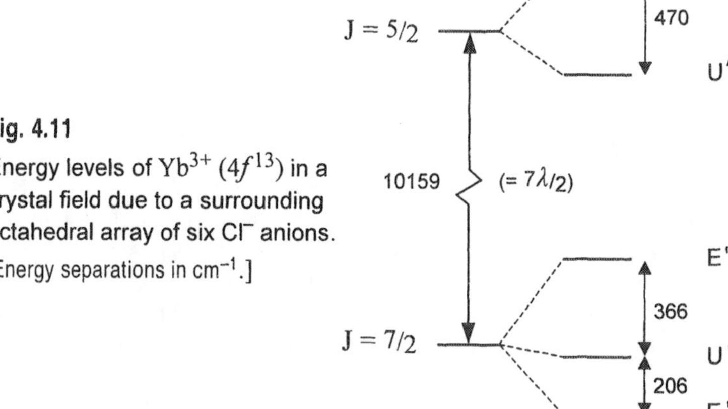

Fig. 4.11
Energy levels of Yb^{3+} ($4f^{13}$) in a crystal field due to a surrounding octahedral array of six Cl^- anions.
[Energy separations in cm^{-1}.]

The variation evident in magnetic moment data in Fig. 4.9 (and also Table 4.4) is due in large part to such crystal field effects.

The energy separations defined in these figures come from the absorption spectra of crystals having the *elpasolite* structure, of stoichiometry $Cs_2NaLnCl_6$, containing a precisely octahedral $[LnCl_6]^{3-}$ complex. The Ce(III) ($4f^1$) compound has an inverted pattern of levels as compared with the Yb(III) compound, with overall splittings of 570 cm^{-1} for the ground multiplet (J = 5/2) and 888 cm^{-1} for the excited multiplet (J = 7/2).

The splittings of the multiplet levels reflect a splitting of the 4*f* orbital energy level by the electrostatic field of the ligands, which here may be simulated quite accurately by an array of point negative charges. The seven *f* orbitals have distinctive spatial characteristics, like the three *p*-orbitals or the five *d*-orbitals [§ 2.1(4)], and their angular distributions will in general differ in relation to the ligand array. For example, one of the *f*-orbitals may have its lobes strongly oriented towards the ligands, while the lobes of another may instead be directed more between the ligands;[c] consequently the two orbitals experience different electrostatic perturbations and are no longer degenerate. (The former *f*-orbital will be the more destabilised.) In general, the ligand field will destroy the degeneracy of the *f*-orbitals, at least in part. The extent of this effect, and the pattern of degeneracies that the perturbed multiplet levels retain, depends principally on the *symmetry* of the ligand array. The higher is the symmetry, the more orbitals can group in sets equivalent with

c See, *e.g.*, Shriver and Atkins (1999) or Huheey *et al.* (1993) for sketches of the *f* orbitals.

[In octahedral symmetry the *f*-orbitals are split into three sets of degeneracy 1, 3 and 3 – usually designated respectively a_2, t_1 and t_2.]

respect to the ligand environment, and hence the less the removal of orbital degeneracy. But the details of this are more easily, and pertinently illustrated for *d*-orbitals in transition metal compounds [§ 5.2] where ligand field effects are considerably greater and often a determinative factor in the chemistry.

Regular octahedral crystal sites or discrete complexes are actually quite rare for Ln^{3+} cations which, being comparatively large, are usually found with higher coordination numbers (CN) ranging from 7 to 9, and occasionally up to 12. (The ions contract in size progressively with increasing nuclear charge as the lanthanide series is traversed, such that the higher order coordination polyhedra are encountered mainly in the first half of the series).[a] For the most part, the ligand geometries are of symmetry lower than octahedral, with the consequence that the destruction of degeneracy is more extensive than shown in Figs. 4.11 & 4.12. In a lower symmetry the U' levels will be split into two doublets (E) but no further.[b]

The loss of degeneracy of the *f*-orbitals, wholly or in part, results in a commensurate degree of *quenching* of orbital angular momentum [§ 2.3(4)] which will be expected normally to reduce the first order magnetic moment. However, for $4f^N$ configurations with $N < 7$, where **L** and **S** are coupled in opposition in the ground multiplet of the free ion, the quenching of $\boldsymbol{\ell}$ might conceivably *enhance* the first order magnetic moment by permitting an increased contribution from electron spin. The outcome in any particular case depends on the *f*-electron configuration, the symmetry of the crystal field, and the relative strength of the electrostatic perturbation and the spin-orbit interaction. There is, in effect, a competition between the crystal field and the spin-orbit coupling, leading to a substantial readjustment in the make-up of electronic states as compared with the free ion. The antagonists are of similar strength early in the lanthanide series, but the spin-orbit interaction gains ascendancy towards its end.[c] Other things being equal, the spin-orbit coupling tends to resist the effects of the crystal field.

Whatever the effect of the ligand field on the *first-order* component of m_{eff}, the splitting induced in the ground multiplet naturally leaves the opportunity for a possibly large *second-order* Zeeman effect, which may compensate to some extent for the quenching of the orbital contribution. This is unquestionably the case for $[YbCl_6]^{3-}$ (Fig. 4.11), with a separation of merely 206 cm^{-1} between ground (E') and first excited state (U').

No simple generalisation can be made, except that *both* the first and second order components of the magnetic moment will typically be temperature dependent in the vicinity of 300 K due to a changing Boltzmann distribution among the accessible crystal field levels. Thus, the relative populations of the ground and first excited state of the $[YbCl_6]^{3-}$ complex must obviously vary with temperature – except at low temperatures when, of course, only the ground state will be occupied. (The third crystal field level (E"), 572 cm^{-1} above the ground level, is not appreciably populated at "room" temperature.)

It should be pointed out, before going into further detail, that *crystal field effects will be unimportant, at least as a first approximation, in the cases of the* $4f^7$ *and* $4f^6$ *configurations* – exemplified most prominently by Gd^{3+} and Eu^{3+}, respectively. First, Gd^{3+} (also Eu^{2+}) is not vulnerable to a crystal field

Fig. 4.12

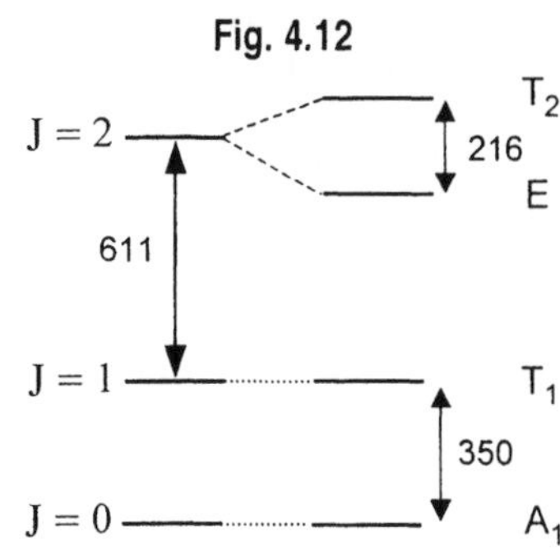

The lower multiplet levels of Eu^{3+} ($4f^6$) in an octahedral crystal field created by six Cl^- anions.
[Energies in cm^{-1}; *cf.* Fig. 4.10.]

[a] CN = 9, which involves commonly a tricapped trigonal prismatic ligand array, is observed in, for instance, $LnCl_3$ (Ce – Gd) and in the hydrated ions $[Ln(OH_2)_9]^{3+}$. CN = 12 is found in $Ln_2Mg_3(NO_3)_{12}.24H_2O$ (Ce – Eu), which contains the icosohedral ion $[Ln(NO_3)_6]^{3-}$ with a bidentate $[NO_3]^-$ ligand.

[b] There is a famous theorem due to Kramers (1930), according to which the ground state of an *odd*-electron system must possess a degeneracy of at least *two*.

[c] Thus, in $[CeCl_6]^{3-}$ ζ_{4f} = 623 cm^{-1} and the overall crystal field splitting of $^2F_{7/2}$ is 888 cm^{-1}; the corresponding figures for $[YbCl_6]^{3-}$ are 2903 and 572 cm^{-1}, respectively.

The paramagnetism of compounds involving $4f^7$ and $4f^6$ ions is comparatively unaffected by crystal fields – somewhat paradoxically, perhaps, in the latter case.

splitting because the ground term ^8S, that of a half-filled subshell, is an orbital singlet (L = 0). Nor, secondly, is Eu^{3+} (or Sm^{2+}) with a singlet, J = 0 ground state (Fig. 4.12), as determined by the spin-orbit coupling. The level having J = 1 will suffer a splitting in crystal fields of lower than octahedral symmetry but, as pointed out in the previous section, the excited multiplet has mainly a second-order role in $4f^6$ ions (at ordinary temperatures and lower) and this function is little affected by modest crystal field splittings.

The more complicated behaviour of lanthanide compounds at large is conveniently illustrated by reference to the Yb^{3+} ($4f^{13}$) case. Consider first, the $[YbCl_6]^{3-}$ system, already characterised in part, where m_{eff} is apparently 4.47 μ_B at ordinary temperatures (as in $YbCl_3$ – *cf.* Table 4.3), in excellent agreement with the Hund-Landé expectation of $g_J[J(J+1)]^{\frac{1}{2}}\mu_B = 4.50\ \mu_B$ for the *free* Yb^{3+} ion.

However, the EPR spectrum[a] of the complex yields a g-value of 2.584, indicating a first order magnetic moment of $g(\sqrt{3}/2)\mu_B = 2.24\ \mu_B$ in the doubly-degenerate ground level (E') with $J = \frac{1}{2}$. Analysis of the temperature-dependence of the paramagnetism reveals that the apparent $m_{eff} = 4.47\ \mu_B$ is due largely to the second-order terms which at 300 K contribute some 72% to the susceptibility. This second-order contribution is due principally to the mixing of the U' excited state into the E' ground state (by the applied magnetic field), this having a T-dependence at 300 K because U' is thermally populated. At low temperatures, the U' excited level becomes depopulated: thus at 20 K the *apparent* $m_{eff} = 2.40\ \mu_B$, consistent with a first order $m_{eff} = 2.24\ \mu_B$ (see above) boosted slightly by virtue of the U' – E' second order (TIP) term.[b]

[a] Lanthanide compounds normally display EPR spectra only at temperatures sufficiently low that no excited crystal field state is populated.

[b] The apparent magnetic moment in the *ground* state is given by

$$m_{eff} = \sqrt{\tfrac{3}{4}g^2 + 3kT\alpha'}\ \mu_B$$

(Recall eqn. 4.13.) The second order component shrinks in comparison with the first order term as the temperature is lowered.

A more straightforward case, perhaps, is that of the Yb^{3+} in $Yb_3Ga_5O_{12}$, a *garnet*, where the lanthanide ion occupies a cubic site with 8-coordination by oxide ions. Not surprisingly, the pattern of the crystal field splitting of the ground multiplet differs from that in an octahedral field; it turns out in fact to be precisely the inverse of that shown in Fig. 4.11, with instead an E" ground state. The U' exited state lies some 550 cm^{-1} above and is barely populated for $T \leqslant 300$ K. (The crystal field level E' has an even higher energy of about 800 cm^{-1}.) The experimental bulk susceptibility per mole of Yb^{3+} can be fitted satisfactorily to the expression $\chi = (1360/T + 5.00) \times 10^{-8}\ m^3\ mol^{-1}$, while EPR measurements on Yb^{3+} doped into the garnet $Y_3Ga_5O_{12}$ (a.k.a. *YGG*) reveal a g-value of 3.420 for the ground doublet.[c] The latter suggests a first order magnetic moment of 2.96 μ_B (assuming again $J = \frac{1}{2}$), in good agreement with the figure of 2.94 μ_B derived from the Curie term in the susceptibility. The *apparent* magnetic moment of 4.27 μ_B at 300 K is due in almost equal measure to the first and second order Zeeman effects.

[c] Very similar g-values are obtained from EPR studies of Yb^{3+} doped into ThO_2 and CeO_2, both of which have the *fluorite* (CaF_2) structure – with perfectly cubic, 8-coordination of the metal ion.

Outstanding, of course, is an explanation of the g-values observed for Yb^{3+} in these two situations, and also the magnitudes of the second order contributions to the paramagnetism. But that requires quantum mechanical arguments beyond the scope of this book.

Also out of bounds is the subject of ***anisotropy*** in the susceptibility, and therefore in the paramagnetic moment and the g-factor determined by EPR spectroscopy. The phenomenon is encountered when (as is usually the case) the site symmetry of the lanthanide ion is lower than cubic, and reflects simply the anisotropic features of the crystal field, *i.e.* the non-equivalence of some mutually perpendicular directions. Many compounds involve a *trigonal* coordination environment, when the crystal field along the *three*-fold symmetry axis will be of different strength to that at right angles to the axis. In measurements on single crystals, the susceptibility observed with the external magnetic field applied *parallel* to the symmetry axis will then, in general, be different from that recorded when the field is directed *perpendicular* to the axis (all such directions being equivalent). One such case, namely the Yb(III) compound of type *C* in Table 4.4, is illustrated by Fig. 4.13 opposite – but on "a dump it and run" basis.

Fig. 4.13

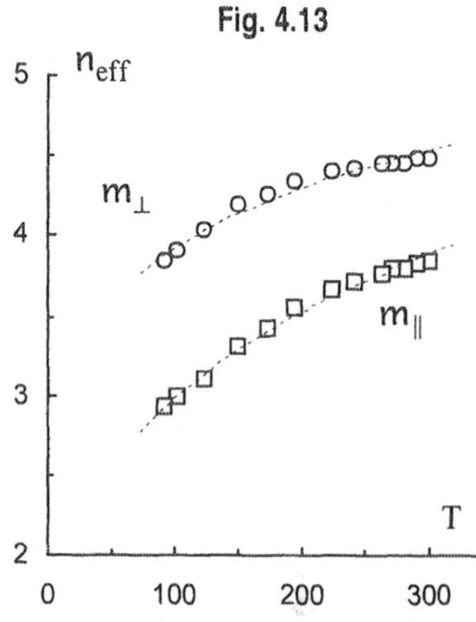

Anisotropy of the paramagnetism of an Yb(III) compound with the metal ion in a trigonal coordination site.

$m_\parallel$ and $m_\perp$ are the *apparent* magnetic moments parallel and perpendicular to the three-fold axis, respectively.

A brief postscript, however, concerning the study of $[YbCl_6]^{3-}$, which provides a salutary lesson. Remarkably, the susceptibility of $Cs_2NaYbCl_6$ down to 45 K can be simulated accurately by the Curie-Weiss law with n_{eff} = 4.47 and θ = 70 K, the effective magnetic moment being in excellent agreement with the Hund-Landé prediction of 4.50 μ_B. However, this processing of the experimental data is arbitrary and potentially misleading, the Curie-Weiss dependence being merely empirical. The large value of the Weiss temperature θ is unrealistic given that the compound remains paramagnetic at 2.5 K. Here, as in many other connections, the fitting of bulk susceptibility data to the Curie-Weiss law may confuse the investigation.#

Recall § 3.2(4)

Both Yb(III) case histories demonstrate the importance of supplementing susceptibility measurements with EPR and optical spectroscopy in order to characterise fully the electronic structure of a compound.

4.4 Compounds of the actinide elements

1. Distinctive electronic structural features

The elements of the actinide series, *thorium* Th to *lawrencium* Lr (with the general symbol An), are closely related to the lanthanide elements, their neutral atoms and cations having electron configurations with, almost invariably, occupancy of the 5*f* (rather than 4*f*) subshell. The actinide ions, most prominently An^{3+} and An^{4+}, all display configurations of the type (Rn) $5f^N$, where (Rn) denotes the radon core. Thus the americium ion Am^{3+}, analogous to Eu^{3+}, has the outer electronic configuration $5f^6$.

There are, however, important differences as regards electronic structure between the two series. The more pertinent of these, (a) – (c) below, are described in order of physical significance.

> In a first reading of this section, the main idea to glean is that ligand field effects are much stronger *early* in the actinide series than they are at any point in the lanthanide series.

(a) In the *lighter* actinide elements (Th – Pu, apparently) *the 5f orbitals are more exposed to the influence of neighbouring atoms*, a property illustrated by the plots in Fig. 4.14 (overleaf).

While the 4*f* electrons reside essentially within the xenon core from the very beginning of the lanthanide series, the 5*f* electrons of the actinide elements have less contracted radial distributions such that, *early* in the series, the 5*f* orbitals protrude substantially from the radon core. The 5*f* orbital however contracts with increasing nuclear charge across the actinide series and retreats into the atomic core by americium or (certainly) curium. All this is well authenticated by quantum theory, and also by the results of both optical (electronic) absorption spectroscopy and measurements of the

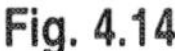
Fig. 4.14

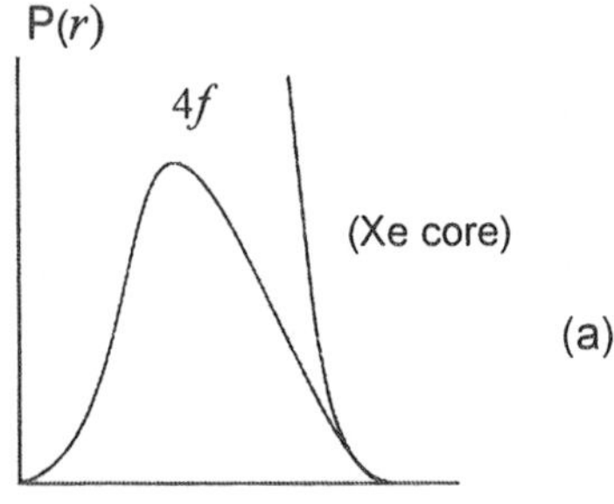

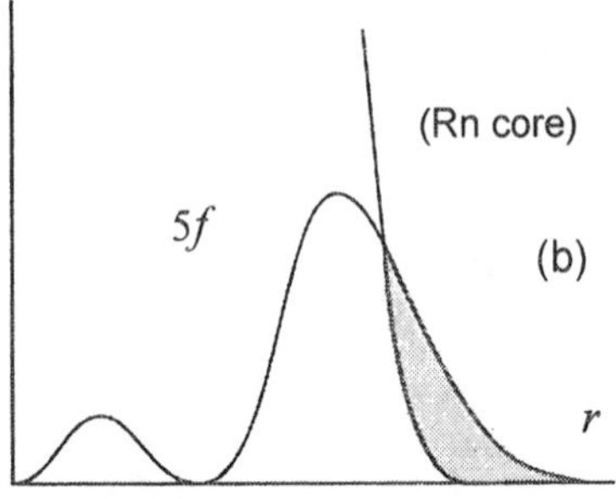

Sketches of the radial distribution functions P(r) for valence shell f-orbitals set against the dwindling electron densities of the cores.

(a) Any lanthanide ion, *e.g.*, Nd^{3+}. [*cf.* Fig. 4.5].

(b) An actinide ion *early* in the series, *e.g.* U^{3+}. The shaded area represents the extension of the $5f$ electron density beyond the radon core.

paramagnetism observed in actinide compounds. The greater spatial extent of the $5f$ orbitals *in relation to the core*, as compared with the $4f$ orbitals in the lanthanides, results in large part from the $5f$ distributions responding more freely to the centrifugal force [recall § 2.5(3)], though it is also partly due to relativistic effects. An analogous effect is well established for the d-electron distributions in the transition elements (Fig. 5.25).

As a result of the protrusion of the $5f$ probability density [Fig. 4.14(b)], the *early* actinide ions experience stronger crystal field effects than the corresponding lanthanide ions. There is, moreover, *greater scope for covalency attributable to overlap of the 5f orbitals with ligand valence orbitals.*

(b) Spin-orbit coupling is stronger *by 50 % or more* in the $5f$ subshell of an actinide element than in the $4f$ subshell of the corresponding lanthanide element. This is due in part to the larger nuclear charge of the actinide, and partly to an increased $\langle r^{-3} \rangle$ for the more penetrating $5f$ radial distribution [*cf.* § 2.5(3), and eqn. 2.34 in particular]. The one-electron parameter ζ_{5f} increases progressively across the actinide series, like ζ_{4f} in the lanthanide elements (Fig. 4.7). It is about 2600 cm^{-1} in Am^{3+}.

As an obvious consequence, multiplet splittings are greatly increased in actinide ions ($5f^N$) as compared with lanthanide ions ($4f^N$). At the same time, the second order (mixing) effects of the spin-orbit interaction are more pronounced, indeed sufficiently so that an *intermediate* coupling scheme is required [§ 2.5(5)]. The mixing of J states (of different term parentage, of course) is enhanced also by a reduction in term separations associated with reduced repulsion between electrons in the more diffuse $5f$ orbitals.

Whereas this mixing "contaminates" the ground multiplets of lanthanide ions by no more that 3 %, the ground multiplets of actinide ions may have admixtures of 10 – 30 % (sometimes even more) of the equivalent multiplets of excited terms. Thus, according to *ab initio* calculations on U^{4+} ($5f^2$), the Russell-Saunders ground state $^3H_4$ acquires a 10 % admixture of the excited term ^1G (with uniquely J = 4). Applied to Am^{3+} ($5f^6$), with a nominal ground state $^7F_0$ (*cf.* Eu^{3+} $4f^6$), such calculations indicate that the ground multiplet (J = 0) has some 38 % $^5D_0$ and 13 % $^3P_0$ in its make-up.

Intermediate coupling affects the atomic magnetic moment because the admixed excited multiplets invariably have different values of the g-factor g_J (in eqn. 4.3) as compared with the unperturbed ground multiplet; these g-values are invariably larger before the half-filled subshell, but smaller afterwards. In general, this leads to increased estimates of the first order (Case 2) m_{eff} for $N = 2 - 5$, but a decrease in the predicted values for $N = 7 - 12$.[a] However, the emendations are not very large, usually $\leqslant$ 10 %.

[a] The estimates for the $5f^6$ case are unchanged because the ground multiplet has J = 0.

(c) Among at least the *early* actinide elements, the $6d$ orbital apparently lies lower in energy relative to $5f$ than does the $5d$ orbital in relation to $4f$ in the lanthanide series.[b]

[b] Note the electron configurations adopted in certain neutral atoms:

Ce $4f^1 5d^1 6s^2$ Th $6d^2 7s^2$
Nd $4f^4 6s^2$ U $5f^3 6d^1 7s^2$

There is accordingly the prospect of a more significant mixing of the states of the ground configuration nf^N with those of the excited configuration $nf^{N-1}(n+1)d^1$ – a (theoretical) effect known as *configuration interaction.* This may greatly complicate the description of the electronic structure of compounds of the very *early* actinide elements, especially for cations of low external charge (as in a compound such as US). However, the $5f-6d$ energy separation increases sharply with both external charge on the atom (and therefore, usually, with oxidation state) and also nuclear charge as the series is traversed. There is little evidence that configurational mixing exerts a significant influence on magnetic properties beyond uranium.

The agency responsible for configuration interaction is the interelectronic repulsion. The "effect" compensates for the shortcomings of the orbital approximation.

In summary, the actinides have distinctly more complicated electronic structures than the lanthanide elements, especially the lighter elements, Th – Pu. As regards the latter, it is likely that *the effects of the ligand field dominate the energetics*. It is less clear how the interplay of crystal field effects and spin-orbit interaction is resolved for the subsequent elements (Am *etc.*), where the $5f$ orbital has contracted essentially into the core.

2. Paramagnetic properties of actinide compounds

Proceeding to the facts, "room" temperature magnetic moments reported for the oxides An_2O_3 and AnO_2 are collected in Table 4.6, where they are compared with data for the lanthanide oxides. The data available for the halides AnX_3 and AnX_4, and the pnictides AnY (Y = N, P *etc.*), paint essentially the same picture. The information concerning the transuranic elements (Np *etc.*) was hard won by.

Table 4.6 The apparent magnetic moments m_{eff} ($/\mu_B$) observed at ordinary temperatures for the actinide oxides, An_2O_3 and AnO_2, compared with the data for the corresponding lanthanide oxides.

N	Ln	Ln_2O_3	An	An_2O_3	AnO_2	
1	Ce	2.35	Th	#	PaO_2	1.0^a
2	Pr	3.59	Pa	#	UO_2	3.7
3	Nd	3.66	U	#	NpO_2	3.0
4	Pm	–	Np	#	PuO_2	(1.07)*
5	Sm	1.50	Pu	1.0	AmO_2	1.4
6	Eu	3.64	Am	(1.24)*	CmO_2	$(1.6\ ?+)^b$
7	Gd	7.90	Cm	7.7	BkO_2	$7.9(?-)^a$
8	Tb	9.67	Bk	9.1	CfO_2	9.1
9	Dy	10.5	Cf	9.95		
10	Ho	10.4	Es	10.1		

It is difficult to prepare these sesquioxides, at least in an adequate state of purity. Some have been reported to be *metallic*.

* Am_2O_3 and PuO_2 display TIP.

a The lanthanide compounds equivalent to PaO_2 and BkO_2 – *viz.* PrO_2 and TbO_2 – have n_{eff} values of, respectively, 2.47 and 7.90. [*cf.* Table 4.5(b)]

b The CmO_2 data may be unreliable due to impurities (*e.g.* Cm^{3+}). The compound is expected to show TIP, like Am_2O_3.

No worthwhile generalisation can be made regarding paramagnetism in the compounds of the *early* actinides (except that ligand field effects are likely to be particularly strong). Additional data for compounds with ions having the $5f^2$ configuration provided in Table 4.7 confirm the expectation that the ligand field is a major determinative factor. The magnetic moments recorded there show a variation of some 50 %, and cannot be adequately rationalised on the basis of either the Hund-Landé equation or the spin-only formula. The data cannot even be employed to identify reliably the oxidation state of the metal concerned, where it is unknown.

Table 4.7 Apparent m_{eff} values at 300 K (/μ_B) for compounds involving the $5f^2$ configuration.

UO_2	3.7	$[NpO_2]^+$	3.00
UCl_4	3.29	$[NpOCl_5]^{2-}$	2.30
$(cp)_4U$	2.76	$[PuO_2]^{2+}$	3.56
$(cot)_2U$	2.39	PuF_6	2.75
$[UCl_6]^{2-}$	(2.22)*		
$[U(NCS)_8]^{4-}$	2.8		
$[U(NO_3)_6]^{2-}$	2.42*		

* $[UCl_6]^{2-}$ displays TIP, like most U(IV) compounds with octahedral coordination. *cot* = octagonal $[C_8H_8]^{2-}$.

Later in the series, however, the experimental magnetic moments are consistent with the prediction that, *as a result of intermediate coupling*, they should be distinctly less than the Hund-Landé values. Thus Bk_2O_3 and BkF_3 display magnetic moments of 9.1 μ_B and 9.4 μ_B, respectively, as compared with the somewhat naïve prediction of 9.72 μ_B. The corresponding lanthanide compounds, Tb_2O_3 and TbF_3, have magnetic moments of (respectively) 9.67 μ_B and 9.69 μ_B. But the rationalisation of the Bk(III) data enjoys, almost certainly, the same fortuitous elements as in the case of Tb(III), and most other lanthanide compounds – namely that the first order magnetic moment is partly quenched by the crystal field, the loss being made up to a considerable extent by the second order Zeeman effect.

Further to the contents of Table 4.6, the influence of an increased spin-orbit interaction is evident also in Am_2O_3, which exhibits TIP in the range 5 – 300 K with a susceptibility of 0.80 x 10^{-8} m^3 mol^{-1}, equivalent to m_{eff} = 1.24 μ_B at 300 K. In contrast, the analogous Eu_2O_3 shows a mixture of first and second order paramagnetism at normal temperatures (recall section 4 earlier). The paramagnetism of Am_2O_3 – and, indeed, all other Am(III) compounds that have been investigated (*e.g.* the halides AmX_3 and AmN) – is exclusively of the T-independent kind because the energy separation of the ground (J = 0) and first excited multiplet (J = 1) has become so large, *ca.* 450 cm^{-1} or more, as to preclude a significant population of the latter for T $\leqslant$ 300 K. Eu(III) compounds display *pure* TIP only at low temperatures.

A great deal is understood about the consequences of ligand field splittings, even in compounds of the early actinides, but the details are beyond the scope of the present text. However, one particular aspect of this subject, where very specific ligand field effects can be discerned *qualitatively*, is perhaps worth mention. This concerns ions possessing the $5f^2$ and $5f^4$ configurations with coordination sites of *cubic* symmetry. Both configurations have a J = 4 ground multiplet but invariably the splitting pattern in one case is inverted as compared with the other.

Consider first PuO_2 ($5f^4$), which adopts the fluorite structure with cubic 8-coordination of the Pu^{4+} cation. The ground multiplet suffers a crystal field splitting such that (so it turns out) the ground state is non-degenerate (A_1) and the first excited state of too high an energy to be thermally populated to a significant extent. This explains the TIP observed in PuO_2 (Table 4.6). However, the splitting pattern of the J = 4 ground multiplet in the analogous U(IV) compound, UO_2 ($5f^2$), proves to be the inverse of that observed in PuO_2, the ground state being instead triply-degenerate (T_2). UO_2 accordingly shows a temperature-*dependent* paramagnetism.

In an *octahedral* ligand field, *e.g.* the halide complexes $[MX_6]^{2-}$, the sequence of crystal field levels is found# to be inverted in relation to that characteristic of cubic 8-coordination. Therefore, it is $[UCl_6]^{2-}$ (for instance) that displays TIP (Table 4.7), while $[PuCl_6]^{2-}$ exhibits typical paramagnetic behaviour with m_{eff} = 1.92 – 1.95 μ_B, depending on the counterion.

\# The ligand field splittings observed in such cases are readily explicable by quantum mechanics.

4.5 Metallic materials involving the lanthanide and actinide elements

1. The elemental metals

The lanthanide *metals* are remarkable in each showing a paramagnetism at ordinary temperatures that is of similar magnitude to that observed in simple *compounds* (Table 4.8, overleaf).[a] Moreover (and Gd included), the susceptibility typically conforms to the Curie-Weiss law or shows the temperature dependence $\chi = C/T + D$ above the temperature, T_C or T_N, at which the metal orders magnetically. Most lanthanide metals evidently have three electrons per atom involved in metallic bonding, the surplus being retained by the $4f$ subshell. The itinerant electrons contribute a net TIP, but this is usually swamped by the T-dependent paramagnetism due to the localised $4f$ electrons of the Ln^{3+} "ion core". The paramagnetism observed in these metals is accordingly often termed by physicists *local-moment* magnetism.[b]

a Gadolinium is exceptional in being magnetically ordered (actually ferromagnetic) below $T_C = 293$ K.

The end members of the lanthanide series, ytterbium and lutetium, show TIP – like lanthanum – due solely to their itinerant electrons.

b It should be appreciated that the *transition metals* are very different as regards their magnetic properties.

Table 4.8 The *apparent* magnetic moments m_{eff} ($/\mu_B$) observed at ordinary temperatures for the lanthanide and actinide metals, and for the generally metallic sulphides LnS and hydrides LnH_2.
[Compare the data given earlier in Tables 4.4 – 4.6.]

Z−57	Ln	metal	LnS	LnH_2	An	metal
0	La	(0.49)	??	(0.38)	Ac	??
1	Ce	2.54	2.78	2.50	Th	(0.44)
2	Pr	3.56	3.59	3.69	Pa	(0.80)
3	Nd	3.33	3.62	3.36	U	(0.97)
4	Pm	–	–	–	Np	(1.15)
5	Sm	1.74	[3.60]*	1.625	Pu	(1.11)
6	Eu	7.94	[7.49]*	[7.0(?)]*	Am	(1.33)✠
7	Gd	7.95	8.16	7.7	Cm	7.6
8	Tb	9.77	9.62	9.8	Bk	9.7
9	Dy	10.67	10.4	10.8	Cf	9.7
10	Ho	10.8	10.3	9.9	Es	11.3(?–)
11	Er	9.8	9.5	9.8		
12	Tm	7.6	7.0	7.6		
13	Yb	(0.4)	[??]*	[0.2]*		
14	Lu	(0.2)	–	–		

The left-hand column of Table 4.8 refers to the lanthanide elements.

Parentheses (....) indicate TIP. The ✠ for Am signifies that the TIP is mainly of the local-moment kind, unlike the earlier actinide metals.

[....]* identifies compounds which are *semiconductors* rather than metals.

[It should be borne in mind that many of these m_{eff} data will have been obtained *via* spurious fits of the susceptibility $\chi(T)$ to the Curie-Weiss law.]

There are no well characterised An(II) compounds comparable to those of Ln(II).

While the Lanthanide metals are normally "trivalent", like lanthanum itself, two of their number – namely, *europium* and *ytterbium* – are instead "divalent", contributing just *two* electrons per atom to the pool for metallic bonding. Europium metal thus displays a magnetic moment characteristic of the $4f^7$ configuration (*cf.* Table 4.5(a)), while ytterbium metal shows only a small TIP consistent with a filled $4f$ subshell $4f^{14}$ (corresponding to Yb^{2+}). The distinctive properties of these two metals reflect the especially high third ionisation energies of their atoms – due in the Yb instance to the very high

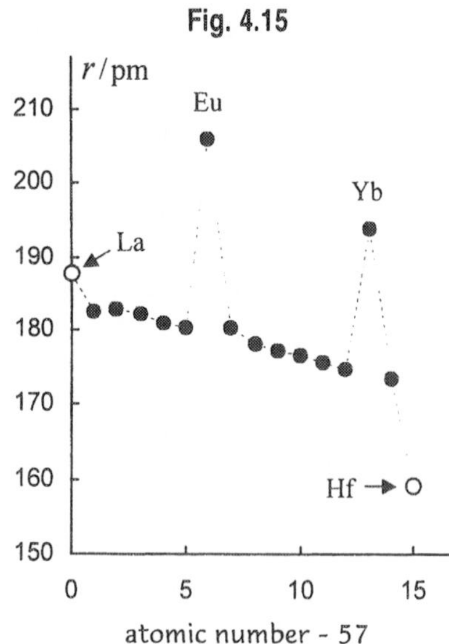

Metallic radii of the lanthanide elements Ce – Lu.

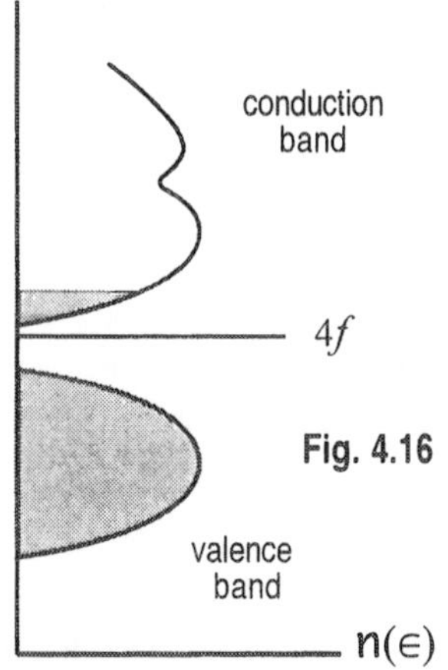

Simple picture of the electronic energy levels (*density of states*) for a *metallic* hydride LnH_2.

Note the partial occupancy of the conduction band. The population $N-1$ of the localised $4f$ level will naturally vary with the element Ln.

nuclear charge at the end of the lanthanide series and, in the case of Eu, to the stabilisation of the half-filled subshell $4f^7$ through spin correlation [recall § 2.2(3,4)]. As is well-known, the Ln^{2+} ion cores make the observed atomic (*i.e.* metallic) radii of Eu and Yb large in relation to the radii of the other Lanthanide elements (Fig. 4.15); moreover, the sublimation energies of the metals are, in a corresponding fashion, unusually *low* for Eu and Yb.

The early actinide elements, Th – Pu, are quite unlike their lanthanide counterparts, displaying almost constant paramagnetism (TIP) down to very low temperatures, there being no indication of magnetic ordering even at *ca.* 1 K. The evidently complete delocalisation of the $5f$ electrons is a further manifestation of the diffuse nature of the $5f$ orbitals in the lighter actinides. There is apparently a change in magnetic type, however, at americium: this metal also shows TIP but of a type consistent with a localised Am^{3+} ($5f^6$) aspect – this conveying TIP, as in Am(III) compounds [see previous section, 6(b)], in addition to that due to the itinerant electrons. Moreover, the metal orders antiferromagnetically below 4 K. The switch to local moment magnetism, with An^{3+} ion cores, is entirely clear beyond americium, the $5f$ orbitals having contracted into the radon core in response to the increasing nuclear charge. (Curium, the element that follows americium, shows magnetic order below $T_N \sim 60$ K.)

2. Metallic compounds of the lanthanide elements

Also listed in Table 4.8 are data concerning the sulphides and dihydrides formed by the lanthanide elements, and by lanthanum. With the exceptions noted in the table, the compounds are *metallic*, both with regard to electrical conductivity and appearance. The metal oxidation state in such compounds is *formally* +2 (with, say, N valence electrons) but their magnetic properties suggest that, for the most part, they involve Ln^{3+} ion-cores ($4f^{N-1}$), one electron per atom having been transferred to the conduction band – in other words, delocalised throughout the solid (Fig. 4.16). The lattice orbitals that make up the conduction band are expected to be mainly of $5d$ character in its lower reaches. Correspondingly, the valence band portrayed in Fig. 4.16 (for LnH_2) is principally of hydrogen $1s$ complexion. (In the LnS case the upper part of the valence band will be largely of sulphur $3p$ composition.)

Unsurprisingly, in Table 4.8, the sulphides and dihydrides of Eu and Yb are exceptional in retaining their Ln^{2+} ion-cores ($4f^7$ and $4f^{14}$, respectively); so also is SmS, with evidently the localised $4f^6$ configuration (of Sm^{2+}).

The dihalides $LnCl_2$ *etc.* are apparently similar in their properties.

4.6 Bibliography

General references

R G Denning, ch. 5 of *Solid State Chemistry: Techniques*, ed. A K Cheetham and P Day, Oxford, 1987

E A V Ebsworth, D W H Rankin and S Cradock, *Structural Methods in Inorganic Chemistry*, Blackwell, Oxford, 1991

T C Gibb, *Principles of Mössbauer Spectroscopy*, Chapman and Hall, 1976

J S Griffith, *The Theory of Transition-metal Ions*, Cambridge, 1961

J E Huheey *et al.*, *Inorganic Chemistry*, HarperCollins, 1993

N Kaltsoyannis and P Scott, *The f Elements*, Oxford, 1999 [OCP no. 76]

D F Shriver and P W Atkins, *Inorganic Chemistry*, Oxford, 1999

J H Van Vleck, *The Theory of the Electric and Magnetic Susceptibilities*, Oxford, 1932

J E Wertz and J R Bolton, *Electron Spin Resonance, Elementary Theory and Practical Applications*, McGraw-Hill, New York, 1972

Lanthanide and actinide elements [a]

The more recent data tabulated above come mainly from articles in

Handbook on the Physics and Chemistry of the Rare Earths, K A Gschneider Jr and L Eyring (eds.), North Holland, vol.1, 1978 – vol. 21, 1995

Handbook on the Physics and Chemistry of the Actinides, A J Freeman and G H Lander (eds.), North Holland, vol. 1, 1984 – vol. 6, 1991

The Chemistry of the Actinide Elements, J J Katz, G T Seaborg and L R Morse (eds.), vols. 1 & 2, 2nd edn., Chapman & Hall, 1986

[a] See also Boudreaux & Mulay (1976) and Carlin (1986), both cited in § 3.

4.7 Exercises

1. Expand and organise the scalar product **m.J**. Hence verify the formula $m_{eff} = g_J\sqrt{[J(J+1)]}\mu_B$ (eqn. 4.9) for the first order magnetic moment of an ensemble of atoms having a thermally isolated ground multiplet.[b]

2. Interpret the bulk susceptibility of 4.44×10^{-8} m^3 (g atom Pr)$^{-1}$ shown by the compound Pr_6O_{11} at 300 K.[c]

3. Confirm that $m_{eff} = 2.00$ μ_B for gaseous NO in the high temperature limit (Fig. 3.6). [*Hint*: adapt eqn. 4.8 appropriately.]

4. A certain lanthanide metal **A** is a temperature-independent paramagnet at SATP. At very high pressures (*e.g.* 40 kbar) it may be converted to a denser form displaying a strong paramagnetism that varies with temperature, also an increased electrical conductivity. Which of the Lanthanide elements must this be? Justify your conclusion.

5. A lanthanide element **B** adopts an f.c.c. crystal structure at SATP with a cell dimension $a = 5.14$ Å; it shows a substantial paramagnetism approximately of the simple Curie type. At 109 K, and 1 atm. pressure, the metal undergoes a phase transition to an alternative f.c.c. form having $a = 4.82$ Å, which displays only a weak TIP, but a greater electrical conductivity. The latter allotrope may be formed at higher temperatures by modest application of pressure. Identify the metallic element **B**, explaining the observations.

6. The *aluminides* $LaFe_2Al_{10}$ and YFe_2Al_{10} display Pauli paramagnetism. Of the lanthanide analogues $LnFe_2Al_{10}$, none ordering magnetically above *ca.* 15 K, the following are essentially Curie paramagnets, at least above 50 K, with the m_{eff} values indicated ($\pm$ 0.1 μ_B):

Pr 3.6; Nd 3.7; Gd 7.9; Tb 10.0; Dy 10.6; Ho 10.6; Er 9.5; Tm 7.9.

The susceptibilities of the Ce, Sm and Yb aluminides do *not* remotely obey the Curie law. Attempts to synthesise the Eu analogue have failed.

$YbFe_2Al_{10}$ is also anomalous in that its cell volume is larger than for the Tm and Lu compounds. The apparent magnetic moment of $SmFe_2Al_{10}$ at room temperature is *ca.* 1.5 μ_B. Comment on these observations.

[b] This derivation is more in tune with modern quantum mechanics than that based on the precessing vector model in section 1.

[c] $\dfrac{\mu_0 N_A \mu_B^2}{3k} = 1.57 \times 10^{-6}\ m^3\ mol^{-1}\ K$

7. The lanthanide oxides PrO, NdO and SmO all exhibit paramagnetism varying with temperature. The apparent n_{eff} at 300 K of PrO and NdO are 3.58 and 3.62, respectively; that of SmO is 2.7. Interpret these observations.

8. The compound SmS is a semiconductor at SATP with $m_{eff} = 3.6\ \mu_B$. Subjected to a pressure of 6 kbar at 300 K, it converts to a metallic form with $m_{eff} = 0.94\ \mu_B$. The boride SmB_6 has a similar magnetic moment at SATP. Comment on these observations.

9. The metals lanthanum and cerium, and their hydrides MH_x are reported to have the following molar susceptibilities (in 10^{-6} cgs/emu) at ordinary temperatures:

[To convert molar susceptibility data given in cgs/emu to the values in SI units ($m^3\ mol^{-1}$) multiply by $4\pi/10^6$.]

La	95	Ce	2500
LaH_2	59	CeH_2	2500
LaH_3	−20	$CeH_{2.9}$	2300

It is also known that the dihydrides display metallic conductivity, like the elements themselves, while LaH_3 and $CeH_{2.9}$ are semiconductors. Interpret these observations.

[*Some hints*:
(11) the f^7 configuration includes ^6P among its excited terms;
(13) the f^5 configuration, with ground term ^6H, has also a ^4G excited term.]

10. Paramagnetic YbH_2 shows only a small TIP (≡ 0.19 μ_B at 300 K). It reacts with H_2 gas under pressure, incorporating hydrogen to form a series of paramagnetic compounds $YbH_{2+\delta}$. At 300 K, $YbH_{2.56}$ has an apparent $m_{eff} = 3.16\ \mu_B$ while $YbH_{2.96}$ has $m_{eff} = 4.37\ \mu_B$. Comment on these data.

11. Explain why compounds of curium(III) and berkelium(IV) invariably display values of $m_{eff} < 7.94\ \mu_B$.

12. $(cp)_3Am$ displays TIP with a molar susceptibility of *ca.* 0.90 x $10^{-8}\ m^3\ mol^{-1}$. Calculate the *apparent* magnetic moment m_{eff} at: (i) 300 K; and also (ii) 10 K. Explain why the paramagnetism of this molecular compound does not show a variation with temperature. [Answers: 1.31 μ_B & 0.24 μ_B]

13. (a) Suggest reasons why compounds of plutonium(III) and americium-(IV) have smaller m_{eff} values at ordinary temperatures than compounds of samarium(III).#

\# For example, Pu_2O_3 has n_{eff} = 1.0, AmO_2 *ca.* 1.4 and Sm_2O_3 1.50 (recall Table 4.6).

(b) Why is the magnetic moment of AmO_2 greater than that of Pu_2O_3?

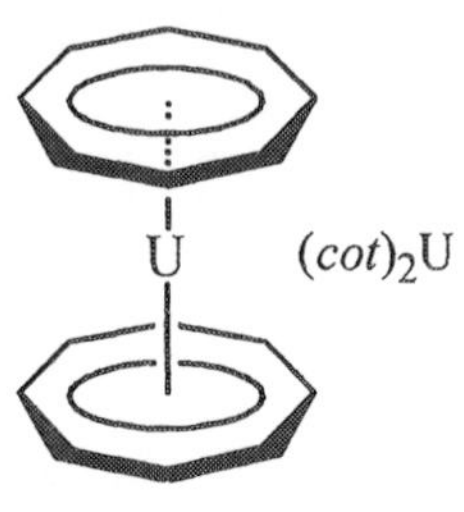

14. The magnetic moments reported for the molecular sandwich compounds $(cot)_2U$ and $(cot)_2Np$ at ordinary temperatures are 2.43 μ_B and 1.77 μ_B, respectively. But $(cot)_2Pu$ is diamagnetic, as is $(cot)_2Th$. Discuss.
[$cot = \eta^8$– cyclo-octatetraenide, $[C_8H_8]^{2-}$.]

15. The vapours of most lanthanide elements react with permethyl-benzene to form sandwich complexes $(bz^*)_2Ln$, where $bz^* = \eta^6\text{–}C_6Me_6$. The m_{eff} reported for some of these are (in μ_B):

Gd 8.75; Tb 10.69; Dy 11.20; Ho 10.10; Er 11.00; & Lu 1.69

(The Tm and Yb complexes are unstable; so is that of Eu.)

The EPR spectra of $(bz^*)_2Lu$, and the scandium and yttrium analogues, are consistent (g being ~ 2) with the presence of just one unpaired electron. What can be inferred about the electronic structures of such compounds?

Analogous complexes of the transition elements are discussed in § 5.5(2d).

[If flummoxed, ask your tutor.]

5 Paramagnetism – part II

Compounds of the transition elements

5.1 General matters

The transition metals form an enormous variety of compounds displaying a remarkable diversity of magnetic properties. This is eloquently witnessed by the contents of Table 3.1 – concerning just one element, iron – though this chapter, like that immediately preceding, is concerned exclusively with *paramagnetic* phases. Comprehension of the more subtle aspects of paramagnetic behaviour, invariably the most revealing, demands the background provided by earlier chapters.

Paramagnetism is a rich source of information about electronic structure. In the last century, the contemplation of paramagnetic properties played a crucial role in the development of now commonplace ideas concerning the electronic structures of transition metal compounds (and, not infrequently, their *raison d'étre*). Together these concepts form the basis of what is called **ligand field theory**. That's history; but magnetic measurements, of the susceptibility and EPR spectrum in particular, remain of fundamental importance as regards the characterisation of a newly synthesised (open-shell) compound.

Table 5.1 The transition elements

group	3	4	5	6	7	8	9	10	11
1st row	Sc	Ti	V	Cr	Mn	Fe	Co	Ni	Cu
2nd row	Y	Zr	Nb	Mo	Tc	Ru	Rh	Pd	Ag
3rd row	La	Hf	Ta	W	Re	Os	Ir	Pt	Au

The group 3 elements are not usually regarded as members of the transition metal block#, while those of group 11 are sometimes otherwise termed *post-transition elements*, along with the group 12 elements (Zn, Cd & Hg) *etc.*

The criterion applied here is simply occupancy of *d* orbitals in the valence shell of the free metal atom.

1. The nature of the valence electrons

Cations of the transition elements have outer electronic configurations of the type d^N, at least when the external charge on the atom exceeds +1. The *neutral* atoms typically display occupancy of the $(n+1)s$ orbital – as in V $3d^3 4s^2$, Cr $3d^5 4s^1$ and Mn $3d^5 4s^2$, for example – but the chemically familiar ions M^{2+}, M^{3+} *etc.* have pure d^N configurations, *e.g.* Cr^{2+} $3d^4$, in their free state. In their *compounds* the formal species M^+ and M^0 (and even M^-), also have essentially d^N configurations because their ligand environments invariably destabilise the diffuse $(n+1)s$ orbital in relation to nd. Thus $Cr(CO)_6$ and $(\eta^6\text{-}C_6H_6)_2Cr$, for instance, may be viewed in simple terms as complexes

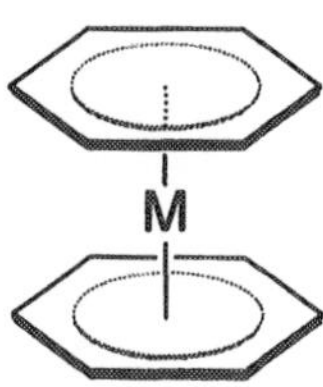

Sandwich structure adopted by the *bis*-benzene complexes $(\eta^6\text{-}C_6H_6)_2M$.

[The eclipsed conformation is shown, but the staggered form is invariably the more stable.]

of Cr^0 $3d^6$, more commonly denoted Cr(0), and the paramagnetic compounds $V(CO)_6$ and $(\eta^6\text{-}C_6H_6)_2V$ (each with one unpaired electron) as complexes of $V(0) \equiv V^0$ $3d^5$. In vital respects, they are akin electronically to compounds of, for example, Fe(II) and Fe(III).

2. Spin-only paramagnetism

The 5-fold degeneracy of the *d*–orbitals characteristic of spherical symmetry is unsustainable in compounds of the transition elements due to the discriminating effects of the ligand environment [section 2(1) below]. Accompanying the loss of degeneracy is a substantial quenching of the orbital angular momentum of the *d* electrons [§ 2.3(4)] such that, as a first approximation at least, the effective magnetic moments observed at ordinary temperatures may be explained by the spin-only formula

$$m_{eff} = g_e[S(S+1)]^{\frac{1}{2}}\mu_B \qquad (5.1)$$

Table 5.2 The spin-only values $n_{eff} = g_e[S(S+1)]^{\frac{1}{2}} = m_{eff}/\mu_B$.

S	n_{eff}
1/2	1.73
1	2.83
3/2	3.88
2	4.90
5/2	5.92

$g_e = 2.00232$

(a restatement of eqn. 3.3). For the most part (though not *always*) the spin state of a transition metal compound may safely be deduced from its m_{eff}. Table 5.3 gives an indication of the reliability of the formula for compounds involving ions with either the d^3 or the d^5 configuration. The expected moments are 3.88 μ_B or 5.92 μ_B, respectively (Table 5.2). Although the data listed do not conform *exactly* with the spin-only expectation, there would be no uncertainty in the characterisation of the spin state of any such compound were it unknown. A more exotic example is $(N^nBu_4)_2[MnMe_6]$ with $m_{eff} =$ 3.9 μ_B; this clearly contains Mn(IV) $3d^3$.

Furthermore, the sandwich compounds $(\eta^5\text{-}C_5H_5)_2V$ (*vanadocene*), with $m_{eff} = 3.78$ μ_B, and $(\eta^5\text{-}C_5H_5)_2Mn$ (*manganocene*) with $m_{eff} = 5.86$ μ_B, may confidently be identified as complexes of V(II) $3d^3$ and Mn(II) $3d^5$, respectively, with the cyclo-pentadienide ligand $[C_5H_5]^-$ (or *cp*), which is electronically similar to benzene as regards its π-orbital structure.

The prototype compound is *ferrocene*, $(\eta^5\text{-}C_5H_5)_2Fe$, a closed shell species involving Fe(II) d^6.

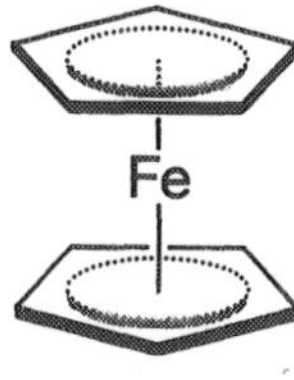

Elsewhere, however, there may be serious ambiguities in the deduction of spin state from magnetic moments. Sometimes, especially with compounds of the heavier transition elements, this is due to magnetic interaction between the open-shell ions. In other words, *the compounds may not be sufficiently dilute magnetically*. Thus many compounds of rhenium(IV) $5d^3$ have values of m_{eff} of 3.35 μ_B or less. It would not be clear, on the magnetic evidence alone, whether such compounds had two or three unpaired electrons (recall Table 5.2). Simulations of the susceptibility data in terms of the Curie-Weiss law [§ 3.2(1)] require quite large (negative) values of the Weiss temperature – *e.g.* $\theta = -88$K for $K_2[ReCl_6]$ with $m_{eff} = 3.25$ μ_B – suggesting that the negative deviation from the spin-only value, $m_{eff} = 3.88$ μ_B, is due in large part to antiferromagnetic interaction between the Re(IV) sites.# (However, the strength of the interaction falls far short of that needed to impose magnetic ordering in the vicinity of room temperature.)

EPR yields a g-factor of 1.816, which suggests $m_{eff} = 3.52$ μ_B – close to the value obtained when the $[ReCl_6]^{2-}$ anion is partnered by large cations.

Thus $(pyH)_2[ReCl_6]$ has $\theta = -14$ K with $m_{eff} = 3.58$ μ_B.

3. Orbital contributions to the magnetic moment

Otherwise, the characterisation of spin state may be confounded by the errors inherent in the spin-only approximation, if these are not understood and borne in mind. Consider, for example, the Co(II) $3d^7$ compounds, $CoCl_2$ and Cs_3CoCl_5, which have effective magnetic moments at 300 K of 5.47 μ_B and

4.67 μ_B, respectively. Naïve application of the spin-only formula to these compounds would indicate S = 2 rather than S = 3/2, the true value. (Considered in isolation, the moment of $CoCl_2$ might even to be taken possibly to imply S = 5/2.) The large positive deviation from the spin-only expectation, m_{eff} = 3.88 μ_B, is due to remanent, *i.e.* incompletely quenched, *orbital* angular momentum of the *d* electrons.

Cs_3CoCl_5 contains tetrahedral $[CoCl_4]^{2-}$ while $CoCl_2$ involves octahedral coordination of Co^{2+} by Cl^-.

Table 5.3 n_{eff} data (~ 300 K) for selected compounds of d^3 and d^5 ions.

d^3	$CrCl_3$	3.90	$K_3[Cr(ox)_3].3H_2O$	3.62
	$[Cr(NH_3)_6]Br_3$	3.77	$KCr(SO_4)_2.12H_2O$	3.84
	$[Cr(en)_3]Br_3$	3.82	$K_3[MoCl_6]$	3.79
	$[Cr(bpy)_3]Cl_3$	3.81	$K_2[MnCl_6]$	3.84
	$K_3[Cr(CN)_6]$	3.87	$[V(en)_3]Br_2$	3.81
	$K_3[Cr(NCS)_6].4H_2O$	3.79	$[V(bpy)_3]Cl_2$	3.67
	$K_3[Mo(NCS)_6].4H_2O$	3.70	$[Mo(bpy)_3]Cl_3$	3.66
	$(N^nBu_4)_3[Cr(N_3)_3]_6$	3.76	$K_4[V(CN)_6]$	3.78
d^5	$MnCl_2$	5.79	$FeCl_3$	5.73
	$MnBr_2$	5.82	$(Et_4N)[FeCl_4]$	5.88
	$(NH_4)_2Mn(SO_4)_2.6H_2O$	5.88	$(NH_4)Fe(SO_4)_2.12H_2O$	5.89
	$[Mn(NH_3)_6]Cl_2$	5.92	$K_3[Fe(ox)_3].3H_2O$	5.90
	$(Et_4N)_2[MnCl_4]$	5.94		

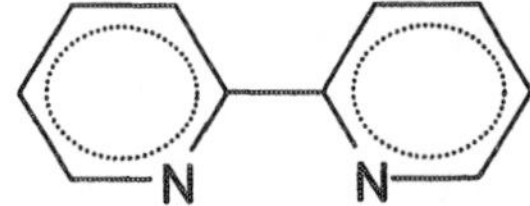

bpy = 2,2'–bipyridine

en = 1,2-diaminoethane (or ethylene diamine) & *ox* = oxalate ion, $[C_2O_4]^{2-}$. Like *bpy*, these are bidentate ligands, and may form *tris*–chelates (pictured below) with many transition metal ions.

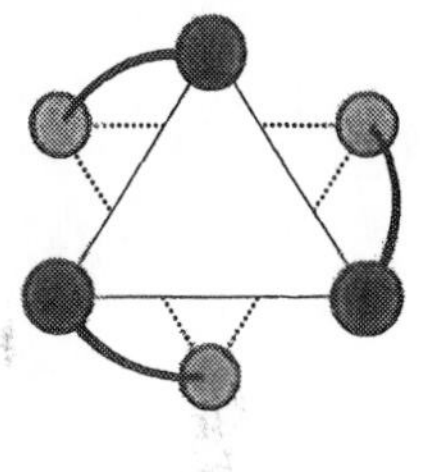

The d^5 configuration is a special case in that, assuming the maximal S = 5/2, the orbital angular momentum will fail to contribute to the magnetic moment, regardless of ligand field effects. The ion has the ground term ^6S (recall Fig. 2.15). With one electron in each of the *d*-orbitals (the spins being parallel) $\Sigma m_l = 0$, whence L = 0 in the free ion and also in any ligand environment allowing S = 5/2. The discrepancy between the observed m_{eff} and the spin-only prediction of 5.92 μ_B is probably due to experimental error and/or infelicitous processing of susceptibility data as a function of temperature when significant antiferromagnetic coupling is present. (Note the case of $[ReCl_6]^{2-}$ above.)

The EPR spectra of high-spin d^5 complexes show g-factors essentially equal to the free spin value g_e.

The invariably *positive* nature of the orbital contribution to m_{eff} in Co(II) compounds reflects the spin-orbit coupling in the ground state. When the *d* subshell is *more* than half-occupied the spin-orbit interaction tends to couple the orbital angular momentum in concert with that due to spin.# Hence, in a d^7 system, the magnetic moment arising from any residual orbital angular momentum will *augment* the spin moment. As illustrated by Fig. 5.1, the effect is almost general. Also discernible in the figure is a tendency, though less pronounced, for compounds with the *d* subshell *less* than half-filled (d^1 – d^4) to display magnetic moments at ordinary temperatures which are *reduced* in relation to the spin-only value. In this latter category, the ground state has the spin and orbital angular momenta coupled in opposition by the

Recall § 2.5(4) and especially Hund's third rule, NB **J**.

spin-orbit interaction. Negative deviations from the spin-only prediction are also evident in almost all of the data for d^3 compounds in Table 5.3.

The subject of orbital contributions to m_{eff} is taken up in more detail in section 6 later. It will seen there that the magnetic moment m_{eff} may often be dependent on temperature,# such that its particular value at ordinary temperatures can be misleading.

However, this is not the case for the data of Table 5.3 (except perhaps at very low temperatures).

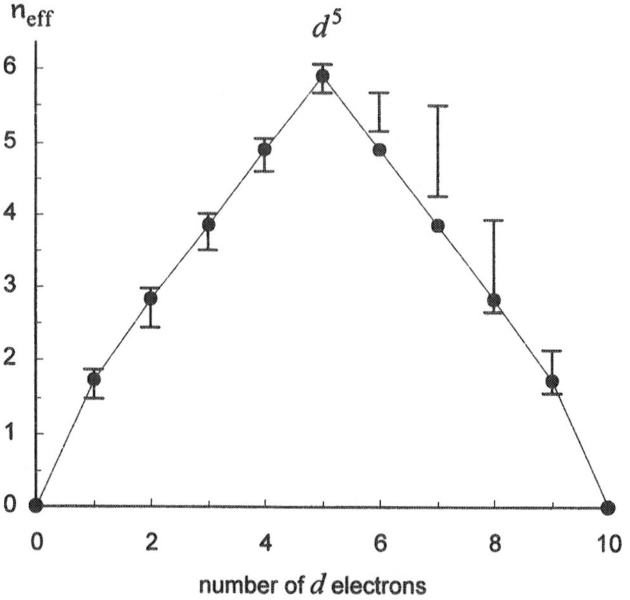

Fig 5.1 The scope of the spin-only formula (●) as applied to n_{eff} data (*ca.* 300 K) for high-spin compounds of the first transition series.

The data spread shown for each d^N configuration is restricted to magnetically dilute compounds. The survey is also confined to compounds in which the transition metal ions have the same spin configuration as in their free state. These are termed "high-spin" compounds. (See section 3.)

[Contrast Figure 4.9]

As indicated by the Co(II) data introduced above, the m_{eff} observed for a particular d^N configuration may also depend critically on the symmetry of the ligand field (which influences the form of the orbital splitting). The influence of the coordination number and disposition of ligands is illustrated further in Table 5.4, devoted to complexes of Ni(II) with S = 1 or 0.

Table 5.4

The range of m_{eff} values (/μ_B) observed in complexes of Ni(II) $3d^8$ with different geometries, at *ca.* 300 K.

octahedral	2.9 – 3.4
tetrahedral*	3.2 – 4.0
trigonal bipyramidal*	3.2 – 3.8
square pyramidal	3.2 – 3.4
square planar	0

* m_{eff} invariably decreases as the temperature is lowered.

Many *tbp* and *sqp* complexes of Ni(II) otherwise have closed-shell (S = 0) ground states.

4. Identification of ligand field geometry

Despite the limitations of the spin-only approximation, and the frequently complex variation in experimental magnetic moments, individual systems are sufficiently well characterised that measurements of m_{eff} may provide structural information on a purely empirical (fingerprinting) basis, especially when the oxidation state of the metal has been independently established. In Co(II) chemistry, for example, the magnetic moments m_{eff} lie typically in the range 4.4 – 4.8 μ_B for tetrahedral complexes, but are invariably larger, 4.8 – 5.3 μ_B, in the case of octahedral complexes; moreover, in the latter case m_{eff} commonly falls substantially with decrease of temperature below 300 K [section 6(5)]. There is usually little difficulty, therefore, in distinguishing those alternative geometries. The paramagnetic properties of a newly synthesised Ni(II) complex may similarly provide a strong indication of ligand field symmetry (note Table 5.4), though certain alternative geometries may not be distinguishable without additional evidence – the EPR spectrum, for example. There is no substitute, however, for structure determination by diffraction methods if a suitable sample is available

5.2 The orbital splitting in cubic ligand fields

The term *cubic* embraces the octahedron and tetrahedron (both of which may be inscribed within a cube), as well as the cube itself. Among compounds of the transition metals, octahedral complexes ML_6 are more prevalent than tetrahedral ML_4, the latter being rare in the second and third transition series. Eight coordination is comparatively unusual, and essentially confined to the larger metal ions of the heavier elements; moreover, very few *discrete* ML_8 complexes are actually cubic,# the alternative ligand dispositions corresponding to the *square antiprism* or *trigonal* (a.k.a. *triangulated*) *dodecahedron* being preferred due to reduced repulsion between the ligands.

– and even fewer are paramagnetic.

e.g. $[Mo(CN)_8]^{x-}$, where x = 3 or 4.

1. The point charge/dipole model – crystal field theory

The orbital splitting is described first in terms of the semi-classical crystal field theory in which the ligands are treated as point charges or dipoles.

(a) Octahedral complexes

Complexes ML_6 of octahedral symmetry are conveniently discussed using the (right-handed) coordinate system of Fig. 5.2.

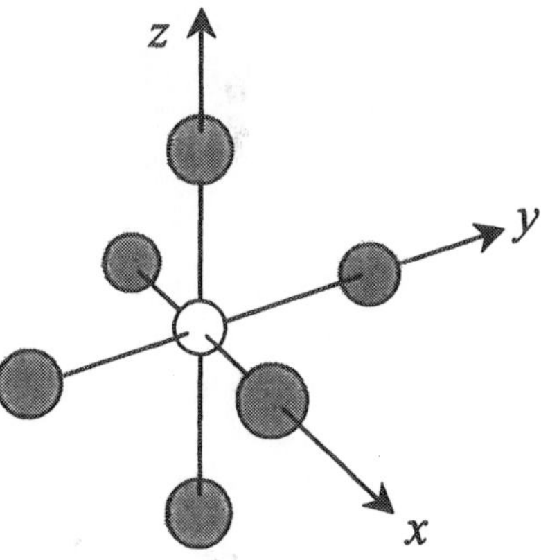

Fig. 5.2 Customary coordinate system for an octahedral complex ML_6.

Note that the ligands are taken to lie on the cartesian axes.

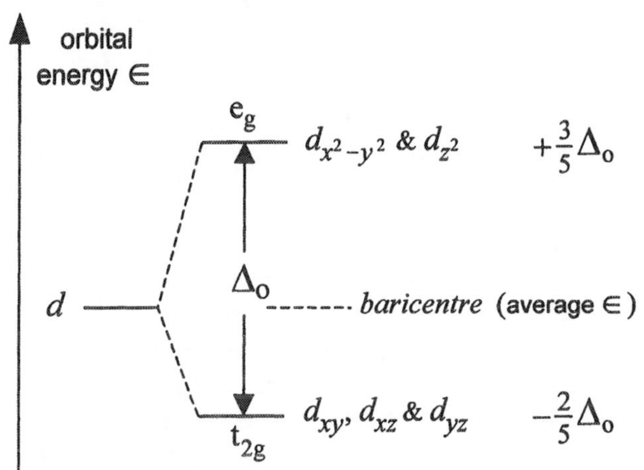

Fig. 5.3 Splitting of the metal *d*-orbitals in an octahedral transition metal complex.

The energies ∈ of the two sets of orbitals that result, e_g and t_{2g}, are specified in relation to their centroid of energy or *baricentre*.

The electrostatic field of the ligands induces a splitting of the atomic *d* level into two, well separated levels designated e_g and t_{2g}, respectively (Fig. 5.3). The particular form of the splitting can be understood by reference to Fig. 5.4. Each of the *d*-orbitals is destabilised by the ligand field, but some more so than others. Thus the $d_{x^2-y^2}$ orbital suffers a greater perturbation than d_{xy} because its lobes of electron density are oriented directly towards the ligands [Fig. 5(4(a)] while those of d_{xy} point between [Fig. 5.4(b)]. Now, the orbitals d_{xz} and d_{yz} are equivalent to d_{xy} with respect to the ligand array

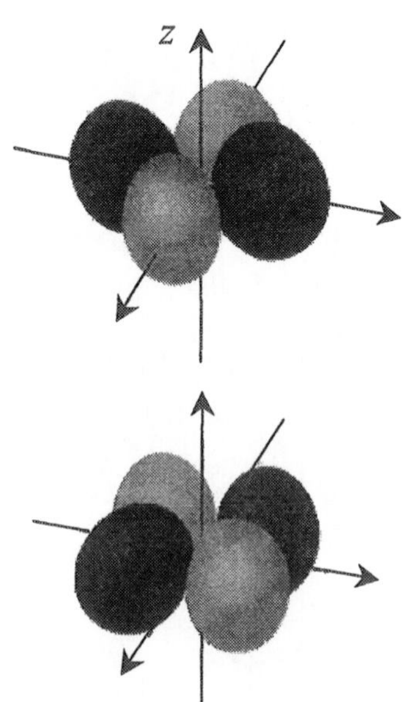

3D picture of the boundary surfaces of the $d_{x^2-y^2}$ and d_{xy} orbitals.
The relative phases of ψ in the lobes are here distinguished by shading.

and therefore these three orbitals remain of the same energy, constituting a three-fold degenerate or *t* set. Less obvious, at first sight, is the effect of the ligand field on the d_{z^2} orbital. However, recalling eqn. 2.6, the latter orbital is simply a linear superposition of two extraneous orbitals,

$$d_{z^2} \sim d_{z^2-x^2} + d_{z^2-y^2}$$

each of which, in an octahedral ligand field, is equivalent in all essential respects to the $d_{x^2-y^2}$ orbital. Given this make-up, the d_{z^2} orbital must experience the same destabilisation as $d_{x^2-y^2}$ and so together they form a doubly-degenerate or *e* set of orbitals.#

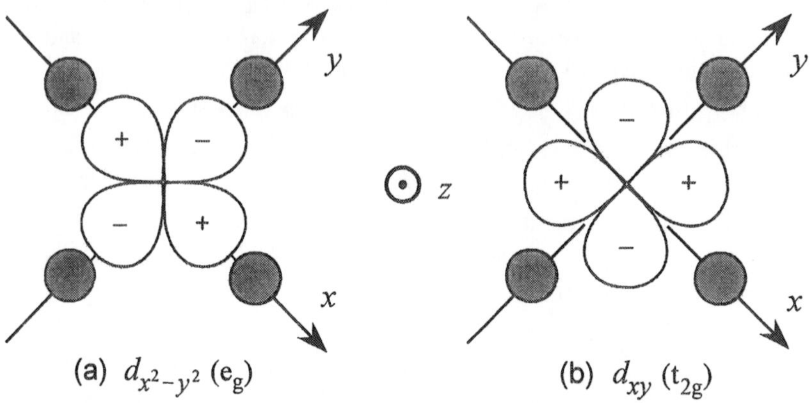

Fig. 5.4 Projections of the *d*-orbitals, $d_{x^2-y^2}$ and d_{xy}, onto the *xy*-plane, adopting the axes defined in Fig. 5.2. [*cf.* Fig. 2.6]

Symmetry labels (an explanatory *aside*): As elsewhere, the symbols *e* and *t*, attributable to Mulliken, simply denote orbital degeneracies of *two* and *three*, respectively. Non-degenerate orbitals are labelled *a* or *b*.

It so happens that cubic symmetry supports *two* kinds of triply-degenerate orbital sets, distinguished as t_1 and t_2. The three *d*-orbitals, d_{xy}, d_{xz} & d_{yz}, belong to the t_2 category. The alternative t_1 species is represented, in the case of octahedral complexes, by the three *p*-orbitals of the transition metal atom, which are not differentiated by the ligand field. The subscript *g* (for German *gerade* = *even*) identifies the *d*-orbital wave-functions as *symmetric* with respect to the operation of inversion through the centre of symmetry of octahedral ML_6. The set of *p*-orbitals is designated t_{1u} (*e.g.* Fig. 5.18 below), where the subscript is *u* (*ungerade* or *odd*) because the *p* wave functions are instead *antisymmetric* – that is, they change sign on inversion through the centre of symmetry.

Symmetry arguments show that the degeneracy of the $d_{x^2-y^2}$ and d_{z^2} orbitals applies in any *cubic* complex regardless of the physical model that is invoked.

See almost any account of applied *Group Theory* in undergraduate textbooks.

Returning to Fig. 5.3, the energy separation of the e_g and t_{2g} orbital sets,

$$\Delta_o = \in(e_g) - \in(t_{2g}) \qquad (5.2)$$

Δ is commonly denoted by 10Dq in the older literature.

is called the *orbital splitting parameter*; the subscript to Δ identifies the ligand field as octahedral. The one-electron energies ∈ are orbital energies *in the field of the atomic core* – $(Ar)^{x+}$ in the case of transition elements of the *first* row.#

These orbital energies are not the same as those identified with *minus* the ionisation energy, which include the electron repulsion within the *d*-subshell.

The average energy ∈ of the split *d*-orbitals, often termed the *baricentre*, is weighted according to the degeneracies of the e_g and t_{2g} sets. It corresponds to the energy of a hypothetical *s*-orbital having the same radial wave function as the *d*-orbitals, and lies high in relation to the *d* orbital energy of the free metal ion.

(b) Tetrahedral complexes

A suitable choice of coordinate system for complexes ML_4 of tetrahedral geometry is shown in Fig. 5.5, and the splitting of the perturbed *d*-orbitals in Fig. 5.6. The pattern of the splitting is the inverse of that observed for octahedral ligand fields.

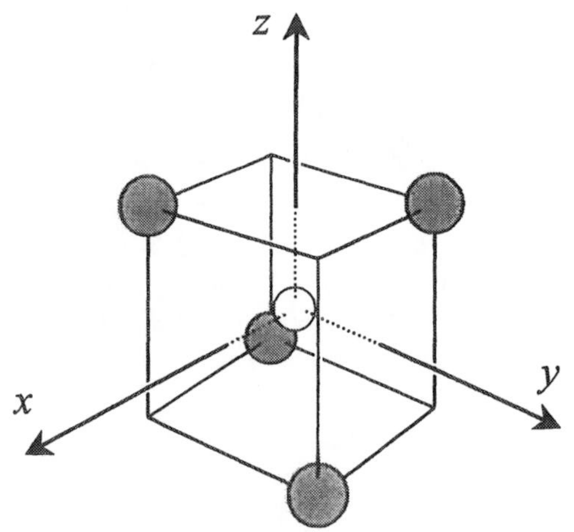

Fig. 5.5 Coordinate system for a tetrahedral complex ML_4 located in its exserted cube.

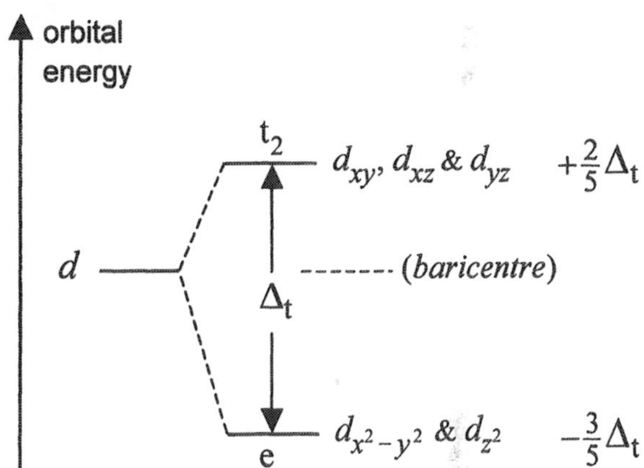

Fig. 5.6 Splitting of the metal *d*-orbitals in a *tetrahedral* transition metal complex.

None of the metal *d*-orbitals points directly at the ligands but each lobe of d_{xy} comes close to one ligand [Fig. 5.7(a), overleaf] while $d_{x^2-y^2}$ has its lobes precisely oriented between ligands [Fig. 5.7(b)], "avoiding" the ligands to best effect. Accordingly, the d_{xy} orbital is destabilised more than $d_{x^2-y^2}$ by a tetrahedral field.

As in octahedral symmetry, d_{xz} and d_{yz} are clearly equivalent to d_{xy} regarding their disposition with respect to the ligands. Moreover, the d_{z^2} orbital, expressible as a simple combination of the alternative orbitals $d_{z^2-x^2}$ and $d_{z^2-y^2}$ (as above), both equivalent to $d_{x^2-y^2}$, must remain degenerate with the latter. Hence the pattern of orbital energy levels shown in Fig. 5.6. Note that the subscript g is no longer appropriate because the tetrahedron lacks a centre of symmetry.

The form of the *d*-orbital splitting in tetrahedral complexes may alternatively be understood by considering first a cubic system ML_8. A splitting with $\in(t_{2g}) > \in(e_g)$ is not difficult to discern. If four ligands are then removed from alternate corners of the cube (and with them the centre of inversion) the splitting pattern of Fig. 5.6 results.

See, *e.g.*, Huheey (1993), § 11.

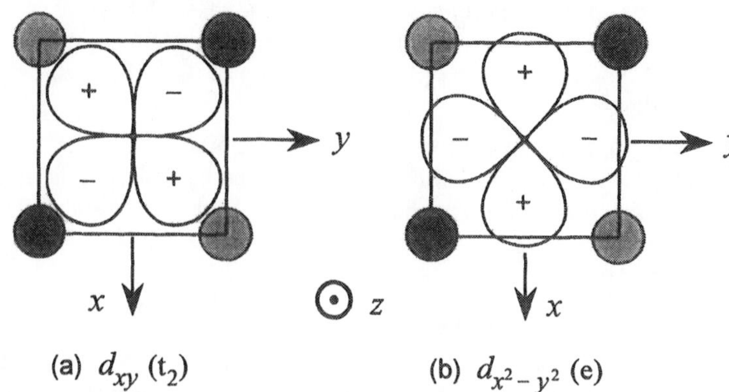

(a) d_{xy} (t_2) (b) $d_{x^2-y^2}$ (e)

Fig. 5.7 Plan view of the orbitals d_{xy} and $d_{x^2-y^2}$ in relation to the tetrahedral array of ligands

2. The variation of Δ

The orbital splitting parameter, as determined from visible/UV absorption spectra, varies almost 20-fold among transition metal complexes – from as little as 2500 cm^{-1} for low coordination numbers (*e.g.* 2650 cm^{-1} in tetrahedral $[CoI_4]^{2-}$) to well in excess of 40,000 cm^{-1} in cases involving "strong" ligands such as cyanide ion (*e.g.* 44,900 cm^{-1} in $[Rh(CN)_6]^{3-}$).

[Anticipating the analysis of magnetic properties, it should be noted that the orbital splitting parameter Δ is always at least an order of magnitude greater than kT at ordinary temperatures, so normally there can be no question of a significant Boltzmann population of *excited* electronic states.]

The more obvious factors influencing the magnitude of Δ are as follows.

(a) The coordination number of the metal ion

First, the size of Δ naturally reflects the number of ligands perturbing the *d*-orbitals. Thus, for a given metal ion and ligand L, and a constant M–L distance, the orbital splitting that occurs in a tetrahedral complex ML_4 will be less than in an octahedral complex ML_6. According to the elementary point charge/dipole model, the splittings are related by

$$\Delta_t = -\tfrac{4}{9}\Delta_o \tag{5.3}$$

In practice, Δ_t is found to be $\approx -\tfrac{1}{2}\Delta_o$.[a]

[a] *e.g.* the Δ values (in cm^{-1})

VCl_4	7900	$[VCl_6]^{2-}$	15,400
$[NiCl_4]^{2-}$	3850	$[NiCl_6]^{4-}$	7,700

(b) The formal charge on the metal ion

For a given transition element and a particular ligand, Δ is observed to increase with the oxidation state of the metal.[b]

[b]

$[VCl_6]^{4-}$	9,350	$[V(OH_2)_6]^{2+}$	12,350
$[VCl_6]^{3-}$	12,100	$[V(OH_2)_6]^{3+}$	16,900
$[VCl_6]^{2-}$	15,400		

(c) The location of the transition element in its subgroup

For any particular coordination type, *e.g.* ML_6, with a given ligand species and constant formal charge on the metal ions, Δ is found to increase significantly (by 75 % or more) on going down any column in the transition metal block.[c] The increase is particularly large (*ca.* 50 %) between the first and second transition series. Thus, other things being equal, the trend in the splitting parameter is

$$\Delta:\quad 3d \ll 4d < 5d$$

In both the point charge/dipole theory and also any covalent model, this observation may be attributed to an increasing protrusion of the *d* electron density from the core (not merely an expansion in absolute terms) on descending a transition element subgroup [see later, section 4(4)].

[c]

$[Co(NH_3)_6]^{3+}$	22,870 cm^{-1}
$[Rh(NH_3)_6]^{3+}$	34,100
$[Ir(NH_3)_6]^{3+}$	41,200
$[Fe(OH_2)_6]^{3+}$	14,000
$[Ru(OH_2)_6]^{3+}$	28,600

(d) The nature of the ligand

The empirical values of Δ from $d-d$ spectra also naturally depend upon the ligand involved. It has been established by comprehensive survey that the relative magnitude of Δ varies with ligand according to the sequence

$$\begin{aligned}&CO > CN^- > H^- \sim [CH_3]^- \sim [C_6H_5]^- > PPh_3 > [\underline{N}O_2]^- \\ &> bpy \sim \text{o-}phen > en > NH_3 \sim py > [\underline{N}CS]^- > OH_2 \\ &> O^{2-} \sim [ox]^{2-} > OH^- > F^- \sim [N_3]^- > [N\underline{O}_3]^- > Cl^- \\ &> [\underline{S}CN]^- > S^{2-} > Br^- > I^-\end{aligned}$$

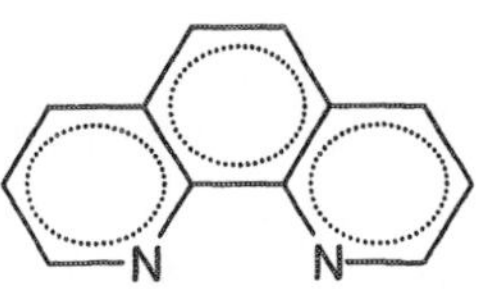

o-*phen* = 1,10–phenanthroline

This is the famous **spectrochemical series.** Ligands may be described as *strong* or *weak* according to their position in this series. Of course, no single metal ion forms a complete set of complexes (*e.g.* ML_6) with all of these ligands; the series has been pieced together from overlapping sequences involving different ions.

Where a ligand is *ambidentate*, the coordinated atom is underlined.

Some of the ligands are *bidentate*, *e.g.* o-*phen* (above). The position of a chelating ligand in the spectrochemical series conveys its effect per coordinating atom.

The spectrochemical series has many intriguing features, certain of which cannot be rationalised on the basis of the point charge/dipole model. In particular, it makes no sense that dipolar ligands such as OH_2 and NH_3 produce a larger orbital splitting than monatomic anions such as $[OH]^-$, F^- and other halide ions.# The position of CO at the head of the series (despite a minuscule dipole moment of *ca.* 0.1 D) is especially bizarre in relation to an ionic picture of transition metal compounds. While simple electrostatic effects undoubtedly contribute, it has long been clear that

Nor, to make a finer point, is it explicable why NH_3, with a smaller electric dipole moment than OH_2, should cause the larger splitting.

> The orbital splitting Δ is due principally to covalent bonding between the metal atom and ligands.

A covalent model is also needed to account for observation (b) above – that Δ invariably increases with the formal charge of the metal ion. If Δ is primarily the result of covalency then, whatever the detail, it should naturally increase with the metal oxidation state – if only because the metal-ligand distance decreases while the metal centre becomes concomitantly a stronger electron acceptor.

It will be shown later in section 4 that the description of metal-ligand covalency in terms of *molecular orbital theory* provides a remarkably successful explanation of the variation of Δ encompassing the whole panoply of transition metal compounds, from the comparatively ionic, containing units such as $[NiF_6]^{4-}$, to the highly covalent, *e.g.* $Ni(PF_3)_4$ and $(\eta^5\text{-}C_5H_5)_2Ni$.

3. Ligand field stabilisation energies

An important consequence of the ligand field splitting, not yet mentioned, is the stabilisation of the metal centre in comparison with the hypothetical situation in which there was no splitting (as with a partly filled *p*-subshell in cubic symmetry). If the metal ion has $N = a + b$ electrons, distributed in the ground state as $(t_{2g})^a(e_g)^b$, its net orbital energy is changed by an amount

$$U_{LF} = -\frac{\Delta}{5}(2a-3b) \tag{5.4}$$

relative to the baricentre (*cf.* Fig. 5.3). The quantity $-U_{LF}$ is called the *ligand field stabilisation energy* (LFSE).

U_{LF} is to be added to any internal energy U (lattice energies *etc.*) calculated within the framework of the ionic model, or its elaborations. It is naturally zero for d^{10} ions (and also for those with the *high*-spin d^5 configuration, when $a = 3$ and $b = 2$; *vide infra*). In the case of *low*-spin complexes U_{LF} must be adjusted to allow for the spin-pairing energy Π [3(4)].

The LFSE can have a significant influence on the energetics of transition metal complexes, apparent in both thermodynamic and kinetic properties. Its precise effects prove highly dependent on the spin configuration adopted in the ground state. The latter is itself highly sensitive to Δ. Among its diverse consequences, the LFSE may determine the geometry of a complex or, when a transition metal cation has alternative sites within a structure, its more stable location – its *site preference* so-called.[a]

[a] For example, the distribution of cations between the alternative octahedral and tetrahedral sites in the spinel lattice [§ 3.6(4b)] is largely dictated by the relative LFSEs at the two locations. Thus, Fe_3O_4 is an inverse spinel because, while high-spin Fe^{3+} (d^5) has a zero LFSE in both tetrahedral and octahedral sites, the LFSE for high-spin Fe^{2+} (d^6) is more negative in the octahedral site.

5.3 High-spin and low-spin complexes

1. Exchange (spin correlation) effects

Recalling § 2.2(3), when two electrons have their spins aligned parallel, the repulsion between them is reduced in relation to the semi-classical expectation, *i.e.* simply the coulomb energy J. This is attributable to *spin correlation* or *exchange*, a consequence of the Pauli principle (in its general form). In the *ground* state of a many-electron system the pair-wise repulsion terms R may usually be distinguished as

$$\text{(i)}\ \uparrow\uparrow\ \mathrm{R = J - K} \quad \text{or} \quad \text{(ii)}\ \uparrow\downarrow\ \mathrm{R = J} \qquad (5.5)$$

according to the orientation of spins, K being the *exchange integral*. This is illustrated by Fig. 5.8. (The J + K repulsion term for an electron pair with *anti*-parallel spins, introduced in eqn. 2.13 for the *two*-electron system, is a special case relating to the *excited* state. This term does not normally arise in *ground* states.)

Fig. 5.8

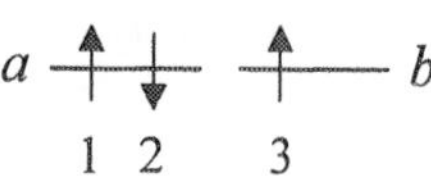

$R_{12} = J_{aa} = J_{bb}$

$R_{13} = J_{ab} - K_{ab}$

$R_{23} = J_{ab}$

The individual electron repulsion terms R within an e^3 configuration. The degenerate orbitals involved are arbitrarily labelled *a* and *b*.

The configuration above might be the $(e_g)^3$ component of an octahedral Cu(II) complex (d^9).

2. Octahedral complexes

As a result of spin correlation, octahedral complexes involving the particular electron configurations d^4, d^5, d^6 and d^7 have *alternative ground states* depending, for any given cation, on the strength of the ligand field. In relatively *weak* ligand fields such complexes may be more stable if the e_g orbitals are populated sufficiently to secure the same optimal spin configuration (with maximum S) as in the free ion. The energy needed to promote one or two electrons from the t_{2g} level to e_g may be more than recovered through the reduction in inter-electronic repulsion occasioned by increasing the number $n(\uparrow\uparrow)$ of pairs of electrons with parallel spins. Those complexes of $d^4 - d^7$ ions that display the maximum possible spin S in their ground states are described as *high-spin*. The d^5 complexes cited in Table 6.3, involving Mn(II) and Fe(III), are prominent examples (while the more credible m_{eff} data for *all* high-spin compounds was included in the survey on which Fig. 5.1 is based).

However, the arithmetic can be different in *strong* ligand fields. Thus the following octahedral d^5 complexes, listed with the m_{eff} values they display at ordinary temperatures,

$[Fe(CN)_6]^{3-}$ 2.25 μ_B $\qquad$ $[Fe(en)_3]^{3+}$ 2.40 μ_B

$[Mn(CN)_6]^{4-}$ 2.18 μ_B

belong clearly to a different category to those in Table 5.3 (all of which have m_{eff} not very different from 5.92 μ_B, the spin-only value for S = 5/2). They represent alternative *low-spin* complexes having the configuration $(t_{2g})^5$ with S = 1/2 (Fig. 5.9)[b] Other examples of low-spin complexes include

[b] Another instance is $V(CO)_6$, with m_{eff} = 1.85 μ_B (at ~ 300K).

The poor agreement with the spin-only formula – which, for S = 1/2, would predict 1.73 μ_B – is due to orbital contributions (section 6).

d^4 :	$[Cr(bpy)_3]^{2+}$	$[Cr(CN)_6]^{4-}$	$[Mn(CN)_6]^{3-}$	t_{2g}^4	$S = 1$
d^6 :	$[Fe(CN)_6]^{4-}$	$[Co(NH_3)_6]^{3+}$	$Cr(CO)_6$	t_{2g}^6	$S = 0$
d^7 :	$[Co(diars)_3]^{2+}$	$[Co(NO_2)_6]^{4-}$	$[NiF_6]^{3-}$	$t_{2g}^6 e_g^1$	$S = \frac{1}{2}$

(where *diars* is short for *diarsine*).#

Having the closed subshell $(t_{2g})^6$, the d^6 complexes cannot display first order paramagnetism and are commonly diamagnetic (or feebly paramagnetic).

\# *diars*

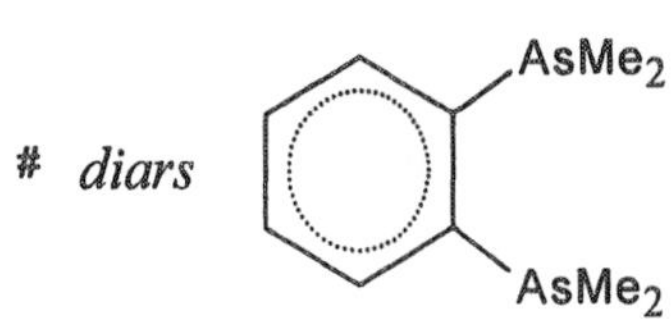

This dichotomy in electronic structure – *high-* vs. *low*-spin states – which proves to have far reaching consequences for both the thermodynamic and kinetic properties of the octahedral complexes concerned, was discovered through magnetic measurements and came to be understood only with the advent of ligand field theory.

Fig. 5.9

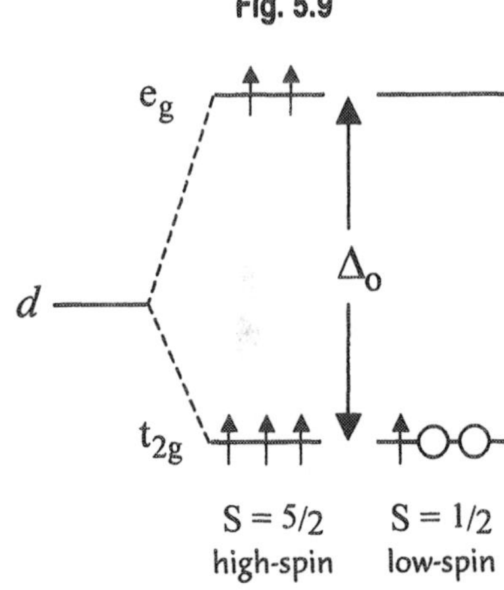

The alternative high- and low-spin ground states of an octahedral d^5 complex. ○ = electron pair

3. The magnetic moments of low-spin octahedral complexes

The variation of m_{eff} at ordinary temperatures among low-spin octahedral complexes from the first transition series is described in Fig. 5.10, which should be viewed in conjunction with Fig. 5.1. The canvass, which is again confined to compounds thought to be magnetically dilute, embraces also related six-coordinate complexes and, in particular, *tris*-chelates such as $[Cr(bpy)_3]^{2+}$ (m_{eff} ~ 3.2 μ_B). The low-spin compounds are placed in the general context, including the (necessarily) high-spin complexes with 4 > *N* > 7. The pattern of experimental m_{eff} for the latter is closely similar to that in Fig. 5.1, which surveys the entirety of high-spin species (from the first transition series), encompassing therefore many of tetrahedral geometry and others of lower than cubic symmetry.

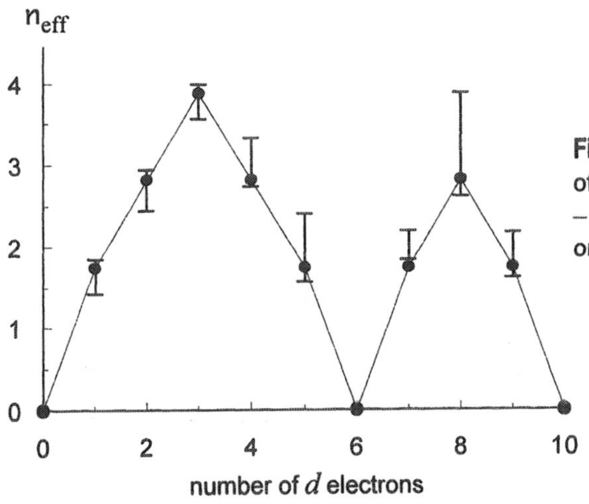

Fig. 5.10 n_{eff} data (300 K) for octahedral complexes of the first transition series, *low* spin in the case of d^4 – d^7, compared with the predictions (●) of the spin-only formula. (Compare Fig. 5.1.)

As already noted, low-spin d^6 complexes show at most a modest second-order paramagnetism (TIP), often swamped by their diamagnetism. The evidence of Fig. 5.10 is that the other low-spin complexes, those having d^4, d^5 and d^7 configurations, usually display *positive*, and commonly large deviations from the spin-only expectation. In the strong ligand fields that are required to induce spin-pairing, $(t_{2g})^N$ configurations behave magnetically like p^N configurations of an atom. When the t_{2g} subshell is *more* than half-filled, the ground state has the spin **S** and orbital angular momentum **L** coupled in unison; hence, for $(t_{2g})^4$ and $(t_{2g})^5$ complexes, any orbital contribution to m_{eff} should increase it beyond the spin-only value.# As regards the $(t_{2g})^4$ species, this contrasts with what is encountered in *high*-spin d^4 complexes. In the low-spin d^7 case, $(t_{2g})^6(e_g)^1$, the positive deviations from the spin-only expectation (*e.g.* $[Co(NO_2)_6]^{4-}$ with $m_{eff} \sim 1.84\ \mu_B$) reflect the normal "parallel" coupling of **S** and **L** in the ground state of a d^N configuration when $N > 5$, as in the case of high-spin complexes (Fig. 5.1).

\# – at least at ordinary temperatures. However, the way in which the spin and orbital moments couple in the ensemble may be more complicated at lower temperatures [see section 6(5)].

4. The factors determining spin state

The increase in the inter-electronic repulsion that militates against a low-spin ground state is termed the *spin-pairing energy*, here denoted Π (but elsewhere often by P). To characterise Π, consider first the simplest case, that of the d^4 configuration (Fig. 5.11) – represented, in particular, by complexes of Cr(II) and Mn(III). The numbers n(||) of pairs of electrons with parallel spins in the alternative states are as detailed in the following scheme.

$$d^4 \begin{cases} \text{high-spin} & t_{2g}^3 e_g^1 & \uparrow\uparrow\uparrow\uparrow & n(\|) = 6 & E_{ex} = 6K \\ \text{low-spin} & t_{2g}^4 & \uparrow\uparrow\uparrow\downarrow & n(\|) = 3 & E_{ex} = 3K \end{cases}$$

Fig. 5.11

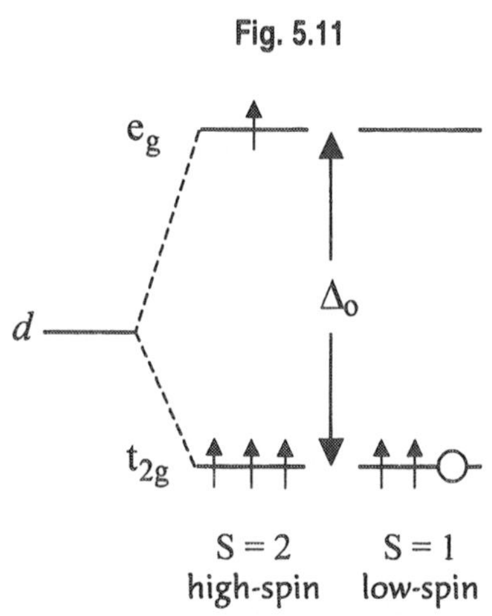

The alternative *high*- and *low*-spin configurations of an octahedral d^4 complex. ○ = electron pair

Also included is the total exchange energy E_{ex} = n(||) x K, it being assumed (as a first approximation) that the exchange integral K is the same for all pairs of d orbitals. If the coulomb integral J is also constant within the d-subshell, then the spin-pairing energy is simply the decrease in exchange energy, *i.e.*

$$\Pi = 6K - 3K = 3K$$

Since the alternative configurations differ in aggregated orbital energy by an amount Δ (the promotion energy), a low-spin ground state may be anticipated when $\Delta > 3K$.

The results of analogous treatment of octahedral complexes involving instead the d^5, d^6 or d^7 configurations are collected in Table 5.5. Notice there that the high-spin states for d^5 and d^6 require the promotion of *two* electrons from the t_{2g} level to e_g.

According to this simple analysis the inherent relative tendencies to adopt a low-spin configuration are: $d^7, d^6 > d^5, d^4$. The d^7 and d^6 cases are not differentiated, nor d^5 from d^4. As a matter of fact, the "ease" of spin-pairing, for comparable values of Δ (and K), appears to vary as

$$d^6 \gtrsim d^7 >> d^4 \gtrsim d^5$$

so that the most commonly encountered low-spin compounds involve the d^6 configuration (where the LFSE is greatest), and the least prevalent d^5 (where Π is maximal). Thus high-spin Co(III) d^6 is known only in the fluoride complexes $[CoF_6]^{3-}$ and perhaps $Co(OH_2)_3F_3$,[a] while the spin-pairing of Fe(III) d^5 requires amines (*en*, for instance) or ligands that lie higher still in the spectrochemical series.

a $[CoCl_6]^{3-}$ *etc.* would be expected to be high-spin also, but apparently cannot be synthesised.

It is worth noting, moreover, that among aquocations of the first transition series, $[Co(OH_2)_6]^{3+}$ is unique in being low-spin.

Table 5.5 High- and low-spin octahedral complexes: the spin-pairing energy Π *vs.* the relative orbital stabilisation energy in the low-spin state.

N = the number of *d*-electrons; n(↑↑) = the number of pairs of electrons with parallel spins.

N	high-spin configuration t_{2g}	e_g	n(↑↑)	low-spin configuration t_{2g}	e_g	n(↑↑)	Π	lowering of orbital energy
4	↑ ↑ ↑	↑	6	↑ ↑ ↑↓		3	3K	Δ_o
5	↑ ↑ ↑	↑ ↑	10	↑ ↑↓ ↑↓		4	6K	$2\Delta_o$
6	↑ ↑ ↑↓	↑ ↑	10	↑↓ ↑↓ ↑↓		6	4K	$2\Delta_o$
7	↑ ↑↓ ↑↓	↑ ↑	11	↑↓ ↑↓ ↑↓	↑	9	2K	Δ_o

Π = Δn(↑↑).K, where Δn(↑↑) is the reduction in the number of pairs of electrons having parallel spins on switching from the high-spin to the low-spin configuration, and K is the average exchange integral for the *d*-subshell (see text).

The observations may be explained by allowing for the variation of both K and J according to the particular *d*-orbitals involved (necessarily *different* orbitals in the case of K).[a] With this refinement, it transpires that the mean pairing energy, *per electron* transferred from t_{2g} to e_g (call it π), has the trend

$$\pi: \quad d^5 > d^4 > d^7 > d^6$$

However, the treatment remains oversimplified and, in particular, because the low-spin complex invariably possesses *shorter* metal-ligand bonds (typically by 0.1 – 0.2 Å) than its high-spin counterpart and has therefore a *larger* orbital splitting Δ_o.[b] Thus the switch to the low-spin configuration occurs at a somewhat lower value of Δ_o than the simple theory would predict. Invoking the point charge/dipole model, the contraction of the metal-ligand bonds may be understood in terms of the depopulation of the e_g orbitals, the consequent reduction in *d* electron density directed at the ligands allowing their closer approach.

[a] The situation is more complicated than Fig. 5.8. The integrals J_{ab} and K_{ab} are not quite the same within the e and t_2 subshells, and they differ again for the e – t_2 repulsion terms.

[b] There may well be a concomitant increase in the pairing energy Π in the effectively smaller metal ion, but this proves less important.

Compounds of the heavier transition elements are straightforward in the present context. Δ_o is greatly increased as compared with compounds involving $3d^N$ cations [section 2(2c)], while the spin-pairing energy Π is diminished (by *ca.* 30 % or more) because the 4*d* and 5*d* valence orbitals are more diffuse and inter-electronic repulsion effects correspondingly reduced. As a result, where the distinction can be made ($d^4 - d^7$ cations),

All known octahedral complexes of the $4d^N$ and $5d^N$ ions have low-spin ground configurations.

5. Tetrahedral complexes

Complexes of tetrahedral geometry differ from octahedral complexes in that *low*-spin states are conceivable for the configurations $d^3 - d^6$, rather than $d^4 - d^7$. The alternative high- and low-spin configurations anticipated in the d^3 case are pictured in Fig. 5.12, and the necessarily high-spin ground state of a tetrahedral d^7 complex in Fig. 5.13.# The latter case is commonplace in the chemistry of cobalt(II), which exhibits a greater variety of tetrahedral complexes than any other transition element.

\# The ground states of both octahedral and tetrahedral complexes are characterised later in Table 5.13.

However, tetrahedral coordination is otherwise relatively uncommon, not least among paramagnetic compounds involving the configurations $d^3 - d^6$, while (recalling section 2(2a) earlier) the ligand field splitting Δ_t is generally small; hence

Low-spin tetrahedral complexes are rare.

The paramagnetic properties of the vast majority of tetrahedral complexes at ordinary temperatures therefore conform to the pattern of Fig. 5.1.

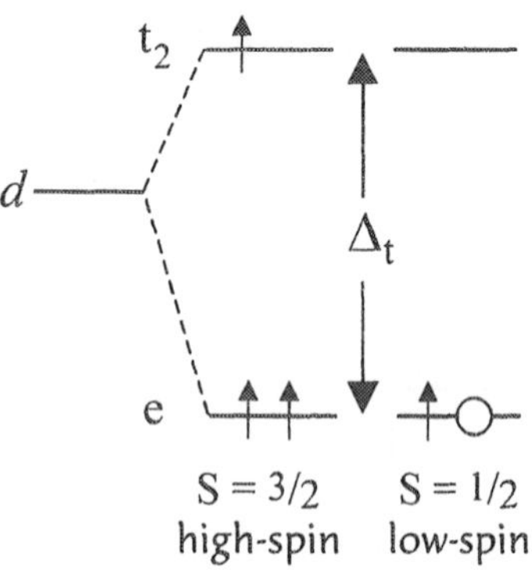

Fig. 5.12 Alternative *high*- and *low*-spin forms of a tetrahedral d^3 complex.

The high-spin complex is realised in $K_3[FeO_4]$, with m_{eff} = 3.71 μ_B; the low-spin form is known in (*e.g.*) the Re(IV) complex Re(o-tolyl)$_4$, with m_{eff} = 1.31 μ_B.

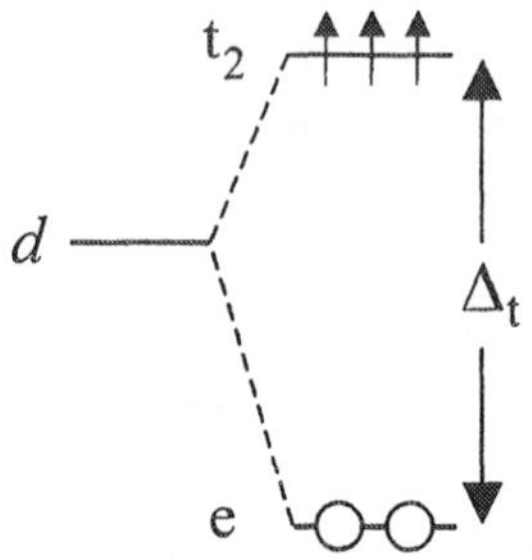

Fig. 5.13 The (unique) electron configuration of a tetrahedral d^7 complex, with S = 3/2.

This is well-known in the chemistry of Co(II) – *e.g.* $[CoCl_4]^{2-}$, with $m_{eff} \approx$ 4.72 μ_B (depending on the cation).

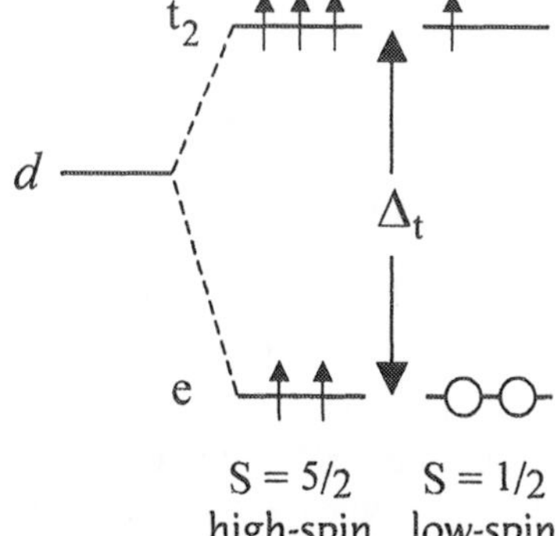

Fig. 5.14 The *high*- and *low*-spin configurations of a tetrahedral d^5 complex.

The high-spin form is exemplified by $(NEt_4)[FeCl_4]$, with m_{eff} = 5.88 μ_B, while the low-spin form is known as the compound (1-norbornyl)$_4$Co.

○ = electron pair

Where a low-spin configuration is feasible its stabilisation requires a combination of a "strong" ligand [one high in the spectrochemical series, *cf.* section 2(2d)] with a metal species in a high oxidation state [2(2b)]. Hence a rare example of a low-spin d^5 system (Fig. 5.14, RHS) furnished by the molecular compound (1-*nor*)$_4$Co, where 1-*nor* = 1-norbornyl, formally a complex of Co(IV) with an *alkide* ligand R$^-$, which induces a large orbital splitting Δ_t (comparable with that expected for $[CH_3]^-$). The complex has m_{eff} = 1.89 μ_B, constant between 299 and 2 K. Remarkably, the Co(III) d^6 complex anion [(1-*nor*)$_4$Co]$^-$ is also low-spin, with ground configuration $(e)^4(t_2)^2$ (S = 1), despite the expected reduction in the strength of the ligand field; it displays an m_{eff} = 3.18 μ_B at ~ 300 K.

1-norbornyl [1-*nor*]

(M)

It is not difficult to show that, in a tetrahedral field, the configuration for which it should be "easiest" to induce spin-pairing is d^4. The ground state would then have the closed subshell $(e)^4$, reminiscent of $(t_{2g})^6$ in the case of octahedral d^6 complexes, and the complex would be diamagnetic (or show TIP). Unsurprisingly, in the light of the (1-*nor*)$_4$Co case, both the Co(V) d^4 complex cation [(1-*nor*)$_4$Co]$^+$ and the isoelectronic Fe(IV) molecular species

$(1\text{-}nor)_4Fe$ have $(e)^4$ ground states; however, such complexes are extremely rare, at least for elements in the first transition series.# The stability of low-spin $(1\text{-}nor)_4Fe$ may be marginal since its Mn(IV) d^3 analogue $(1\text{-}nor)_4Mn$ has the high-spin $(e)^2(t_2)^1$ ground configuration, evidenced by $m_{eff} = 3.78$ μ_B (~ 300 K). The Mn(III) d^4 complex $[(1\text{-}nor)_4Mn]^-$ is naturally also high-spin, with the $(e)^2(t_2)^2$ ground configuration

Another example, of an electronically different kind, is the pyramidal molecule $Cr(NO)(N^iPr_2)_3$.

Low-spin $(e)^4$ complexes have a stronger presence among those of the heavier transition elements – where Δ_t is inherently larger and the spin-pairing energy smaller. Well established examples include a variety of MR_4 species from group 8 (M = Ru and Os), the ligands R being of a bulky nature, *e.g.* cyclohexyl or alkyl-substituted phenyls such as $2,6\text{-}C_6H_3Me_2$.

However, there are comparatively few tetrahedral complexes with open-shell configurations because the bigger metal cations tend ordinarily to adopt ligand coordination numbers that are higher than four. Voluminous ligands are *essential* not only to occlude additional ligands but also to prevent the formation of polynuclear species involving metal-metal bonding.

There exists a populous class of 4-coordinate complexes of d^8 cations, exemplified by $[PdCl_4]^{2-}$ & $[AuBr_4]^-$. However, for reasons very particular to the d^8 configuration, such complexes are invariably *square-planar* rather than tetrahedral. [See section 5(2a).]

6. High-spin → low-spin transitions (spin "cross-over")

Occasionally, the internal energy difference between the high- and low-spin configurations of a complex is sufficiently small for a transition, commonly referred to as a *spin cross-over*, to occur between the alternative electronic states when the temperature is varied. In practice, the high-spin state is invariably that of higher energy and the spin transition is observed on lowering the temperature. Since the metal-ligand distance contracts in the low-spin state such transitions are often facilitated by an increase of pressure.

(a) Spin equilibria

When the difference in internal energy is comparable with kT the alternative forms of the complex may co-exist as *electronic isomers*, in a relative proportion depending on temperature through the Boltzmann law (at a given pressure). This phenomenon, known as *spin equilibrium*, has been encountered in a considerable variety of compounds, especially complexes of Fe(II) and Fe(III).

One copiously documented category consists of the *tris*-dithiocarbamates, $Fe(S_2CNR_2)_3$, of Fe(III) d^5, where spin cross-over occurs for $\Delta_o \sim 16{,}500$ cm^{-1} in the high-spin complex. The sigmoidal dependence of the effective magnetic moment on temperature for the case R = CH_3 (Me) is shown in Fig. 5.15. At very low temperatures m_{eff} is characteristic of the $(t_{2g})^5$ *low*-spin configuration (*i.e.* about 2.25 μ_B), but increases gradually and continuously with temperature to more than 4 μ_B above 300 K. The limiting value of m_{eff} at high-temperatures, when the amount of the high-spin isomer approaches that of its low-spin counterpart, is theoretically 4.47 μ_B, the root-mean square of the moments for the individual species. (It is the moments *squared* that must be averaged because $\chi \propto$ the square of m_{eff}, the susceptibilities χ being additive.) Essentially similar results are obtained whether the measurements are carried out on the solid materials or on their solutions in inert solvents.

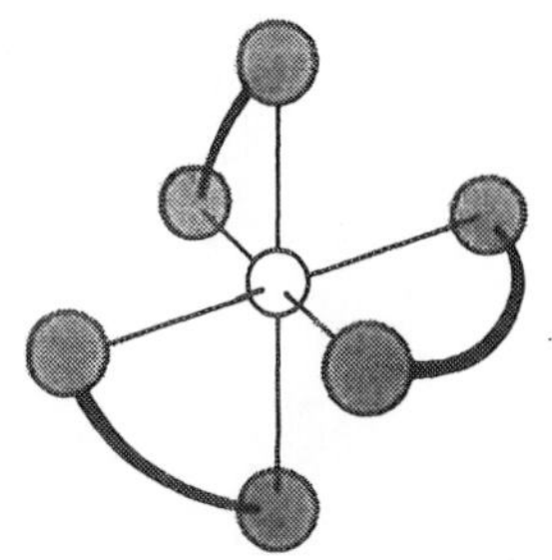

A *tris*-chelate such as $Fe(S_2CNR_2)_3$

Fig. 5.15

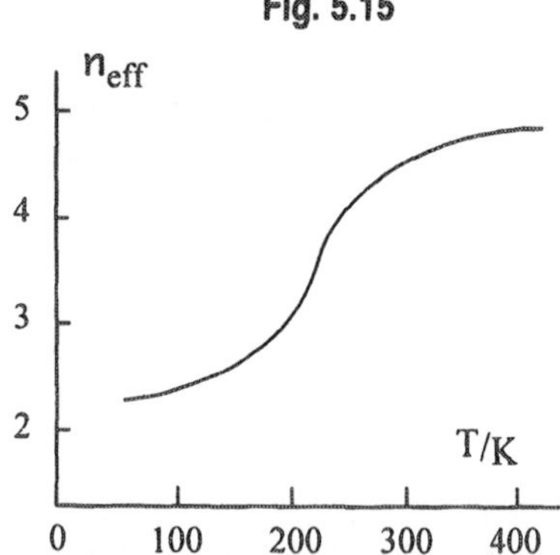

Variation of the magnetic moment of $Fe(S_2CNMe_2)_3$ with temperature.
– an example of *spin equilibrium*.

\# *pyrrolidyl* = —N
(a.k.a. *pyrrolidino*)

Detailed analysis of the data of Fig. 5.15 indicates that the energy difference between the spin isomers is (in spectroscopic units) *ca.* 390 cm^{-1}. The existence of spin equilibrium depending on a very marginal energy term, quite minor changes in the ligand can have a dramatic effect on the position of the equilibrium. Thus, when NMe_2 is replaced by *pyrrolidyl*,# the complex is purely high-spin (at least for $T \gtrsim 80$ K).

(b) More complex spin transitions

Among compounds that display spin cross-over, simple equilibrium between spin isomers is the exception rather than the rule, especially in the solid state where the interactions between molecules, including the mutual interference of vibrational motions, may hinder the conversion and sometimes trigger a *phase* transition. This is the more likely the greater are the stuctural changes accompanying the spin transition.[a] A common outcome, not necessarily involving a phase change, is very *sudden* spin-pairing as the temperature is gradually lowered. In many such instances, however, the spin cross-over proves to be incomplete, the limiting magnetic moment at low temperatures corresponding to a particular "frozen" mixture of the alternative spin-states.[b] Both effects are evident in Fig. 5.16, which is illustrative of many complexes from the family based on Fe(*phen*)$_2$XY and Fe(*bpy*)$_2$XY, where X and Y are certain anionic ligands of moderate field strength. The broken curve in the figure describes the hypothetical behaviour of the system were there a simple spin equilibrium (*cf.* Fig. 5.15).

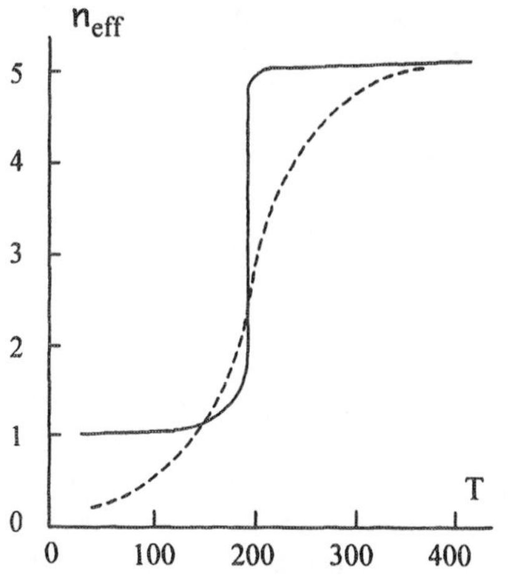

Fig. 5.16

Magnetic moment *vs.* temperature for Fe(II) d^6 complexes of the type Fe(*phen*)$_2$XY showing an abrupt, but incomplete spin cross-over.
(Somewhat schematic – see text.)

One intensively studied example, Fe(*phen*)$_2$(NCS)$_2$, shows a spin transition at about 174 K, m_{eff} falling suddenly from 5.0 μ_B to 1.5 μ_B. Otherwise, m_{eff} increases to 5.20 μ_B at 440 K, while its limiting value at low temperatures is 0.65 μ_B (rather than ~ 0). The orbital splitting Δ_o is known to be 11,900 and 16,300 cm^{-1} in the high- and low-spin states, respectively, the (one-electron) spin pairing energy π (here = $\Pi/2$) being about 11,300 cm^{-1}. (Δ_o represents the average effect of the two ligands, o-*phen* and $[NCS]^-$.)

The precise form of the spin cross-over may be fine-tuned by introducing substituents on the *phen* or *bpy* rings or by *minor* variation of the field strength of the ligands X and Y. However, as with the *tris*-chelates in section

[a] The contraction in metal-ligand bond length occurring on spin-pairing in the tris-chelates discussed above is apparently no more than 0.08 Å, but may be as large as 0.25 Å elsewhere.

[b] In other cases, a limiting hybrid state may be generated in transitions from low- to high-spin states as the temperature is raised. Also, there are many systems in which hysteresis is observed on cycling through a spin cross-over (by first lowering the temperature and then raising it again).

(a), even a moderate change in the average Δ may instead produce exclusively low- or high-spin ground states. Thus, ligands X and Y situated around or above amines NR_3 in the spectrochemical series [section 2(2d)] induce spin-pairing, while those below F^- yield purely high-spin configurations.

Both smooth and discontinuous spin cross-overs are well established elsewhere – among compounds of Co(II) and Cr(II), for example.

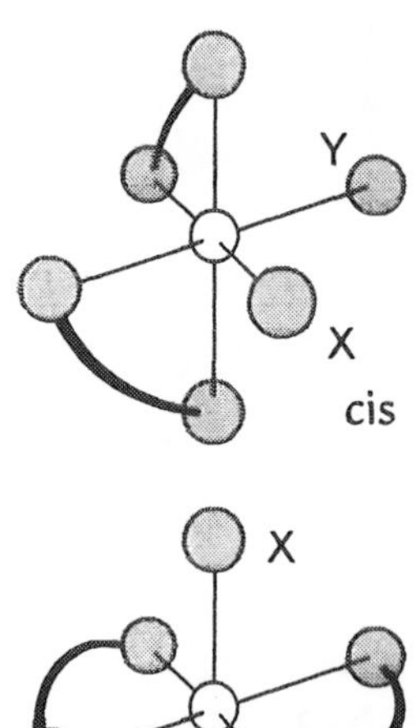

Complexes of the type Fe(*phen*)$_2$XY or Fe(*bpy*)$_2$XY.

7. Intermediate spin-states

In the particular case of octahedral complexes involving the d^5 and d^6 configurations there is superficially the possibility of ground terms having *intermediate* spin states – *viz.* S = 3/2 for $(t_{2g})^4(e_g)^1$ and S = 1 for $(t_{2g})^5(e_g)^1$ – in which the e_g orbital is occupied by only one electron. However, the theory of spin-pairing [section 3(4)] shows that, if Δ_o is large enough to induce the switch from the high to the intermediate spin state, it is at the same time more than sufficient to demote a second e_g electron and impose the low-spin state. Thus, invoking the arguments on which Table 5.4 is based, it is easy to show in the d^6 case that the switch to the intermediate spin configuration (S = 1) requires $\Delta_o > 3K$; but, as shown in the table, complete spin-pairing to form the $(t_{2g})^6$ (S = 0) state is effected when $\Delta_o > 2K$.[a]

> A state of *intermediate* spin can never be the ground state in an octahedral ligand field.

A similar conclusion is reached for tetrahedral complexes, where intermediate spin arrangements are conceivable for the d^4 and d^5 configurations.

The ground state may however involve an intermediate spin configuration in certain complexes of *lower symmetry* than cubic, where the energy balance between alternative spin states can be slightly different. This is encountered in, for example, complex spin cross-over in some compounds of the type Fe(*phen*)$_2$XY, an S = 1 spin isomer joining the medley at intermediate or low temperatures. An unadulterated triplet ground state is also known in some complexes of this type, *e.g.* one crystalline form of Fe(*bpy*)$_2$(NCS)$_2$, and in Fe(*phen*)$_2$(*ox*).[b]

[a] In the case of d^5 complexes, formation of the intermediate spin state (S = 3/2) requires $\Delta_o > 4K$, while $\Delta_o > 3K$ is sufficient to impose a low-spin ground state.

[b] Triplet Fe^{2+} is otherwise well known in square-planar and square pyramidal symmetry, *e.g.* Fe(*dtc*)$_2$Cl.

5.4 The origin of the orbital splitting Δ

The variation in Δ observed among transition metal complexes was surveyed in section 2(2). The aim of the present section is to make sense of all that.

1. The LCAO-MO viewpoint

It is well established that the splitting of the *d*-orbitals is due principally (~ 80% or more) to covalent bonding between the metal centre and the ligands. This bonding is most conveniently described by means of molecular orbital (MO) theory, simulating the MOs by linear combinations of atomic orbitals (the LCAO-MO model).[c] The perturbed metal *d*-orbitals of crystal field theory, polarised simply by the electrostatic field of a point-charge/-dipole array, are supplanted by (generally) *anti*-bonding polycentric MOs that, while made up principally of the metal *d*-orbitals, have also significant ligand valence orbital character. In general, the ligands possess valence orbitals of both σ and π type relative to the metal-ligand axes; the bonding is thus mediated by both σ and π orbital overlap, the former being apparently the larger in most situations.

[c] It is assumed that the reader has previous knowledge of elementary MO theory, also the characterisation of molecular symmetry.

Fig. 5.17

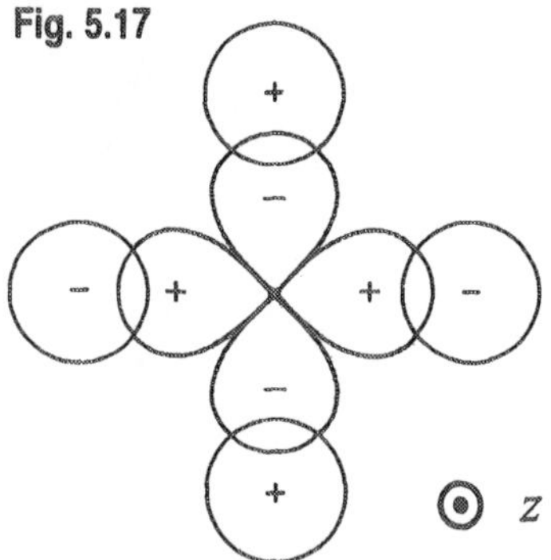

Section in the *xy*-plane of one of the two e_g σ *anti*-bonding MOs of an octahedral hydride $[MH_6]^{n-}$, *e.g.* the Fe(II) complex $[FeH_6]^{4-}$.

The zero net overlap of a t_{2g} *d* orbital with a ligand σ orbital

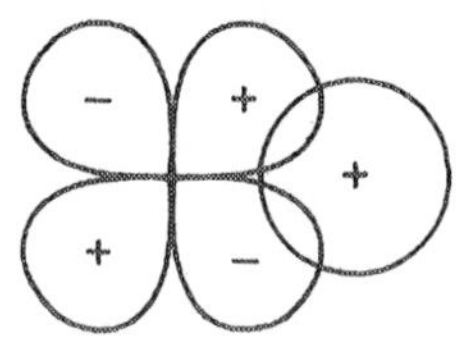

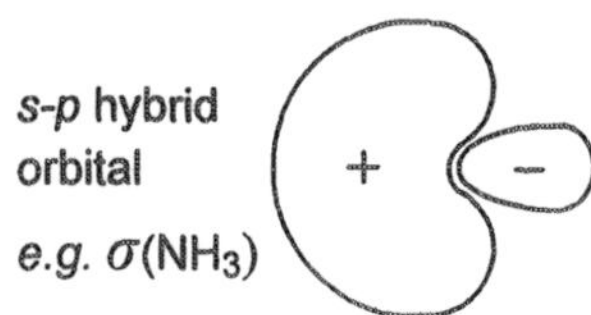

Of the metal valence AOs, the *d*-orbitals have the dominant role in metal-ligand bonding.

Consider, first of all, the σ aspect of the bonding, a simple context in which some general principles may conveniently be introduced. The $e_g(\sigma^*)$ MO of an homoleptic hydride complex $[MH_6]^{n-}$, derived from the $d_{x^2-y^2}$ orbital through σ-overlap with hydrogen 1*s* orbitals, is sketched in Fig. 5.17. The negative phase of the orbital overlap shows it to be *anti*-bonding. The orbitals have positive overlap in the corresponding bonding MO, for which the wave-functions of the metal and ligand orbitals are in phase. The degenerate partner MO based on the d_{z^2} orbital is naturally of a different appearance and involves also the ligand H atoms on the *z*-axis.

The ligand *s*-orbitals can participate only in σ overlap, so the t_{2g} frontier orbitals of an octahedral $[MH_6]^{n-}$ complex are not engaged in the bonding. (A t_{2g} *d* orbital has no net, "broadside" overlap with a ligand σ orbital.)# Described as *non*-bonding, they are completely localised *d*-orbitals (d_{xy} *etc.*), as in crystal field theory. The MO *energy level diagram* for such a complex is shown in Fig. 5.18, which may also be invoked for hexammine complexes, $[M(NH_3)_6]^{n+}$, if it is assumed that only the lone-pair orbital on nitrogen is involved in the metal-ligand bonding. The $e_g(\sigma^*)$ MO is then built from the appropriate *s-p* hybrid AO of the nitrogen atom (sp^3 approximately), rather than an *s* orbital (as in Fig. 5.17).

Notice in Fig. 5.18 that the *bonding* MOs (which are of mainly ligand composition) are exactly filled by electrons formally contributed by the free ligand (H^-, NH_3 *etc.*). The electrons of the metal cation, d^3 as illustrated, are fed into the frontier orbitals, t_{2g}(n) and $e_g(\sigma^*)$. The occupancy of the anti-bonding MOs being less than that of the bonding MOs, metal-ligand covalency results in the transfer of electron density from the ligand σ orbitals to the metal centre (σ donation).

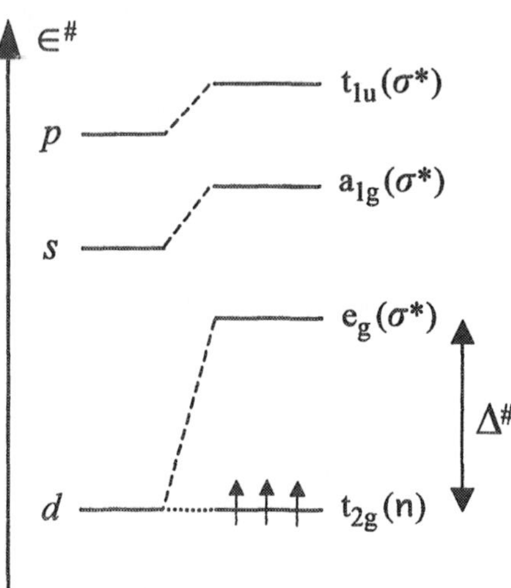

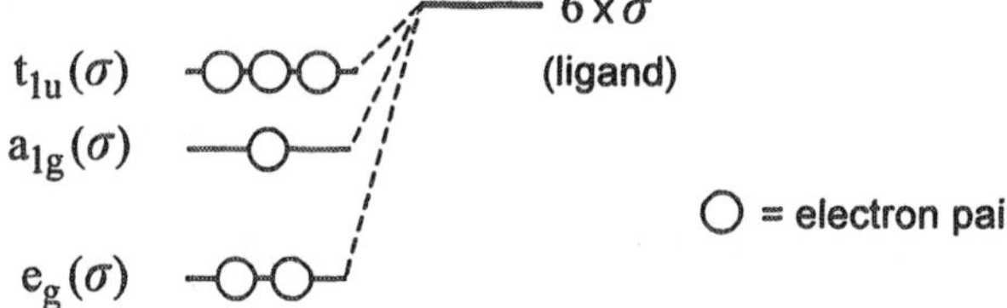

Fig. 5.18 Simple LCAO-MO energy level scheme for an octahedral complex involving only σ bonding (see text).

The occupancy of the frontier orbitals is that for a d^3 cation – *e.g.* $[Cr(NH_3)_6]^{3+}$, with m_{eff} = *ca.* 3.8 μ_B depending on the counter-anion.

While in practice the orbital energy separation $\Delta^\#$ in Fig. 5.18 has the same trend as Δ_o of crystal field theory (Fig. 5.3), the two quantities are identical (or very nearly so) only when *both* frontier orbitals are occupied. This is because *the orbital energy* $\epsilon^\#$ *includes the inter-electronic repulsion within the frontier orbitals* and, at the simplest level of approximation, is to be interpreted as *minus* the ionisation energy (IE) when a molecular orbital is occupied (Koopmans' theorem) but, instead, as the *electron affinity* in the case of an empty MO, *i.e. minus* the orbital energy in the negative ion formed on capture of an electron by that orbital. It is not difficult to show that, for a ground configuration $(t_{2g})^a$,

$$\Delta^\# \approx \Delta_o + J \tag{5.6}$$

where J is again the coulomb integral, so that $\Delta^\#$ is very much bigger than Δ_o in such a case. [As pointed out in connection with eqn. 5.2, the ϵ of the point charge/dipole model are simply the orbital energies relative to the atomic core.] $\Delta^\#$ and Δ_o do apparently correlate well, however.

In a rudimentary MO treatment, adequate for present purposes, $\Delta^\#$ is given approximately by

$$\Delta^\# \propto S^2 / [\epsilon^\#(d) - \epsilon^\#(\sigma)] \tag{5.7}$$

where S is the overlap integral between the $d_{x^2-y^2}$ orbital (or d_{z^2}) and the appropriate combination of ligand σ orbitals (*cf.* Fig. 5.17). The expression is borrowed from the analysis of an elementary heteronuclear system, pictured in Fig. 5.19.$

Fig. 5.19

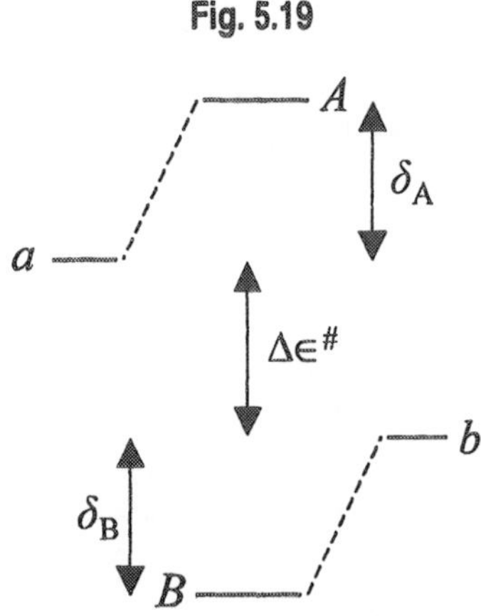

Energetics of bicentric MOs in the heteronuclear case:

$$\delta_A \geq \delta_B \propto \frac{S^2}{\Delta\epsilon^\#}$$

B and *A* are respectively the bonding and anti-bonding MOs simulated by a superposition of AOs *a* and *b*. *B* is made up largely of AO *b*; *A* is correspondingly of mainly *a* character.

$ The bonding MOs $e_g(\sigma)$ lie at an energy *ca.* $\Delta^\#$ below the ligand σ orbitals from which they are largely constructed.

2. The different categories of ligand

The ligand species encountered in transition metal chemistry are essentially of *three* types:

(a) purely σ donor ligands (as described above);

(b) ligands, such as F^- and OH_2, which are both σ and π donors; and

(c) π acid (acceptor) ligands, such as CN^-, *bpy*, PR_3 and C_6H_6.

In categories (b) and (c), the ligands have valence orbitals of both σ and π type with respect to the metal-ligand axes, and the metal-ligand bonding is accordingly more complicated. The frontier energy levels for octahedral complexes in the three different cases are contrasted in Fig. 5.20. Aspect (a) of the figure corresponds, of course, to Fig. 5.18. The form of the frontier orbitals in complexes of category (b) ligands is illustrated by Fig. 5.21, appropriate to monatomic anions such as halide or O^{2-}.

π acceptor ligands also have donor properties: thus CN^- is a strong σ donor, while C_6H_6 and $[C_5H_5]^-$ are notably also π donors (better as such than as σ donors).

[See section 5(2d)]

The $e_g(\sigma^*)$ MO pictured in Fig. 5.21(a) is analogous to that in Fig. 5.17. The major difference between (*e.g.*) $[FeF_6]^{4-}$ and $[FeH_6]^{4-}$ is obviously that, in the fluoride complex, the t_{2g} frontier orbital is anti-bonding (π^*) rather than non-bonding. Now the sideways π overlap pictured in Fig. 5.21(b) proves invariably to be smaller than the more directed σ overlap of metal d and ligand p_σ orbitals. Therefore, by recalling eqn. 5.7, and assuming $\epsilon^\#(\pi) \sim \epsilon^\#(\sigma)$, the $t_{2g}(\pi^*)$ MO must be of lower energy, *i.e.* it is less antibonding, than $e_g(\sigma^*)$. The ligand field splitting, $\Delta^\# = \epsilon^\#(e_g) - \epsilon^\#(t_{2g})$, may be expressed as the difference in antibonding effects,

$$\Delta^{\#} = \Delta^{\#}(\sigma) - \Delta^{\#}(\pi) \quad (5.8)$$

where $\Delta^{\#}(\sigma)$ are $\Delta^{\#}(\pi)$ are the displacements in energy of the $e_g(\sigma^*)$ and $t_{2g}(\pi^*)$ MOs, respectively, in relation to the unperturbed *d* level.

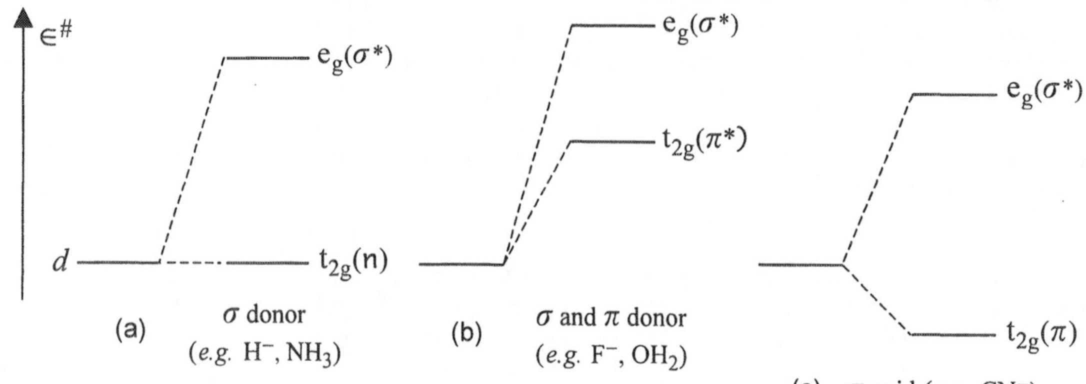

Fig. 5.20 The relative energies of the frontier orbitals of octahedral complexes with different types of ligand.

Covalent bonding with ligands of type (b) involves the transfer of both σ and π electron density to the metal. [F^- and O^{2-} are particularly effective at stabilising high oxidation states by relieving the potential electron deficiency of the metal centre.]

Fig. 5.21 The antibonding frontier MOs in octahedral complexes of monatomic ligands with *p* valence orbitals ("exploded" view).

The MOs are projected onto the *xy* plane. There is no contribution from the ligands above and below the

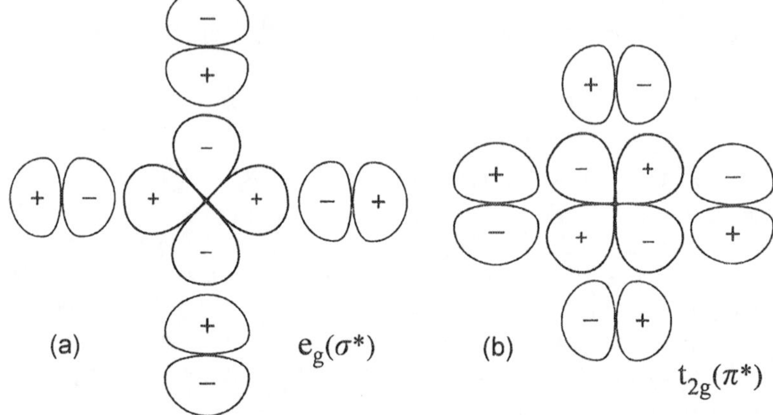

Common examples include $[Cr(CN)_6]^{3-}$ and $V(CO)_6$, with the electron configurations $(t_{2g})^3$ and $(t_{2g})^5$, respectively.

$\in^{\#}(p)$ has the trend: $p_C > p_N > p_O$.

Feature (c) of Fig. 5.20 depicts the frontier energy levels in complexes with ligands such as cyanide ion or CO (both of which bind through carbon). The valence orbitals include π and σ bonding MOs, the latter composed principally of an *s-p* hybrid orbital on carbon. As with the *p* orbitals of F^- *etc.*, both MOs are fully occupied in the free ligand. The ligands possess also empty low-lying π^* orbitals, though with $\in^{\#}(\pi^*) > \in^{\#}(d)$, which prove an important ingedient in the synthesis of the frontier MOs. A distinctive outcome of π bonding *within* such ligands is that the bonding π MO is strongly biased towards the N or O atom, the greater nuclear charge of the latter making their *p* valence AOs more stable than those of the C atom (*cf.* Fig.

5.19). As Fig. 5.22 shows, the antibonding partner (π^*) "leans" commensurately towards the carbon atom.

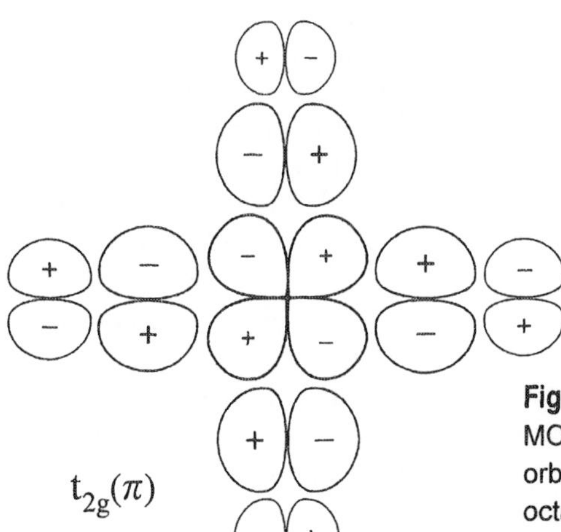

Fig. 5.23 The blending of the π^* MOs of the ligand into the t_{2g} *d*-orbitals in the t_{2g} frontier MO of an octahedral complex $[M(CN)_6]^{n-}$, or of a hexacarbonyl species.

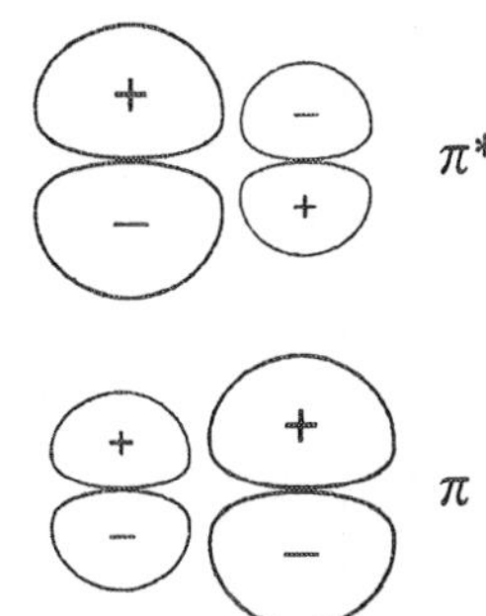

Fig. 5.22 The π MOs of diatomic ligands such as CN^- and CO (also NO). [Each is doubly degenerate; there is an equivalent set of MOs oriented perpendicular to the page.]

As a consequence of the asymmetry of the ligand MOs, the t_{2g} *d* orbitals overlap more strongly with the π^* MO than with its bonding counterpart. The admixture of π^* (Fig. 5.23) therefore dominates the ligand contribution to the t_{2g} frontier MOs and such that, as portrayed by Fig. 5.20(c), the t_{2g} MOs are usually *bonding* in nature.

The metal-ligand π bonding usually results in a net transfer of electron density *from metal to ligand*, often termed *back-donation* (and symbolised $d_\pi \rightarrow \pi^*$). This aspect of the bonding can be crucial to the stability of low oxidation state compounds, which would otherwise have excessive electron density at the metal site. CO is more effective than $[CN]^-$ because it is only a weak σ donor and probably the stronger π acceptor.

The binding of NO is closely similar, provided the molecule coordinates in linear M–N–O fashion, as is *always* the case with CO and CN^-. It may be viewed formally as NO^+ (isoelectronic with CO and CN^-), with its π^* electron transferred to the metal site.

3. Tetrahedral complexes

Covalent bonding in tetrahedral complexes ML_4 differs from that in the octahedral case in that the $t_2(d)$ orbitals can overlap with ligand orbitals of both σ and π symmetry, while the e(*d*) orbitals engage only in π overlap. In an LCAO-MO description, the t_2 frontier MO is therefore invariably more antibonding than the e MO. This is illustrated by Fig. 5.24, constructed with ligands of type (b) in mind. In the case of π acceptor ligands – as in such molecular compounds as $Ni(CO)_4$ or $Pt(PF_3)_4$ – the frontier MOs will be considerably less antibonding, and the e MO conceivably somewhat bonding. When the ligand is simply a σ donor, as is the case for H^- – *e.g.* $[MnH_4]^{2-}$ (in K_3MnH_5) – or approximately so for NH_3 and $[alkyl]^-$, t_2 is a purely σ^* MO while the e MO is non-bonding.

Fig. 5.24

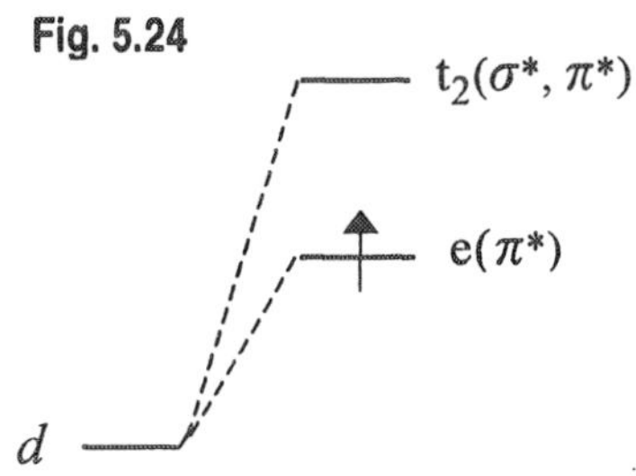

Frontier MO energy levels for a tetrahedral complex such as VCl_4 (d^1).

4. The rationale of the trends in Δ

Since the metal-ligand bonding involves the transfer of electron density, the spectrochemical series [section 2(2d)] can be understood in terms of the σ donor and π donor/acceptor characteristics of the ligands. The substantial ligand field splittings observed for H^- and $[CH_3]^-$ [Fig. 5.20(a)] reflect their

The relative *proton affinities* for a selection of ligands are:

$[CH_3]^- \gtrsim H^- > [OH]^- > [CN]^- > F^- > Cl^- > NMe_3 > py > NH_3 \gg CO$

[a] Complexes with these ligands are invariably *low*-spin – *e.g.* $[FeH_6]^{4-}$, $[Fe(CN)_6]^{4-}$ and $Cr(CO)_6$, all having the $(t_{2g})^6$ ground configuration. So also, usually, are *tetra-alkyls* MR_4 [recall section 3(5)].

[b] – and all the more so in the case of tetrahedral complexes.

very strong (and purely) σ donor properties, correlating with their similar, large proton affinities. The weaker σ donor NH_3 causes a splitting of smaller, but intermediate magnitude.

That $[CN]^-$ produces an even larger splitting than H^- and $[CH_3]^-$ is attributable to strong σ donation (though cyanide ion is not as strong a σ donor) combined with considerable π back-donation of electron density [Figs. 5.23 and 5.20(c)].[a] The isoelectronic ligand CO sits at the very top of the spectrochemical series apparently because it is a much stronger π acceptor than $[CN]^-$, perhaps because the energy separation $\in^\#(\pi^*) - \in^\#(d)$ is significantly smaller.

Octahedral complexes with ligands of type (b), which are simultaneously σ and π donors, display comparatively small ligand field splittings because, recalling eqn. 5.8, the effects of σ and π bonding are subtracted in $\Delta^\#$ (and therefore in Δ). The trend in Δ among such complexes [section 2(2d)] is a subtle issue[b] but, as regards monatomic ligands, it mirrors straightforwardly the variation expected in the metal-ligand bond strength: the stronger the covalent interaction, the greater the differentiation in energy of the $e_g(\sigma^*)$ and $t_{2g}(\pi^*)$ antibonding orbitals.

The magnitude of the d orbital splitting is also highly dependent, of course, on the nature of the metal. The observation [section 2(2b)] that Δ *tends to increase with oxidation state* is explicable by the variation in energy $\in^\#(d)$ of the d orbital, which is stabilised by increase in partial charge on the metal centre. Its spacing in energy from the ligand σ and π orbitals is thereby reduced and the covalent interaction therefore increases. The variation in overlap integrals S (recall eqn. 5.7) is not easily assessed, but is thought to be less important.

On the other hand, the overlap term is probably dominant in determining the increase in Δ observed on scanning down a transition metal subgroup [section 2(2c)]. As illustrated by Fig. 5.25, the metal d orbitals *project increasingly from the core* such that, despite the lengthening of the metal-ligand bond (at least in the second transition series), the overlap integrals become larger. The influence of this increasing overlap is partly countered by the variation in d orbital energies, $\in^\#(3d) < \in^\#(4d) < \in^\#(5d)$, but that trend is strong only for highly charged metal atoms. As judged by ionisation energies, the $\in^\#(d)$ are not very different for low positive charge.

The increasing protrusion of the d electron density from the atomic core, on descending any subgroup, is well established theoretically, and appears to have two causes. First, the more flexible form of the $4d$ and $5d$ orbitals allows a greater relative expansion under the centrifugal force. Secondly, there is a requirement in quantum mechanics that the radial wave functions of the outer d orbitals be *orthogonal* to (*i.e.* have no net overlap with) those of any d orbitals within the core. In the atoms of the second transition series, this causes the $4d$ orbital to expand in relation to $3d$ and, in the third transition series, the $5d$ valence orbital is further expanded to ensure orthogonality with both the $4d$ and $3d$ orbitals of the lutetium core. A similar "effect" contributes to the greater projection of the $5f$ orbitals from the radon core in

Fig. 5.25

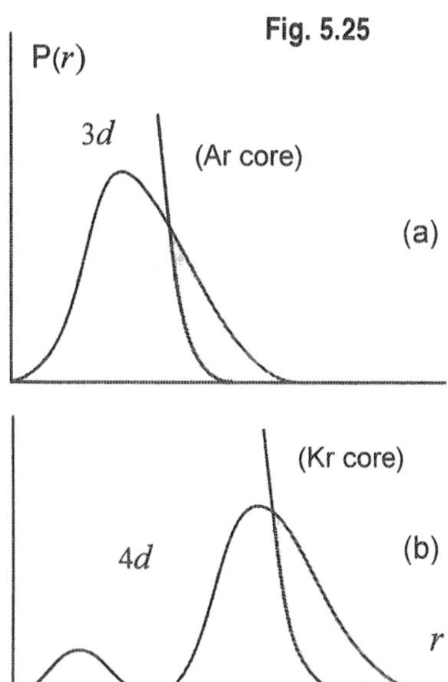

RDFs of d orbitals showing, by the shading, their extension beyond the atomic cores (*cf.* Fig. 4.14).

(a) A metal ion from the first transition series, *e.g.* Cr^{3+}.

(b) The congeneric metal ion of the second transition series, *e.g.* Mo^{3+}.

actinide ions, as compared with that of the $4f$ orbitals from the xenon core in the case of lanthanide ions [§ 4.3].

5. Experimental evidence for metal-ligand covalency

There is abundant evidence, from a variety of sources, for the delocalisation of the d electrons through covalence. The role of electron paramagnetic resonance spectroscopy in this connection is briefly described.

EPR spectroscopy provides an immediate demonstration of delocalisation by the frequent observation of hyperfine structure (sometimes referred to as *superhyperfine*) due to magnetic coupling of the d electron spin with the spin of the ligand nuclei. Fig. 5.26 offers a simple illustration. There the 1:1:1 triplet resonance arising unmistakably from coupling with the I = 1 nucleus of nitrogen (^{14}N), is direct evidence of the presence of d electron density on the NO ligand. The magnitude of the splitting may be used to estimate the extent of delocalisation.

Fig. 5.26

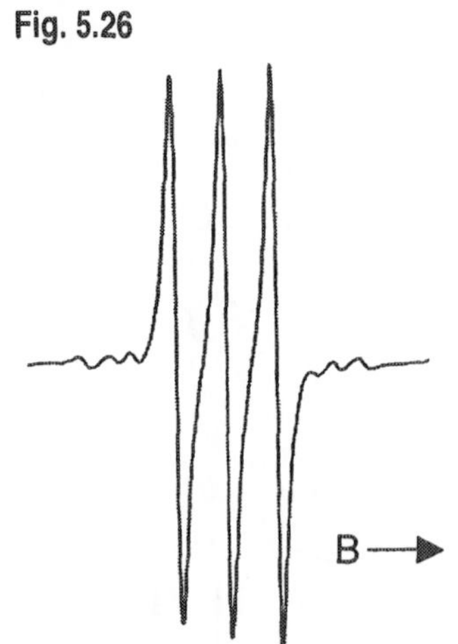

The dominant features in the EPR spectrum of the low-spin Cr(I) d^5 complex $[Cr(CN)_5(NO)]^{3-}$.

The three, equally intense signals are due to hyperfine coupling of the unpaired electron with the ^{14}N (I = 1) nucleus of the nitrosyl ligand. The complex displays a g-value of 1.9945.

The EPR spectrum of $[Cr(CN)_5(NO)]^{3-}$ enriched with ^{13}C (I = ½) also exhibits superhyperfine structure due to the CN^- ligands, interpretable in terms of four equivalent (equatorial) and a unique (axial) ligand. Hyperfine splitting due to interaction with the N atoms of CN^- is not resolved, indicating limited delocalisation beyond the C atom. This is consistent with the picture of covalent bonding advanced in section 4, cyanide ion being a π-acid ligand with both σ-donor and π-acceptor MOs strongly biased towards the carbon atom. (Recall, in particular, Fig. 5.23.)

$[Cr(CN)_5(NO)]^{3-}$ displays m_{eff} = 1.87 μ_B (at 295 K), which establishes a low-spin ground configuration ~ $(t_{2g})^5$ (roughly put). This in turn indicates that, like CO and CN^-, NO binds in linear mode[a] and validates the simple notion that, for the purposes of electron "book-keeping", it may be regarded *formally* as NO^+, the chromium centre being accordingly Cr(I) $(3d)^5$.

Similarly, while tetrahedral $Co(PPh_3)_4$ is paramagnetic with m_{eff} = 1.70 μ_B (at ~ 300 K), with evidently one unpaired electron and suggestive of Co(0) $(3d)^9$, the pyramidal ("pseudo-tetrahedral") complex $Co(NO)(PPh_3)_3$ is diamagnetic, formally incorporating therefore Co(–I) $(3d)^{10}$. The latter complex is electronically analogous to $Ni(PPh_3)_4$.

a However, NO sometimes binds in a non-linear manner with (*e.g.*) ∠ MNO ~ 120°. The metal oxidation state may then be assigned (formally!) by regarding the ligand as NO^- instead of NO^+.

5.5 Complexes of lower than cubic symmetry

Most transition metal complexes have shapes of low symmetry, displaying at most one rotational symmetry axis C_n of order $n \geqslant 3$.[b] The magnetic properties often reflect acutely the limited symmetry of a system and naturally may show, at least, an *anisotropy* in the susceptibility and the EPR spectrum.

b Thus $[Cr(CN)_5(NO)]^{3-}$ shows only a four-fold axis (C_4), coincident with the linear array NC – Co – NO.

1. Distortions from high symmetry

(a) The static Jahn-Teller effect

In general, the integrated d electron "cloud" of a cation situated in a cubic ligand field has an *anisotropic* form – that is to say, its radial extension varies significantly in the different cartesian directions. Exceptions to that statement, among *octahedral* complexes, include those with ions having the ground configurations $(t_{2g})^3$ (S = 3/2), $(t_{2g})^6$ (S = 0) and $(t_{2g})^6(e_g)^2$ (S = 1), also the configuration $(t_{2g})^3(e_g)^2$ (S = 5/2) which is *spherical*, like the filled subshell d^{10}, *i.e.* $(t_{2g})^6(e_g)^4$ (S = 0). But most cations have anisotropic d electron distributions *imposed on them by the ligand field*, these reflecting degeneracies in the disposition of electrons among the d orbitals. As a result, the complexes are prone to *geometrical distortions* [*e.g.* Fig. 5.27(a)] that accommodate the anisotropy, driven by a concomitant lowering in electronic energy. This is a manifestation of what is called the Jahn-Teller effect.

Jahn and Teller (1937) formulated a theorem, based for the most part on symmetry arguments, according to which

> Any non-linear molecule with a symmetrical shape allowing an orbitally degenerate electronic ground state should distort so as to eliminate that degeneracy and thereby lower its electronic energy.

Typically quite small, Jahn-Teller distortions may nonetheless exert a profound influence on paramagnetic behaviour.

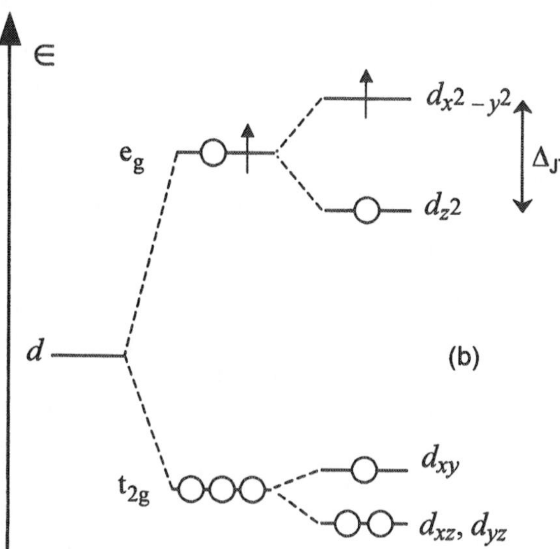

Fig. 5.27 Tetragonal Jahn-Teller distortion of octahedral d^9 and high-spin d^4 complexes:

(a) the usual form of the distorted geometry (somewhat exaggerated);

(b) the associated ligand field levels, with an orbital occupancy relevant to d^9, *e.g.* Cu^{2+}.

[*e.g.* $[Cu(OH_2)_6]^{2+}$ in $K_2Cu(SO_4)_2.6H_2O$ (a *Tutton salt*) with $m_{eff} = 1.93\ \mu_B$ (335 – 1 K) and $m_{\parallel} = 2.15\ \mu_B$ & $m_{\perp} = 1.81\ \mu_B$ at 300 K; or $[CuCl_6]^{4-}$ in $CsCuCl_3$, with $m_{eff} = 1.95\ \mu_B$ for 300 – 90 K, also anisotropic.]

(b) Tetragonal distortions

The basis of the Jahn-Teller distortion is most simply explained by reference, first, to 6-coordinate d^9 complexes (those of Cu^{2+}, in particular) where the anisotropy is associated with the $(e_g)^3$ aspect of the electron configuration. Suppose the odd electron resides in $d_{x^2-y^2}$, the d_{z^2} orbital being doubly occupied, as pictured in Fig. 5.27(b). The d electron cloud will then have a shape corresponding to spherical d^{10} pinched about its waist in the xy-plane, albeit in a "knobbly" fashion (Fig. 5.28). The asymmetry in the d^9 distribution is conducive to a tetragonal distortion# in which the four ligands in the xy-plane approach the metal ion more closely than the two that lie on the z-axis. This distortion causes a further splitting of the d-orbitals, as shown in Fig. 5.27(b), whereby, because the $d_{x^2-y^2}$ and d_{z^2} orbitals have different occupancies, the net orbital energy of the d electrons is lowered (actually by an amount $\frac{1}{2}\Delta_{JT}$). It is assumed that the *average* metal-ligand distance is unchanged and also, therefore, the position of the baricentre.

– *tetragonal* because one of the four-fold symmetry axes (C_4) of the octahedron is preserved.

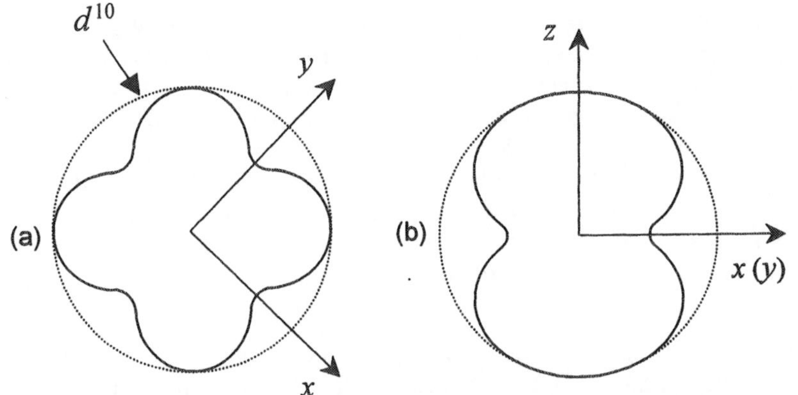

Fig. 5.28 An impression of the anisotropic d electron density characteristic of a d^9 cation in an octahedral field, viewed as a spherical d^{10} distribution with a hole in the $d_{x^2-y^2}$ orbital:
(a) section through boundary surface in the xy-plane;
(b) perpendicular section in the xz or yz-planes.

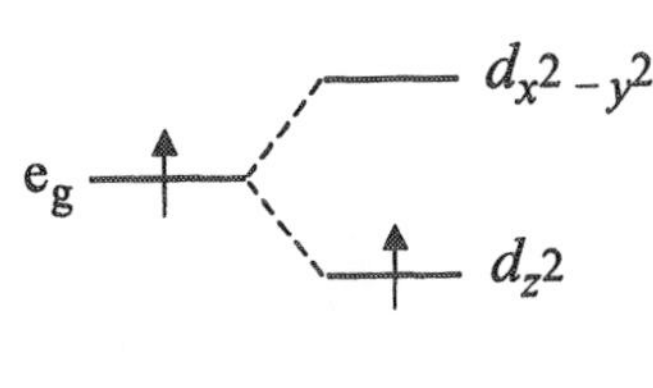

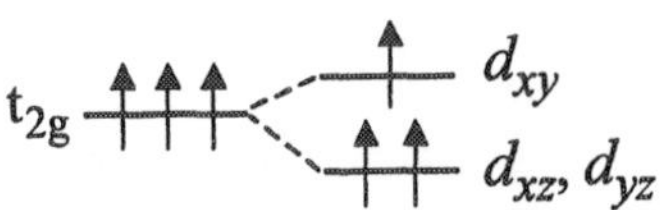

Fig. 5.29 Electronic structure of a high-spin d^4 cation in a tetragonally distorted octahedral environment.

[*e.g.* $[CrCl_6]^{4-}$ in $CrCl_2$, which shows $m_{eff} = 5.10\ \mu_B$ at 300 K, again with a considerable anisotropy.]

Nominally octahedral high-spin d^4 complexes, *e.g.* those of Cr^{2+} and Mn^{3+}, generally display similar structural and electronic properties (*e.g.* Fig. 5.29), as do *low*-spin complexes of Co^{2+} $(t_{2g})^6(e_g)^1$ (S = 1/2), though the latter (*e.g.* $[Co(CNPh)_6]^{2+}$) are comparatively rare. In the former case, the asymmetric d^4 electron distribution can usefully be pictured in terms of a $d_{x^2-y^2}$ hole in the spherical charge cloud of high-spin d^5 (*cf.* Fig. 5.28).

What if, however, the hole were in the d_{z^2} orbital? On this, equally plausible assumption, the asymmetry in the d^4 or d^9 electron densities would invite the inverse distortion, with two shorter bonds to the ligands on the z-axis and four longer ones in the xy-plane. In practice, this latter form of tetragonal distortion is rarely observed.[a] The predisposition towards the distortion pictured in Fig. 5.27(a) is perhaps due to a secondary effect of covalency, namely the mixing of metal $(n+1)s$ and nd_{z^2} orbitals in the lower symmetry, which accentuates the stabilisation of d_{z^2}.

[a] Known instances include the hexafluoride complexes in $Ba_2[CuF_6]$ (d^9) and $K_3[NiF_6]$ (d^7).

The d splitting pattern in Fig. 5.27(b) applies also to complexes of the type $Fe(phen)_2XY$ [*cf.* section 3(6b)], which are inherently tetragonal (or nearly so), if the axial field is weaker than the equatorial. A similar sequence of d energy levels is also encountered in many planar complexes such as metal phthalocyanines [section (2a) below].

Tetrahedral complexes are equally susceptible to Jahn-teller distortions (when the d electron distribution is anisotropic).[b] One well documented case is $[CuCl_4]^{2-}$ $(e)^4(t_2)^5$ (S = 1/2), which undergoes a significant tetragonal distortion involving a flattening of the tetrahedron in a direction perpendicular to an edge (*cf.* Figure 5.34) to form a *dihedron*.[c] [See section 6(e).]

[b] The only open configurations with *isotropic* electron densities are here $(e)^2$ (S = 1) and $(t_2)^3$ (S = 3/2).

The Jahn-Teller distortions often assume other forms, *e.g.* *trigonal* (see below) or *rhombic*, when 6-coordinate complexes show three (or more) different metal-ligand bond lengths.

[c] However, $[CoCl_4]^{2-}$ $(e)^4(t_2)^3$ (S = 3/2) and $[NiCl_4]^{2-}$ $(e)^4(t_2)^4$ (S = 1) both show a slight tetragonal *elongation*. It is not clear why.

Sometimes, Jahn-Teller distortions fail to materialise where anticipated. Frequently, this is due to the metal-ligand vibrations having sufficient amplitude to carry the system between alternative distorted structures though usually, in such instances, the complex does show the expected distortion at low temperatures when the vibrations are dampened – *e.g.* $[Cu(OH_2)_6]^{2+}$ in certain crystals. In other cases, the Jahn-Teller distortions may be restrained when, as is invariably the case, the spin-orbit coupling favours a different form of splitting of the cubic energy levels and is strong enough to pre-empt the situation.# The freedom of a complex to distort may otherwise depend critically on its environment within a crystal lattice.

The Jahn-Teller effect and the spin-orbit interaction are almost always in competition.

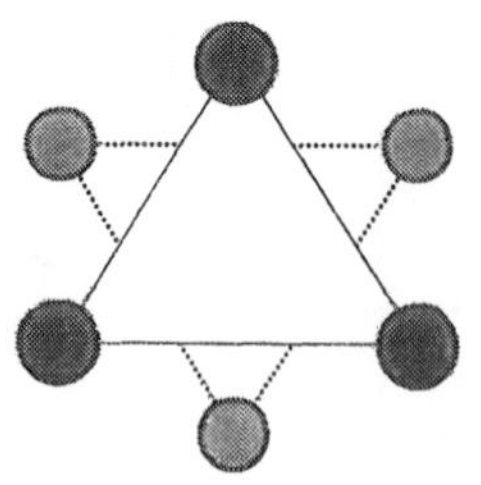

Plan view along a C_3 axis

(c) Trigonal distortions

When the anisotropy in the *d* electron density is due to an asymmetrical occupancy of the t_2 subshell, the Jahn-Teller distortion, if not rhombic, is usually *trigonal* rather than tetragonal. In the case of octahedral complexes the distortion is such as to preserve one of the three-fold symmetry axes, as pictured in Fig. 5.30.

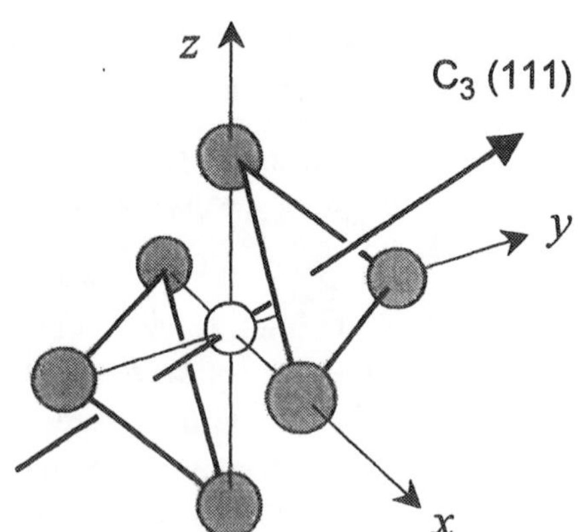

Fig. 5.30 One of the four three-fold axes (C_3) of an octahedral complex highlighted.

In the case of a $(t_{2g})^1$ system a trigonal Jahn-Teller distortion involves compression (or elongation) along such an axis bringing a pair of opposite faces of the octahedron closer together (or spacing them further apart). The other three C_3 symmetry axes are lost in the process.

The resulting *d* energy levels for a $(t_{2g})^1$ complex are shown in Fig. 5.31 (facing). A well characterised instance is $[Ti(OH_2)_6]^{3+}$ in, for example, the *alum* $CsTi(SO_4)_2.12H_2O$ where m_{eff} = 1.80 μ_B at 300 K with a large anisotropy. The EPR spectra are naturally also anisotropic: that reproduced in Fig. 5.32 is attributable to a substantial *compression* of the complex along a C_3 axis (though the analysis involved is beyond the scope of this text). The Jahn-Teller splitting Δ_{JT} in such complexes can amount to several hundred cm^{-1}. One estimate for $[Ti(OH_2)_6]^{3+}$ is 360 cm^{-1}, as compared with Δ_o = 19,000 cm^{-1}.

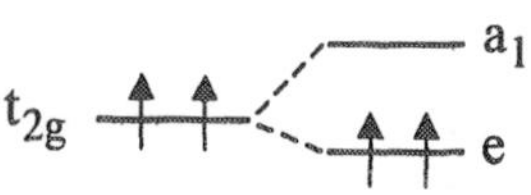

Fig. 5.33 The ground electronic state of $[V(OH_2)_6]^{3+}$ and many other octahedral complexes of vanadium(III) d^2.

The inverse splitting of the t_{2g} level may be observed elsewhere, *e.g.* in the $[V(OH_2)_6]^{3+}$ d^2 cation residing $(NH_4)V(SO_4)_2.12H_2O$, where Δ_{JT} has the (unusually large) value of *ca.* –2000 cm^{-1} (Fig. 5.33) and m_{eff} = 2.80 μ_B at *ca.* 300 K. In such cases, the octahedron is *elongated* in the direction of one of the four C_3 axes.

The splitting of the form in Fig. 5.33 is apparently so large in the intrinsically trigonal Ti(0) complex Ti(*bpy*)$_3$ that the ground state has the closed-shell configuration $(e)^4$.

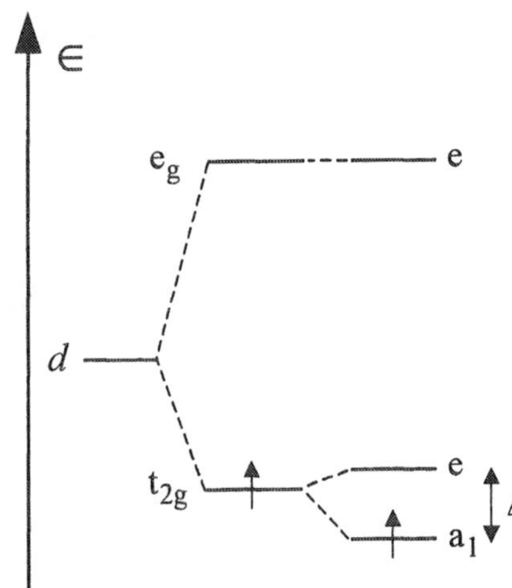

Fig. 5.31 The effect of a trigonal Jahn-Teller distortion on the *d* energy levels of an octahedral d^1 complex.

The complex is compressed symmetrically in the direction of a C_3 axis (Fig. 5.30).

The orbital splitting pattern in d^1 chelates $M(L\text{–}L)_3$ is usually similar in form but the (inherent) $e - a_1$ splitting may be considerably larger – *e.g.* it is some 7,500 cm^{-1} in Ti(*acac*)$_3$,$ with $m_{eff} = 1.73\ \mu_B$.

$ *acac* = [acetylacetonate]$^-$

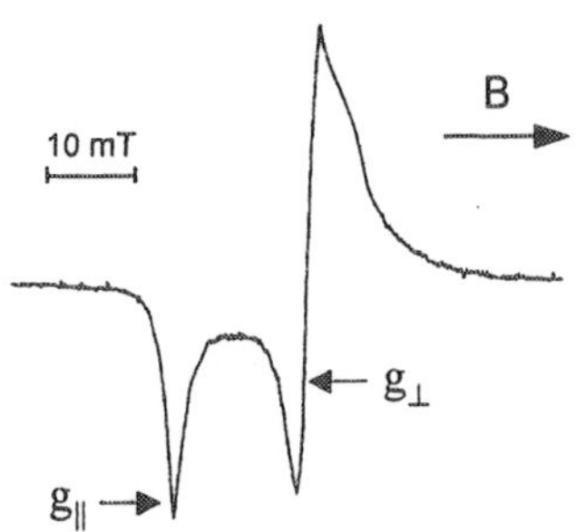

Fig. 5.32 EPR spectrum of the Ti(III) aquocation in a 2-propanol-D_2O matrix at 77 K.

$g_\parallel = 1.994$ and $g_\perp = 1.896$.

2. Complexes of uniaxial symmetry

There is a huge variety of complexes characterised by a *unique* symmetry axis of order ≥ 3,# which allows the limited spatial degeneracy of *two*. The properties of certain categories of such complexes are now reviewed.

– a.k.a. *symmetric tops*, in the context of rotational motion.

(a) Square-planar complexes

Four-coordinate complexes might be expected ordinarily to be tetrahedral, or approximately so, repulsion between the ligands thereby being minimised. However, for certain *d* electron configurations – notably d^8, especially when the ligand field is strong – a preference often exists for planar coordination, *i.e.* a *square*-planar geometry when the ligands are identical. The notional transformation involved is illustrated by Fig. 5.34, and the consequent effect on the *d* energy levels by Fig. 5.35.

d^8 cations show a strong tendency to form planar 4-coordinate complexes. These are invariably closed-shell and *diamagnetic*.

Fig. 5.34

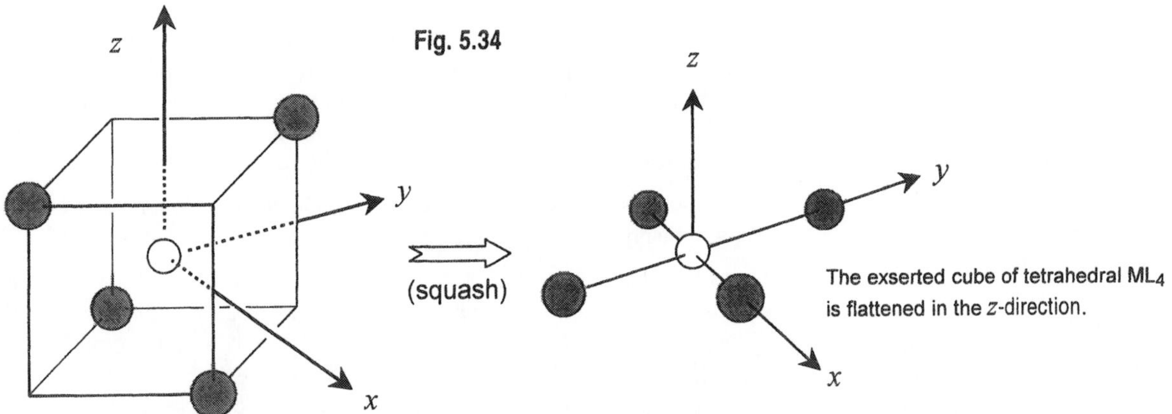

The exserted cube of tetrahedral ML_4 is flattened in the *z*-direction.

A convenient choice of coordinate system for correlating the *d* orbital energies in tetrahedral and square-planar ML_4 complexes.

[The *x*- and *y*-axes are rotated by 45° as compared with Fig. 5.5.]

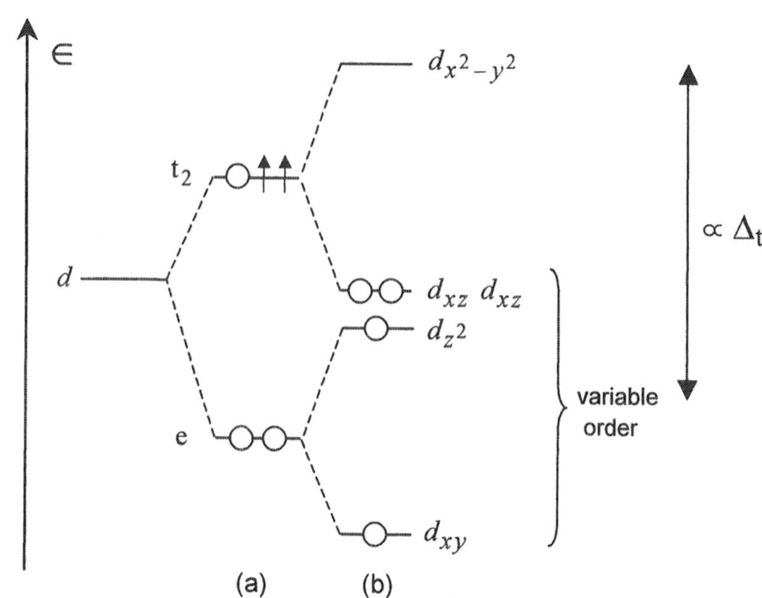

Fig 5.35 Comparison of the d orbital energies for
(a) a tetrahedral ligand field;
(b) a square-planar field of a comparable strength. #

The orbital occupancies shown are those appropriate to the d^8 case.

i.e. the same M–L distances.

○ = electron pair

Most striking in Fig. 5.35 is the sharp destabilisation of the $d_{x^2-y^2}$ orbital in relation to the other four d orbitals, its separation in energy from the average of the latter reflecting the strength of the ligand field.

There is a certain similarity to the d splitting occasioned by Jahn-Teller distortion in octahedral d^9 complexes (Fig. 5.27). However, contrary to many textbooks, the planar shape of most d^8 complexes has almost nothing to do with the Jahn-Teller effect.

$R_2P(=O)-N^-R'$ ⟷ $R_2P(-O^-)=NR'$

(i) d^8 complexes

In sufficiently strong fields the electrons of a d^8 cation will assume the closed-shell configuration shown in Fig. 5.35(b), and hence benefit from a lower net *orbital* energy than in the tetrahedral situation. However, the planar geometry should be preferred only where this orbital stabilisation is large enough to outweigh both the spin-pairing energy (~ K) and the increase in ligand-ligand repulsion (assuming that the metal-ligand distance is unaltered). It turns out, in panoramic view, that

> The vast majority of 4-coordinate d^8 complexes are planar, all of them being therefore *diamagnetic*.

– unless they happen to display TIP, invariably confined to highly coloured materials such as (perhaps) $[PdCl_4]^{2-}$. The generalisation embraces also 4-coordinate complexes lacking a four-fold (C_4) axis – for example, *bis*-chelates $[M(L–L)_2]^{m-}$ (see below) and ML_2X_2 d^8 species.[a]

[a] Apparently there is however at least one class of planar nickel(II) complex which may be open-shell and paramagnetic, namely chelates $Ni(O–N)_2$, where *O–N* is *phosphinic amidato*, the bidentate anionic ligand pictured above. When R = *t*-Bu and R′ = *i*-Pr the complex is reported to display an m_{eff} of 3.40 μ_B at 297 K.

The exceptions are almost entirely complexes of Ni(II)[b] with ligands low in the spectrochemical series, and commonly anions, such as $[NiX_4]^{2-}$ (X = Cl, Br & I), but also complexes with bulky ligands that prevent the adoption of planar geometry. The latter include $[Ni(SPh)_4]^{2-}$ and perhaps pseudo-tetrahedral $Ni(PPh_3)_2Br_2$ (m_{eff} = 3.27 μ_B at ~ 300 K).[c] At the margin, and in non-coordinating solvents (*e.g.* CH_2Cl_2), the tetrahedral and planar isomers

[b] But Ni(II) forms a huge variety of 6-coordinate and (always) paramagnetic complexes, with m_{eff} typically in the range 2.9 – 3.4 μ_B at ~ 300 K. [Recall Table 5.4.]

[c] See note # on facing page.

of many Ni(II) complexes may co-exist in thermal equilibrium, a further category of spin equilibrium [section 3(6a)]. They may even occur together in the same crystalline compound.

Also numerous chelates, $Ni(L\text{–}L)_2$, where L–L is *e.g.*, an N-alkyl substituted *salicylaldiminato* anion (below).

O^-

CH=NR

When R = *t*-Bu, for example, m_{eff} is 3.29 μ_B (at ~ 300 K).

The electronic factors favouring square planar coordination have even greater weight in compounds of the congeneric cations, Pd(II) and Pt(II), where the ligand field is stronger [section 2(2c)] and the spin-pairing energies reduced in comparison with Ni(II); steric hindrance is much less important, moreover. Complexes of Pd(II) and Pt(II) are therefore invariably planar. Furthermore, for reasons that remain unclear, six-coordination is virtually unknown for these ions (though some 5-coordinate, closed-shell complexes have been characterised). A rare instance of six-coordination is PdF_2 with, like NiF_2, the rutile structure and also showing first order paramagnetism.[a]

[a] PdF_2 has $m_{eff} \approx 3.0\ \mu_B$ above 250 K, compared with *ca.* 3.5 μ_B for NiF_2. PdF_2 is weakly ferromagnetic with a T_C of about 217 K, while NiF_2 is prone to antiferromagnetic ordering with T_N ~ 80 K.

Strong ligand fields similarly ensure that four-coordinate complexes of Rh(I) and Ir(I) are almost always of planar geometry, as are the great majority of complexes involving Cu(III), Ag(III) and Au(III) – all also closed-shell d^8, of course. However, "high-spin" Cu(III) is known in the compound $K_3[CuF_6]$ (m_{eff} = 3.01 μ_B) and, remarkably, the analogous Ag(III) species in paramagnetic $KCs_2[AgF_6]$ (m_{eff} ~ 2.6 μ_B). (These compounds are reminiscent of PdF_2.)

The well characterised complexes of Co(I) ($3d^8$), *e.g.* $[Co(PMe_3)_4]^+$ and $CoCl(PPh_3)_3$, are tetrahedral (or nearly so) rather than planar, reflecting comparatively weak ligand fields. The cationic species has m_{eff} = 3.39 μ_B (~ 300 K). Such complexes are analogous to $[NiX_4]^{2-}$ (X = Cl, Br & I).

(ii) d^9 complexes

There is no cost as regards spin-pairing energy for metal ions with the d^9 configuration – Cu(II), most prominently – though the orbital stabilisation in a planar complex is significantly reduced by the unavoidable half-occupation of the high-lying $d_{x^2-y^2}$ orbital. Nonetheless, *the predilection of Cu(II) to form planar, rather than tetrahedral, complexes approaches that of Ni(II).*[b] Ag(II) is apparently similar, though less extensively characterised.[c]

[b] Like Ni(II), however, both Cu(II) and also Co(II) tend preferentially to form 6-coordinate (high-spin) complexes with most ligands.

[c] – also Au(II), as in the remarkable dark-red xenon complex $[AuXe_4]^{2+}$.

Planar complexes of Cu(II) and Ag(II) (S = 1/2) have m_{eff} in the range 1.75 – 2.25 μ_B at ordinary temperatures.[d] In comparison, the magnetic moments observed for six-coordinate Cu(II) complexes range between 1.75 and 2.00 μ_B; five-coordinate complexes of Cu(II) also typically have m_{eff} greater than the spin-only value (1.73 μ_B).

[d] *e.g.* square planar $[Cu(NH_3)_4]^{2+}$ and $[Cu(OH)_4]^{2+}$ have m_{eff} = 1.87 and 1.85 μ_B, respectively (at ~ 300 K).

In weak ligand fields, the net lowering of electronic orbital energy may prove insufficient to stabilise the complex in planar form; for that reason, apparently, the planar geometry is not normally adopted by the tetrahalide species $[CuX_4]^{2-}$ (X = Cl & Br).[e] However, these complexes are intrinsically *dihedral* (flattened tetrahedral) through the operation of the Jahn-Teller effect and, consequently, show anisotropy in their paramagnetic properties. Magnetic measurements alone cannot easily distinguish such cases from the more populous class of planar Cu(II) complexes.

[e] Belonging to the same category are many (non-planar) complexes of the type $Co(PR_3)_4$, $[Ni(PR_3)_4]^+$ and $NiX(PR_3)_3$ (X = Cl, Br & I). In the particular case of $Co(PMe_3)_4$, m_{eff} = 1.70 μ_B (~ 300 K).

(iii) High-spin planar complexes

The lowering of electronic energy can stabilise the planar geometry in relation to tetrahedral for certain other d^N configurations. Most obvious, perhaps, is the high-spin d^4 case (S = 2), *e.g.* Cr(II), where again (as in the d^9 case) no spin-pairing energy is involved; moreover, as in the d^8 case, occupancy of $d_{x^2-y^2}$ is avoided.[e] One paradigm is the square-planar complex $[Cr(NCS)_4]^{2-}$ in $(NBu_4)_2[Cr(NCS)_4]$, which has $m_{eff} = 4.76\ \mu_B$, constant in the range 300 – 85 K. Examples involving Mn(III) include $[MnMe_4]^-$, with a moment ~ 4.8 μ_B (297 K), and $[MnOMe_3]^{2-}$ with m_{eff} ~ 4.9 μ_B (~ 300 K). As elsewhere, preference for planar geometry can be frustrated by steric crowding when the ligands are bulky. Chromium(II) is otherwise apparently unknown as a discrete tetrahedral complex.

[The description "sterically demanding" is often preferred to "bulky" elsewhere in the English-speaking world.]

However, very strong ligands may stabilise the *tetrahedral* complex by imposing the low-spin $(e)^4$ configuration – *e.g.* Fe(1-*nor*)$_4$ and the cationic Co(V) species [Co(1-*nor*)$_4$]$^+$ [recall 3(5)], or "tetrahedral" Cr(II) nitrosyl complexes such as $Cr(CH_2SiMe_3)_2(NO)(N^iPr_2)$ and $Cr(NO)(N^iPr_2)_3$ – when the system will be diamagnetic.# The $(e)^4$ ground configuration is naturally even more stable in complexes of the later transition elements, *e.g.* MPh_4 and M(*cyclohexyl*)$_4$, with M = Ru or Os.

\# – or perhaps display TIP, like many closed-shell transition-metal compounds.

Also belonging to the *high*-spin category, in principle, are 4-coordinate complexes of d^3 cations. But the only well authenticated example of a planar d^3 species would appear to be a V(II) phenoxide anion $[V(OAr)_4]^{2-}$ where Ar = *dipp* [2,6-*bis*(^{i}Pr)phenyl] which, partnered by the cation [Li(*thf*)]$^+$, has $m_{eff} = 3.83\ \mu_B$.

Square coplanar coordination is sometimes encountered in crystal lattices having a primary structure that incidentally provides a suitable site. Thus the planar d^6 unit $[FeO_4]^{6-}$ is stabilised in the mineral *gillespite* $BaFeSi_4O_{10}$; this displays $m_{eff} = 5.01\ \mu_B$ (~ 300 K) confirming the expected high-spin (S = 2) ground state.

(iv) Complexes of low or intermediate spin

In other cases, planar is preferred to tetrahedral geometry when the ligand field is sufficiently strong to impose spin-pairing. For example, Co(II) d^7 forms square-planar [Co(*py*)$_4$]$^{2+}$ (with S = 1/2), analogous to [Cu(*py*)$_4$]$^{2+}$ and [Ag(*py*)$_4$]$^{2+}$,[b] except that the unpaired electron resides in the d_{z^2} orbital rather than $d_{x^2-y^2}$. Co(II) also forms a variety of planar complexes with bidentate ligands such as dithioacetylacetonate (*dt-acac*) and the maleonitriledithiolate (*mnt*) dianion, $[S_2C_2(CN)_2]^{2-}$ (as do many other cations, of course). A selection of magnetic moment data for planar d^7 complexes is provided in Table 5.6. Five- and six-coordinate complexes of low-spin d^7 ions also invariably show $m_{eff} > 1.73\ \mu_B$.

Table 5.6 m_{eff} values (/μ_B) at *ca.* 300 K for some d^7 ML_4 complexes of planar geometry

Complex	m_{eff}
[Co(*py*)$_4$]$^{2+}$	2.10
$[Co(CN)_4]^{2-}$	2.15
[Co(*mnt*)$_2$]$^{2-}$	2.16
[Rh(*mnt*)$_2$]$^{2-}$	1.91
$Rh(S_2CMe_2)_2$	2.08
$[Co\{S_2C_2(CF_3)_2\}_2]^{2-}$	2.06
$[Ni\{S_2C_2(CF_3)_2\}_2]^-$	1.85
Co(*dt-acac*)$_2$	2.26
[Co(*diars*)$_2$]$^{2+}$	2.10
[Ni(*diars*)$_2$]$^{3+}$	1.89
$[Pt(C_6Cl_5)_4]^-$	2.45
$[Rh(C_6Cl_5)_4]^{2-}$	2.10

The data for charged species may vary somewhat with counter-ion. S = 1/2, and the spin-only expectation is $m_{eff} = 1.73\ \mu_B$.

Planar complexes of d^5 cations are also conceivable (providing they are not high-spin!) but, in practice, are exceedingly rare. One instance is the intriguing square-planar $[IrO_4]^{4-}$, involving Ir(IV) ($5d^5$), with an m_{eff} of 2.89 – 3.25 μ_B (~ 300 K), depending on the counter-cation. The ground electronic state is apparently of *intermediate* spin, S = 3/2, the $d_{x^2-y^2}$ orbital being again unoccupied. As in other strongly tetragonal species (the phthalocyanines below, for example), the spacing of the d_{z^2} and d_{xz}, d_{yz} energy levels is ordinarily too close to allow a *low*-spin configuration.

(v) Complexes of macrocyclic ligands

Furthermore, many *tetra*dentate macrocyclic ligands of inflexible planar structure can impose planar four-coordination upon cations. The *phthalocyanine* dianion $[pc]^{2-}$, for example, complexes with (*inter al.*) the cations V(II) through to Cu(II) (Fig. 5.36). Some magnetic moment data for the M(II) phthalocyanines are collected in Table 5.7.

The β-form of Co(*pc*) has an effective magnetic moment which is almost constant (at *ca.* 2.51 μ_B) between 300 and 20 K. Detailed investigation of the system (including EPR measurements) points conclusively to the energy level scheme in Fig. 5.37, with the unpaired electron residing in the d_{z^2} orbital; this appears to be typical of the planar d^7 case, though the relative energies of the occupied d_{xy} and d_{xz}, d_{yz} orbitals may vary. Essentially the same sequence of *d* energy levels applies also to the other M(*pc*) species.

– except perhaps for the Fe(II) compound (d^6) which remains a controversial case.

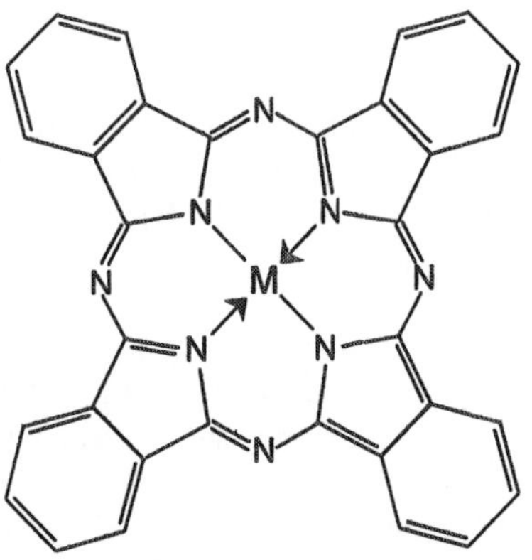

Fig. 5.36 Square-planar coordination of an M^{2+} cation by the phthalocyanine dianion.

Table 5.7

Some m_{eff} values $(/\mu_B)^{\#}$ for phthalocyanine (*pc*) complexes of M(II) cations (at *ca.* 300 K).

Cr	3.49	1
Mn	4.30	3/2
Fe	3.90	1
Co (β)*	2.54	1/2
Ni	0	0
Cu	1.82	1/2
Ag	1.84	1/2

\# These magnetic moments are constant down to at least 100 K.

* The *α*-form of Co(*pc*) has an m_{eff} of 2.38 μ_B.

Fig. 5.37

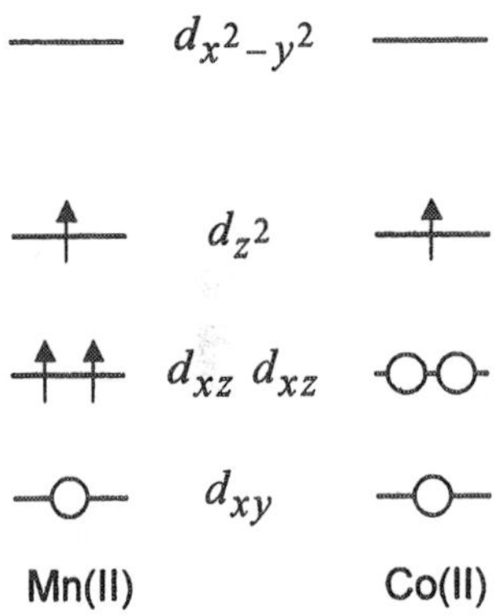

$d_{x^2-y^2}$

d_{z^2}

d_{xz} d_{xz}

d_{xy}

Mn(II) Co(II)

Ground electronic states of Mn(II) and Co(II) phthalocyanines.

All uniaxial complexes display anisotropy in their paramagnetism and this is particularly strong in the phthalocyanines. The principal components of m_{eff} for Co(*pc*) are $m_{\parallel}$ = 1.73 μ_B and $m_{\perp}$ = 2.88 μ_B, varying little between 300 and 90 K. In planar complexes of the type $M(L\text{–}L)_2$, the perpendicular moment $m_{\perp}$ usually has distinguishable *x*- and *y*-components. Thus $Co(dt\text{-}acac)_2$ has the principal magnetic moments (m_{eff} at ~ 300 K) m_z = 1.77 μ_B (≡ $m_{\parallel}$), m_x = 1.89 μ_B and m_y = 2.93 μ_B, displaying a remarkably large in-plane anisotropy.

Complexes of uniaxial symmetry invariably have strongly anisotropic magnetic properties. Thus the EPR spectrum of Co(*pc*) shows principal g-values $g_{\parallel}$ = 1.891 and $g_{\perp}$ = 2.939.

The precise form of the magnetic anisotropy, here and elsewhere, can be an important guide to the make-up of the electronic ground state. Thus, for example, that $\chi_{\perp}$ proves much greater than $\chi_{\parallel}$ in Mn(*pc*) is a strong indicator of the ground configuration pictured in Fig. 5.37. But the theory required cannot be explained here.

(b) Other tetragonal systems

There are a great many tetragonal 6-coordinate complexes of the type ML_5X, *trans*-ML_4XY and (stretching a point) *trans*-$M(L\text{–}L)_2XY$.# Also well characterised, but less abundant, are a variety of *square-pyramidal* complexes, ML_5 or ML_4X, together with approximately tetragonal species $M(L\text{–}L)_2X$ (with the ligand X apical, *i.e.* lying on the principal symmetry axis). The magnetic properties of any such complex are generally intermediate between those of related octahedral and square-planar complexes (where they both exist), and may be *high* or *low*-spin, occasionally of *intermediate* spin, depending on the metal cation and ligands that are involved. The variation in the relative strength of axial and equatorial ligand fields may sometimes embrace spin cross-over.

The latter category includes the complexes *trans*-$Fe(bpy)_2X_2$, *etc.* introduced in 3(6b) to illustrate the spin cross-over phenomenon.

(c) Trigonal complexes

In addition to the *tris*-chelates $M(L\text{–}L)_3$, introduced earlier as approximately octahedral complexes, there is a diversity of *uniaxial* trigonal complexes of lower coordination number.

(i) Trigonal bipyramidal (*tbp*) species, *trans*-ML_3XY

Most transition metal ions form 5-coordinate complexes of general formula ML_3XY, the equivalent ligands L occupying the equatorial sites (Fig. 5.38) defined as lying in the *xy*-plane. Perturbed by such a ligand field, the metal *d* orbitals are expected to have the relative energies

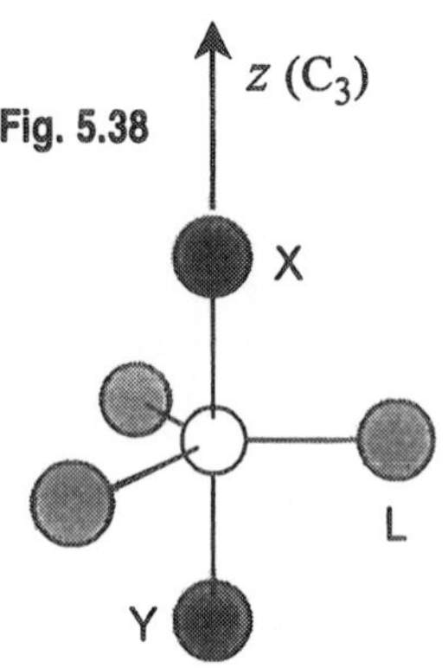

Fig. 5.38 Trigonal bipyramidal ML_3XY

$$d_{z^2} > d_{x^2-y^2} = d_{xy} > d_{xz} = d_{xz}$$

– a very different splitting pattern from that in an octahedral field. The d_{z^2} orbital should be the least stable because it points directly at the axial ligands, X and Y, whilst also being significantly destabilised by the equatorial ligands L (Fig. 5.39). Detailed application of the point-charge/dipole model shows, moreover, that the in-plane orbitals $d_{x^2-y^2}$ & d_{xy} suffer a greater repulsion from the ligands than the out-of-plane d_{xz} & d_{xz}. However, covalency often inverts that energy sequence, especially when π bonding features strongly.

That there are two sets of doubly-degenerate orbitals (Mulliken symbol e) is dictated by the three-fold symmetry. They may conveniently be distinguished here as $e_a(d_{xz}, d_{xz})$ and $e_\beta(d_{x^2-y^2}, d_{xy})$ though, it should be made clear, this is *not* standard notation.$ The orbital sets have π and δ symmetry, respectively, in relation to the three-fold axis.

$ When there is a plane of symmetry, as in *t*-ML_3X_2 (the ML_3 or *xy*-plane), the doubly-degenerate sets are distinguished as e″ and e′, respectively.

The notation identifies the orbitals as antisymmetric or symmetric, respectively, as regards their behaviour on reflection in the symmetry plane.

The two orbital sets become partially hybridised when the plane of symmetry is absent.

The description of the frontier orbitals in LCAO-MO terms is provided in Fig. 5.39, it being assumed there that the ligands are both σ and π donors. If the ligands were purely σ donors then the MOs derived from $d_{x^2-y^2}$ & d_{xy} (e_β) would be simply σ^* antibonding MOs, while the d_{xz} & d_{xz} orbitals (e_a) would be disengaged, *i.e.* non-bonding. Ligands of π-acid character would tend to stabilise both degenerate sets.

Some representative information regarding the magnetic properties of *tbp* complexes is given in Table 5.8. One ubiquitous class of *tbp* complex, especially with M(II) ions of the first transition series, incorporates tetradentate "tripod" ligands such as *tren* $N(CH_2CH_2NH_2)_3$ (and its derivatives) or the ligand QP $P(o\text{-}C_6H_4PPh_2)_3$, equivalent to XL_3 in Fig. 5.38; the remaining coordination site is taken up by (usually) an anionic ligand Y (Fig. 5.40). As

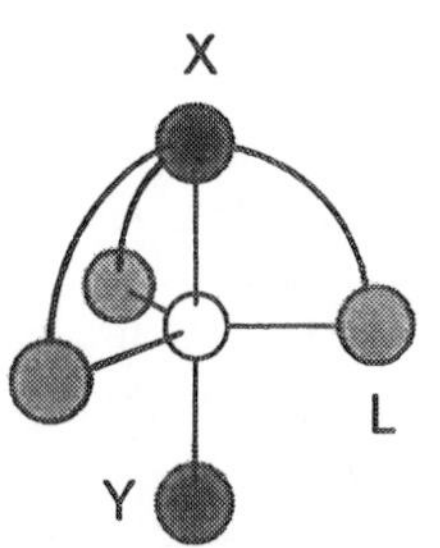

Fig. 5.40 A *tbp* complex $M(XL_3)Y$ involving a tripod ligand XL_3 such as *tren* or QP.

illustrated by Fig. 5.41, M(II) complexes of the type $[M(Me_6\textit{tren})X]^+$, with X = halide, invariably adopt *high*-spin ground configurations. The tripod ligand QP, binding *via* phosphorus rather than nitrogen, exerts a sufficiently stronger field to induce maximal spin-pairing, *e.g.* in $[Fe(QP)Cl]^+$ $3d^6$, the ground configuration of which is pictured in Fig. 5.39.

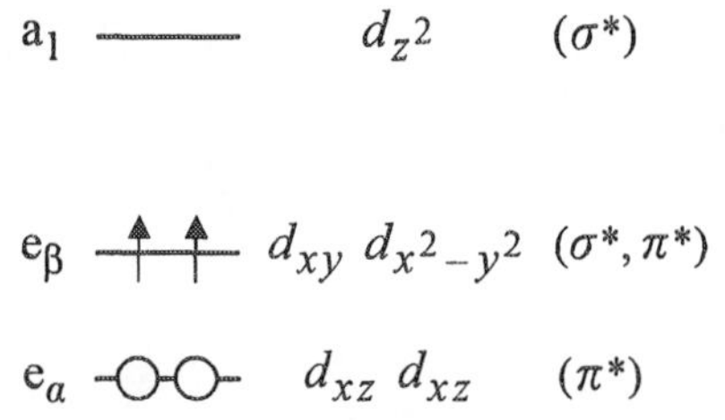

Fig. 5.39 Frontier energy levels of a trigonal bipyramidal complex ML_3XY. The occupancies are those for low-spin $3d^6$ (S = 1), *e.g.* Fe(II) and Co(III). Contrast the energy levels for *tris*-chelates M(L–L)$_3$ in Fig. 5.31.

Table 5.8 m_{eff} data (/μ_B) at *ca.* 300 K for a selection of trigonal bipyramidal complexes.

S	Complex	m_{eff}	Complex	m_{eff}
S = 0	$[Ni(CN)_5]^{3-}$, *t*-$Ni(PMe_3)_3(CN)_2$	0 (closed-shell d^8)		
	$[Mo(NO)(SPh)_4]^-$	0 (closed-shell d^4)		
S = 1/2	*t*-$Ti(NMe_3)_2Cl_3$	1.69 (d^1)	$[VCl_5]^-$	1.70 (d^1)
	t-$Co(PMe_3)_3Br_2$	2.10 (d^7)	$[CuCl_5]^{3-}$	1.89 (d^9)
S = 1	*t*-$V(PMe_3)_2Cl_3$	2.61 (d^2)	*t*-$Co(PMe_3)_2Cl_3$	3.05 (d^6)
	$[Ni(Me_6\textit{tren})Cl]^+$	3.42 (d^8)	$[Fe(QP)Cl]^+$	3.10 (d^6)
S = 3/2	*t*-$Cr(NMe_3)_2Cl_3$	3.88 (d^3)	*t*-$Fe(PMe_3)_2Cl_3$	4.22 (d^5)
	$[Co(Me_6\textit{tren})Cl]^+$	4.45 (d^7)	$[Co(\textit{Opy})_5]^{2+}$	4.56 (d^7)
S = 2	$[Cr(Me_6\textit{tren})Cl]^+$	4.85 (d^4)	*t*-$Mn(PMe_3)_2I_3$	4.80 (d^4)
	$[Fe(Me_6\textit{tren})Br]^+$	5.34 (d^6)		
S = 5/2	$[Mn(Me_6\textit{tren})I]^+$	5.95 (d^5)		

$Me_6\textit{tren}$ = *tris*-(2-dimethylaminomethyl)amine, $N(CH_2CH_2NMe_2)_3$
QP = *tris*-(2-diphenylphosphinophenyl)phosphine, $P(o\text{-}C_6H_4PPh_2)_3$
Opy = pyridine N-oxide, C_5H_5NO

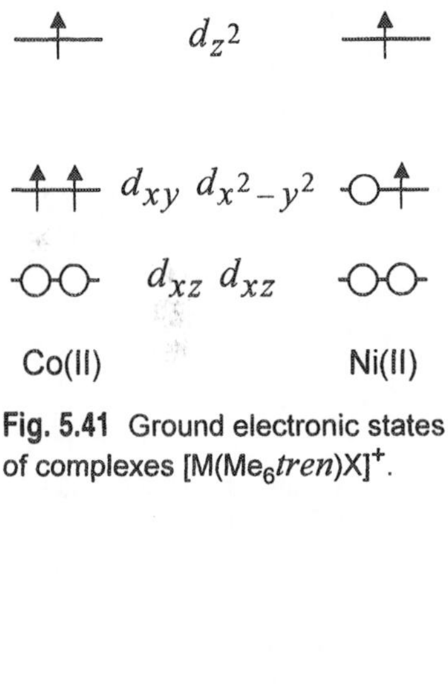

Fig. 5.41 Ground electronic states of complexes $[M(Me_6\textit{tren})X]^+$.

It is a striking consequence of the d orbital splitting pattern in *tbp* systems that the *low*-spin d^6 configuration, $(e_\alpha)^4(e_\beta)^2$, is a spin triplet (S = 1), in contrast to the singlet (S = 0) $(t_{2g})^6$ ground state of octahedral complexes. On the other hand, d^8 cations, with exclusively a triplet (S = 1) ground state in octahedral symmetry, have an alternative low-spin (S = 0) ground state in *tbp* complexes (*e.g.* $[Ni(CN)_5]^{3-}$). Another notable feature of low-spin *tbp* complexes, reflecting the small energy separation of the e_α and e_β orbital sets, is that a singlet $(e_\alpha)^4$ ground state is unusual among d^4 complexes, and even the intermediate spin state (S = 1) is rare. (A similar tendency prevails among the *tetragonal* complexes of d^4 cations.) Thus complexes of the type $[Cr(QP)X]^+$ (X = Cl, Br *etc.*) have m_{eff} in the range 4.80 – 4.89 μ_B at ordinary temperatures (*cf.* Table 5.8). Similarly, *tbp* d^3 complexes seldom display the doublet $(e_\alpha)^3$ ground configuration.

There is a quite different class of trigonal pyramidal ML_3X complex that involves tripod ligands XL_3 (Fig. 5.40) having sufficiently large groups attached to the "feet" L that further coordination is blocked. Much deployed have been *triamidoamine* ligands such as $[(RNCH_2CH_2)_3N]^{3-}$ with basal amido N^-.# Their complexes are typically high-spin: the Fe(III) compound $\{(^tBuMe_2SiNCH_2CH_2)_3N\}Fe$ ($3d^5$), for example, has an m_{eff} of 5.95 μ_B,

commonly symbolised by $[N_3N]^{3-}$.

and its Mn(III) analogue ($3d^4$) 5.01 μ_B. More conventional complexes of this kind include $[Co\{N(CH_2CH_2PPh_2)_3\}]^+$ ($3d^8$) with $m_{eff} = 3.11$ μ_B.

(ii) Trigonal planar complexes ML_3

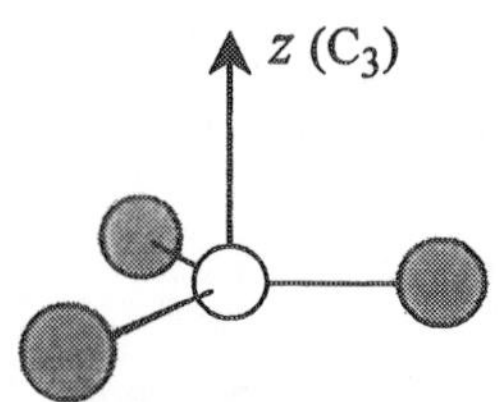

Trigonal planar ML_3

The simplest category is that of *trigonal planar* complexes ML_3, represented by a wide range of amides, alkyls, alkoxides *etc.*, many stabilised by virtue of bulky, occluding substituents. The ligand field being intrinsically weak, such complexes have usually *high*-spin ground states, at least for metal ions of the first transition series. Consider, for example, the m_{eff} data listed in Table 5.9, which characterise the *tris*-amido complexes $M\{N(SiMe_3)_2\}_3$. However, the complexes of the heavier transition elements are sometimes *low*-spin – as are, apparently, all of the remarkable nitride complexes $[MN_3]^{6-}$ of the first row elements (see below).

According to ligand-field theory, the relative energies of the perturbed d orbitals are as shown in Fig. 5.42, with d_{z^2} naturally lowest. (Contrast the energy sequence for trigonal bipyramidal ML_3XY in Fig. 5.39.) However, the energy sequence of d_{z^2} and d_{xz}, d_{xz} (e_a) may be inverted as compared with Fig. 5.42 when σ bonding is much stronger than out-of-plane π bonding.

Table 5.9

m_{eff} values ($/\mu_B$) at ~ 300 K for the trigonal planar amido complexes $M\{N(SiMe_3)_2\}_3$.

Ti	S = 1/2	1.62#
V	1	2.38
Cr	3/2	3.73
Mn	2	5.38
Fe	5/2	5.94
Co	2	4.73

After allowance for TIP; beforehand, 1.73 μ_B.

The Lanthanide elements form analogous compounds, including $Gd\{N(SiMe_3)_2\}_3$ ($4f^7$) with m_{eff} = 7.89 μ_B at *ca.* 300 K.

e_β —— d_{xy} $d_{x^2-y^2}$ (σ^*, π^*)

e_a ↑↑ d_{xz} d_{xz} (π^*)

a_1 ↑ d_{z^2} (σ^*)

Fig. 5.42

The frontier energy levels of trigonal planar complexes ML_3.

The occupancies are those appropriate to high-spin d^3 (S = 3/2), *e.g.*

$Cr(N^iPr_2)_3$ $m_{eff} = 3.80$ μ_B;
$Mo(N^tBu_2)_3$ $m_{eff} = 3.56$ μ_B.

Among *low*-spin complexes of this kind is the diamagnetic tantalum(III) ($4d^2$) siloxide $Ta(OSi^tBu_3)_3$, clearly having the ground configuration $(d_{z^2})^2$. Further examples are the d^6 complexes,[a] $Rh(mes)_3$ and $Ir(mes)_3$ (though, curiously, these are markedly *pyramidal* in shape). On the other hand, equivalent complexes of Mo(III) ($4d^3$) are planar and high-spin (S = 3/2), like their Cr(III) ($3d^3$) counterparts, and have magnetic moments in the range 3.55 – 3.75 μ_B at ordinary temperatures. Comparable complexes of tungsten(III) have not as yet been adequately characterised, but W(IV) analogues of the Ta(III) siloxide above are diamagnetic (and for the same reason).

Also planar with closed-shell ground configurations are the formally M(VI) *imido*[b] complexes $Os(NAr)_3$ and $[Re(NAr)_3]^-$ (both $5d^2$), where Ar (aryl) is a phenyl group with bulky substituents.

A further category of trigonal planar complex comprises the so-called *Hoppe anions* $[MO_3]^{4-}$ where M = Fe, Co & Ni, together with (of very recent vintage) the nitride complexes $[MN_3]^{6-}$ with M = V through to Co. The oxyanions $[MO_3]^{4-}$, formally containing M(II) cations, all have high-spin ground states: the Co(II) species $[CoO_3]^{4-}$ ($3d^7$) in $Na_4[CoO_3]$, for example, has m_{eff} = 4.02 μ_B (at ~ 300 K), corresponding to S = 3/2 and explicable in terms of the energy level scheme in Fig. 5.42.

a *mes* = mesityl, 2,4,6-$C_6H_2Me_3$.

b The imido group is formally RN^{2-}.

The nitrido-anions $[MN_3]^{6-}$, involving M(III) cations, are apparently *low-spin*, however; this presumably reflects the stronger ligand field due to nitride ion N^{3-}, as compared with oxide O^{2-}, and the increase in field strength associated with M(III), as opposed to M(II) cations. The complexes are not as yet completely characterised, but it is clear that $[VN_3]^{6-}$ is diamagnetic with the ground configuration $(d_{z^2})^2$, while $[CoN_3]^{6-}$ $(3d^6)$ is a temperature independent paramagnet. These observations may again be rationalised on the basis of the *d* orbital energy sequence of Fig. 5.42, as can the magnetic moments of 1.85 μ_B and 2.75 μ_B displayed by $[CrN_3]^{6-}$ $(3d^3)$ and $[MnN_3]^{6-}$ $(3d^4)$, respectively, at high temperature. However, there is substantial antiferromagnetic interaction in many of the compounds concerned.[a]

[a] $Ca_3[CrN_3]$ has $T_N \sim 240$ K. The moment quoted for $[CrN_3]^{6-}$ refers to the paramagnetic phase at $T \gtrsim 500$ K.

(iii) Trigonal prismatic complexes

A credible alternative to the octahedral arrangement for six-coordinate complexes, trigonal prismatic coordination is actually rare. It was first established in the enigmatic *tris*-chelate $Re(S_2C_2Ph_2)_3$, involving a 1,2-dithiolene ligand. The complex displays an m_{eff} = 1.79 μ_B (~ 300 K) suggesting a metal oxidation state of +6; however, the hyperfine structure in its EPR spectrum indicates that the unpaired electron resides largely on the ligands.[b]

[b] The same phenomenon is observed in many related complexes, *e.g.* the anions $[M(S_2C_2Ph_2)_2]^-$ (M = Pd and Pt) which may not contain M(III) cations.

Trigonal prismatic coordination is also known in the corresponding *tris*-chelates of (*inter al.*) vanadium, chromium, molybdenum and tungsten. It occurs, moreover, in the hexamethyl complexes $[MMe_6]^{2-}$ (M = Zr & Hf), WMe_6 and $ReMe_6$. The rhenium compound is paramagnetic, formally containing Re(VI) $5d^1$.

(d) Sandwich compounds

Higher orders of axial symmetry are encountered in organometallic species of the sandwich type, referred to earlier [*e.g.* section 1(1,2)]: thus ferrocene $(\eta^5\text{-}C_5H_5)_2Fe$ and $(\eta^6\text{-}C_6H_6)_2Cr$ have rotational symmetry axes of order *five* and *six*, respectively. If the carbocyclic rings are in *eclipsed* conformation the (proper) symmetry axes are signified C_n (C_5 or C_6); when the rings are *staggered* the axes are instead denoted S_n (S_5 or S_6).[c]

[c] A symmetry axis S_n is called an *improper* or *alternating* axis.

(i) Common electronic structural features

The frontier MO energy levels typical of a sandwich complex are shown in Fig. 5.43 – the general conclusion of both magnetic measurements and electronic (including photoelectron) spectroscopy. The ligands C_6H_6 and (formally) $[C_5H_5]^-$ belong to the π acid category of ligand [section 4(2)] but, while weak σ donors, they are evidently quite strong π *donors* (quite unlike CO and CN^-).

The antibonding $e_1(\pi^*)$ frontier orbitals are out-of-phase combinations of d_{xz} and d_{yz} with bonding π MOs of the ligands. The $e_2(\delta)$ orbitals, *bonding* in nature (though probably not strongly) are composed of $d_{x^2-y^2}$ and d_{xy} and an in-phase admixture of empty π^* MOs of matching symmetry (e_2). Remarkably, the $a_1(\sigma)$ frontier orbital, based on d_{z^2}, is comparatively non-bonding because the d_{z^2} orbital proves to have only a small overlap with the a_1 π bonding MOs of the arene ligands. Moreover, mixing of metal *s* and d_{z^2} orbitals under the ligand field tends to lower the energy of $a_1(\sigma)$. The orbital energy sequence $a_1 > e_2$ in Fig. 5.43(b) is that appropriate to $(\eta^6\text{-}C_6H_6)_2Cr$;

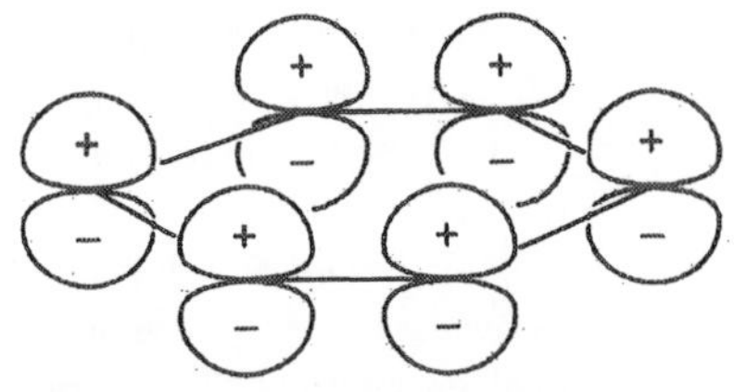

The $a_1(\pi)$ bonding MO of benzene.
At metal-ring distances optimal for π bonding this MO is drawn into the conical nodal surface of the metal d_{z^2} orbital (*cf.* Fig. 2.7) so that the net overlap is small.

the reverse order prevails in $(\eta^5\text{-}C_5H_5)_2Fe$. The relative energies of the combining metal and ring orbitals are such that the δ aspect of the bonding (resulting in $d \to \pi^*$ "back-donation") is typically stronger in the *bis*-arene complexes than in metallocenes.

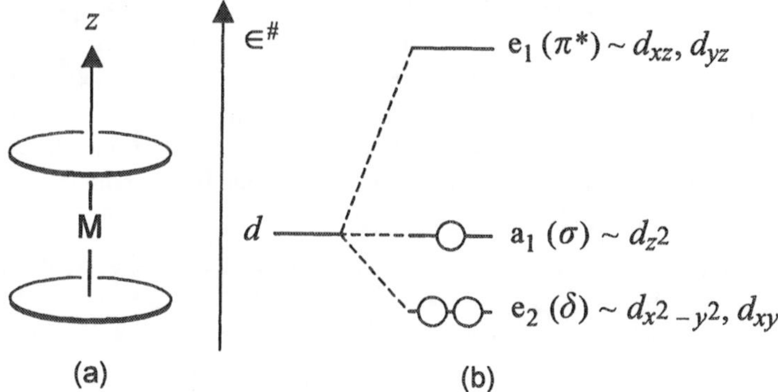

Fig. 5.43 Sandwich compounds such as $(\eta^5\text{-}C_5H_5)_2Fe$ and $(\eta^6\text{-}C_6H_6)_2Cr$: (a) generalised molecular structure; (b) frontier MO energy levels.

The symmetry axes are taken to lie in, and define the z direction.

The orbital occupancies shown are those pertaining to the prototype d^6, 18-electron complexes, which are closed-shell species.

When a molecule has a high-order symmetry axis, C_n or S_n, of order $n \geqslant 5$, alternative doubly-degenerate symmetry types, e_1 and e_2, may be distinguished.

(ii) Survey of magnetic moment data

Much of the m_{eff} data available for sandwich complexes is collected in Tables 5.10 and 5.11.

Table 5.10 The effective magnetic moments m_{eff}/μ_B at *ca.* 300 K of *bis*-cyclopentadienide complexes from the first transition series, and the ground electronic configurations thereby inferred. $[cp = \eta^5\text{-}C_5H_5]$

complex	electron confign. a_1	e_2	e_1	S	m_{eff}/μ_B	
$[(cp)_2Ti]^+$ †		1		1/2	1.68	(1.73) #
$[(cp)_2V]^{2+}$		1		1/2	1.90	(1.73) #
$[(cp)_2V]^+$		2		1	2.86	(2.83)
$(cp)_2V$	1	2		3/2	3.78	(3.88)
$[(cp)_2Cr]^+$	1	2		3/2	3.73	(3.88)
$(cp)_2Cr$	1	3		1	3.20	(2.83) #
$(cp)_2Mn$ $	1	2	2	5/2	5.86	(5.92)
$[(cp)_2Fe]^+$	2	3		1/2	2.54	(1.73) #
$(cp)_2Fe$	2	4		0	0	(0)
$[(cp)_2Co]^+$	2	4		0	0	(0)
$(cp)_2Co$	2	4	1	1/2	1.95	(1.73) #
$[(cp)_2Ni]^+$	2	4	1	1/2	1.82	(1.73) #
$(cp)_2Ni$	2	4	2	1	2.89	(2.83) ‡

Some of the data opposite (also the m_{eff} in Table 5.11) may be in error by as much as ± 0.15 μ_B.

‡ The parenthetical figures are the expected spin-only moments.

† $(cp)_2Ti$ has a dimeric structure while $(cp)_2Cu$ is unknown.

Among the heavier transition metals, only Ru and Os form simple sandwich compounds $(cp)_2M$ stable at ordinary temperatures; they are closed-shell d^6 (18 electron) species like ferrocene.

$ The m_{eff} quoted for $(cp)_2Mn$ relates to its solutions or melt (T > 445 K), or to the solid above 432 K. Solid $(cp)_2Mn$ is polymeric below 432 K.

A dilute (~ 8 %) solution of $(cp)_2Mn$ in solid, molecular $(cp)_2Mg$ obeys the Curie-Weiss law between 432 and 77 K with m_{eff} = 5.94 μ_B. The EPR spectrum gives g ≈ 2.00.

In these cases the paramagnetic properties are likely to be affected by unquenched orbital angular momentum. [See section 6(3b).]

The characterisation of ground configurations detailed in the tables is sometimes dependent on the results of EPR and photoelectron spectroscopy. For example, the techniques agree in identifying the ground state of the ferrocinium cation $[(cp)_2Fe]^+$ as $(a_1)^2(e_2)^3$ (with S = 1/2) but instead as $(e_2)^4(a_1)^1$ (also S = 1/2) in both $[(bz)_2Cr]^+$ and $(bz)_2V$.

(iii) η^5-cyclopentadienyl complexes

There remains, however, some room for argument concerning the assignment of ground configurations, especially as regards early members of the metallocene series up to $(cp)_2Cr$. It may indeed be the case, as implied in Table 5.10, that the e_2 orbital is more stable than a_1 at the beginning of the transition series, as it clearly is in all the *bis*(arene) complexes (Table 5.11). If so, the switch to the energy sequence $e_2 > a_1$, firmly established for $[(cp)_2Fe]^+$, is probably due to destabilisation of the e_2 (δ) orbital, which is expected to become less bonding as the metal *d* orbitals stabilise in response to the increasing nuclear charge.

The *permethyl* analogues $(cp^*)_2M$ ($cp^* = \eta^5$-C_5Me_5) of the metallocenes have also been extensively studied.[a] The sandwich complex decamethyltitanocene $(cp^*)_2Ti$ ($3d^2$), a 14-electron species, is known to be paramagnetic but its full characterisation is impeded by the fact that, at ordinary temperatures, it co-exists in equilibrium with a diamagnetic 16-electron isomer.

[a] μ_{eff} data (*ca.* 300 K) for some of the metallocenes $(cp^*)_2M$

$(cp^*)_2V$	3.69
$(cp^*)_2Cr$	3.20 (? 2.97)
$(cp^*)_2Mn$	2.18
$(cp^*)_2Fe$	0
$(cp^*)_2Co$	1.45 (?)
$(cp^*)_2Ni$	2.93
$[(cp^*)_2Fe]^+$	2.54
$[(cp^*)_2Os]^+$	2.75

Sandwich complexes have generally ground states of *low* or *intermediate* spin but *molecular* $(cp)_2Mn$ and its dimethyl derivative $(\eta^5$-$C_5H_4Me)_2Mn$ ($3d^5$) straddle the spin cross-over point, both displaying spin equilibrium in solution and in the vapour phase. $(cp)_2Mn$ is essentially high-spin, though its low-spin isomer is present at very low (~ 1%) concentration, while the low-spin form predominates in the case of $(\eta^5$-$C_5H_4Me)_2Mn$.[b] The methyl substituent increases the donor strength of the ligand, thereby raising the energy of the e_1 MO in relation to a_1 (Fig. 5.43) and discouraging a high-spin ground state. The permethyl compound $(cp^*)_2Mn$ is decisively low-spin, with μ_{eff} = 2.18 μ_B, having a ground configuration $(a_1)^2(e_2)^3$ equivalent to that of both $[(cp)_2Fe]^+$ and also $[(cp^*)_2Fe]^+$. Its rhenium analogue $(cp^*)_2Re$ ($5d^5$) is also low-spin, displaying μ_{eff} = 1.69 μ_B (constant in the range 300 – 100 K) but with instead the ground configuration $(e_2)^4(a_1)^1$.[c]

[b] In one report, the difference in internal energy between the spin-isomers is some 175 cm^{-1}, equivalent to about 2 kJ mol^{-1}.

[c] However, spectroscopic studies at very low temperatures (including MCD) indicate the alternative, more usual ground configuration $(a_1)^2(e_2)^3$, as they do for $(cp)_2Re$. The low-spin, excited state $(a_1)^1(e_2)^4$ may sometimes have a small thermal population in $[(cp^*)_2Fe]^+$.

Spin equilibrium in sandwich complexes is not confined to manganese. It is observed also in, for example, the *iso*-propyl derivative of chromocene, $(\eta^5$-$C_5H^iPr_4)_2Cr$ ($3d^4$): μ_{eff} = 2.83 μ_B at 15 K, increasing very gradually and smoothly to 4.90 μ_B at 300 K. The system is evidently high-spin at ordinary temperatures, having the ground configuration $(a_1)^1(e_2)^2(e_1)^1$, with S = 2. This is initially puzzling given that chromocene itself is a complex of *intermediate* spin, with S = 1 (Table 5.10). Apparently the repulsion between the ^iPr substituents forces the C_5 rings apart somewhat, reducing the ligand field strength despite the electron-donating alkyl substituents. In support of this interpretation, the manganese analogue $(\eta^5$-$C_5H^iPr_4)_2Mn$ is purely *high*-spin, as is also $(\eta^5$-$C_5H_2{}^tBu_3)_2Mn$.

A similar phenomenon is encountered in related *bis*(η^5-indenyl) complexes. The permethyl compound $(\eta^5$-$C_9Me_7)_2Cr$ displays μ_{eff} = 2.67 μ_B (constant between 300 and 4 K) indicating S = 1, while the (1,3)–*iso*propyl

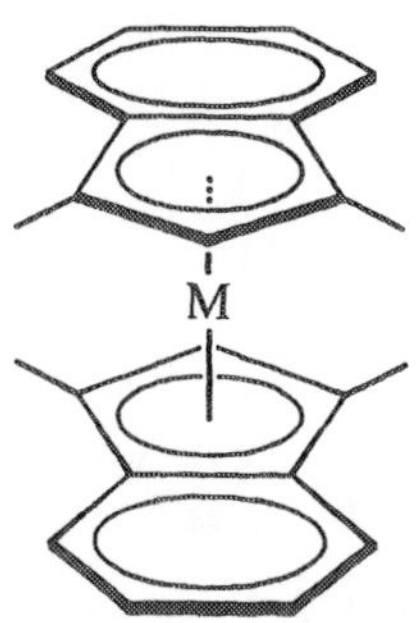

A (1,3)–substituted indenyl complex (η^5-indenyl)$_2$M

derivative (η^5-$C_9{}^iPr_2H_5)_2Cr$ is *high*-spin (S = 2) having m_{eff} = 4.55 μ_B at 300K (and averaging 4.4 μ_B in the temperature range 300 – 20 K). The Cr–C distances in the two compounds are 218 pm and 232 pm, respectively. (The parent complex (η^5-indenyl)$_2$Cr dimerises at ordinary temperatures.)

(iv) *bis*(η^6-arene) complexes

The magnetic data for *bis*-arene complexes (Table 5.11) have several notable features. In particular, it is remarkable that both $3d^5$ and $3d^4$ species are low-spin while, among the metallocenes (Table 5.10), the $3d^5$ complex $(cp)_2Mn$ is high-spin whereas the $3d^4$ and $3d^3$ species have ground configurations of *intermediate* spin. The greater degree of spin pairing in the $(bz)_2M$ complexes is due probably to a larger $a_1 - e_2$ energy spacing (Fig. 5.43),# together with reduced exchange integrals K.

\# The increased stability of the e_2 MO is presumably the secret of the existence of *bis*-arene complexes.

However, methyl substitution of the benzene ring apparently reduces the π acid strength of the ligand such that complexes of intermediate spin can form. Thus, the mesitylene complex $[(\eta^6\text{-}C_6H_3Me_3)_2V]^+$ ($3d^4$) has m_{eff} = 2.80 μ_B (at ~ 300 K), suggestive of a ground configuration $(e_2)^3(a_1)^1$.

Table 5.11 The effective magnetic moments m_{eff}/μ_B at *ca.* 300 K of *bis*-benzene complexes of the first row transition elements, and the ground electronic configurations they imply (see text).

complex†	m_{eff}/μ_B‡		S	electron confign. e_2	a_1	e_1
$(bz)_2Ti$	0	(0)	0	4	0	
$(bz)_2V$	1.68	(1.73)	1/2	4	1	
$[(bz)_2Cr]^+$	1.77	(1.73)	1/2	4	1	
$[(bz)_2V]^-$	0	(0)	0	4	2	
$(bz)_2Cr$	0	(0)	0	4	2	
$(bz^*)_2Cr$	0	(0)	0	4	2	
$[(bz^*)_2Mn]^+$	0	(0)	0	4	2	
$[(bz^*)_2Fe]^{2+}$	0	(0)	0	4	2	
$[(bz^*)_2Fe]^+$	1.89	(1.73)	1/2	4	2	1 #
$[(bz^*)_2Co]^{2+}$	1.73	(1.73)	1/2	4	2	1 #
$(bz^*)_2Fe$	3.08	(2.83)	1	4	2	2
$[(bz^*)_2Co]^+$	2.95	(2.83)	1	4	2	2
$[(bz^*)_2Ni]^{2+}$	3.00	(2.83)	1	4	2	2
$(bz^*)_2Co$	1.86	(1.73)	1/2	4	2	3 #

$bz = \eta^6\text{-}C_6H_6$; $bz^* = \eta^6\text{-}C_6Me_6$

‡ The parenthetical figures are the nearest spin-only values.

† Many analogous complexes have been reported for the heavier transition elements, including the $(bz)_2M$ species with M = Nb, Ta, Mo & W, and also permethyls $(bz^*)_2M$ – the cations $[(bz^*)_2Re]^+$ and $[(bz^*)_2Ru]^{2+}$, for example. Of the above, only the Nb and Ta complexes are paramagnetic.

The cationic Ru(II) ($4d^6$) complex can be reduced to a compound $(bz^*)_2Ru$ ($4d^8$) which, unlike $(cp)_2Ni$ ($3d^8$) (Table 5.10), is diamagnetic, thus denying a *bis*-η^6 sandwich structure. Like $[(bz^*)_2Rh]^+$, involving instead Rh(I) (also $4d^8$), it has the asymmetrical structure

$$(\eta^6\text{-}C_6Me_6)(\eta^4\text{-}C_6Me_6)M$$

and, accordingly, a valence configuration having formally 18 electrons.

The parent compound $(bz)_2Ru$ is a mixture of two isomers, one of the η^6/η^4 form as above, the other (~ 20 %) with the structure

$$(\eta^5\text{-}C_6H_6)_2Ru.$$

The η^4- and η^5-bound rings are non-planar in such complexes.

(v) Other sandwich complexes

There exist a variety of sandwich-type compounds involving other formally aromatic ligands, and many of these have been characterised magnetically. Thus, further to the *bis*(η^6–arene) systems, there are sandwich complexes formed by heterocyclic molecules such as pyridines, one example being (2,6)-dimethyl-pyridine (*dmpy*). Complexes $(dmpy)_2M$ of that ligand display magnetic properties[a] recalling those of the corresponding $(bz)_2M$ species (Table 5.11) though the loss of axial symmetry must be recognised.

a *e.g.* the d^5 complexes (obviously low-spin),

$(dmpy)_2V$	$m_{eff} = 1.75\ \mu_B$
$[(dmpy)_2Cr]^+$	$1.82\ \mu_B$

both constant in the range 300 – 5 K. The d^4 and d^6 complexes are closed-shell (*cf.* Table 5.11).

Also demanding mention here is the large category of "mixed" sandwich complexes, those involving *different* planar carbocyclic rings. Prominent, paramagnetic examples include (*cp*)(*bz*)Cr and (*cp*)(*ch*)V (both formally $3d^5$), where *ch* = planar *cycloheptatrienyl* C_7H_7, η^7-bound. Both are again *low*-spin complexes with m_{eff} values of 1.70 μ_B and 1.69 μ_B (~ 300 K), respectively and, apparently, the $(e_2)^3(a_1)^1$ ground configuration. Another variant of the theme is (*cp*)(*cot*)Ti ($3d^1$), where *cot* denotes η^8-bound *cyclo-octatetraenyl* C_8H_8 (often viewed as aromatic $[C_8H_8]^{2-}$); the complex has $m_{eff} = 1.60\ \mu_B$, sensibly independent of temperature.

5.6 Orbital contributions to the magnetic moment

Beyond the spin-only approximation .....

The spin-only formula for m_{eff} clearly has its limitations regarding data obtained at ordinary temperatures [1(3)]; its deficiencies become the more evident when applied to much of the data relating to low temperatures.

The failure of the spin-only approximation, when dealing with genuinely paramagnetic materials, free of complications such as spin isomerism [recall 3(6)] and incipient magnetic ordering, is due principally to the orbital motion of the *d* electrons contributing to m_{eff}. The purpose of this section is to elucidate the orbital contributions, distinguishing the first and second-order elements. Attention is focussed on cubic complexes.

1. Residual orbital angular momentum in the ligand field

(a) First order orbital momentum

It was explained in § 2.1(5) that a *d* electron (with $l = 2$) can display orbital angular momentum $\ell_z = \pm 2\hbar$ only when the $d_{x^2-y^2}$ and d_{xy} orbitals are degenerate. The removal of their degeneracy by a cubic ligand field (Figs. 5.3 and 5.6) therefore *quenches* these, the major components of the angular momentum (at least in first order).[b]

b The concept of the *quenching* of orbital angular momentum was introduced in § 2.3(4).

However, the d_{xz} and d_{yz} orbitals remain degenerate[c] and the $\pm\hbar$ components of ℓ_z are still well-defined. Now the d_{xy} orbital, decisively separated in energy from $d_{x^2-y^2}$, has $\ell_z = 0$ so that the t_2 set of *d* orbitals is formally similar as regards angular momentum to the set of atomic *p* orbitals. Close scrutiny of this connection by the algebra of quantum mechanics reveals that the angular momenta of the two sets of orbitals are precisely related by

c – as they are in any axially symmetric complex (the principal axis defining the *z*-direction).

$$\boldsymbol{\ell}(t_2\, d) = -\boldsymbol{\ell}(p) \tag{5.9}$$

This correspondence is known as the $t_2 - p$ isomorphism. Note the change in phase in the above equation.[d]

d – it means (rather oddly, perhaps) that *in cubic symmetry* the orbitals $d_{\pm 1}$ have the components of angular momentum $\ell_z = \mp\hbar$.

Ordinarily, covalency tends to reduce $|\boldsymbol{\ell}|$ somewhat below the atomic value since it is essentially quenched in the admixed ligand valence orbitals; this is known as the **orbital reduction effect**. It may be allowed for in an elaboration of (5.9),

$$\boldsymbol{\ell}(t_2\,d) = -\kappa\boldsymbol{\ell}(p) \tag{5.10}$$

where κ is a constant ($\leqslant 1$), characteristic of the cubic complex considered, called the **orbital reduction factor**.[a] The parameter κ, an inverse measure of the metal-ligand covalency, may often be determined from EPR and susceptibility data.

[a] Not to be confused with the quantum mechanical mixing coefficient [§ 2.6(1)]. κ is, unfortunately, another overworked symbol.

Delocalisation onto the ligands also leads usually to a reduction in the spin-orbit coupling constant, often therefore denoted ζ_{eff}.

Both e *d*-orbitals have $\ell_z = 0$, $d_{x^2-y^2}$ because it has been prised from d_{xy}, while d_{z^2} is identical with d_0. Electrons occupying the e subshell cannot make an orbital contribution to the magnetic moment (except through second order effects).

In ligand fields of lower symmetry the d_{xy} orbital is separated in energy from d_{xz} and d_{yz} and the *p*-like angular momentum characteristic of the t_2 *d* set (eqn. 5.8) is extinguished. However, if there is present a rotational symmetry axis C_n (or S_n) with $n \geqslant 3$, the d_{xz} and d_{yz} orbitals, which are of π type with respect to the axis, remain degenerate and may together display a well-defined angular momentum $\ell_z(e\,\pi) \sim \pm\hbar$. With the exception, in particular, of tetragonal complexes [5(1b),(2a,b)], the $d_{x^2-y^2}$ and d_{xy} (d_δ) orbitals are also degenerate with, correspondingly, the residual angular momentum $\ell_z(e\,\delta) \sim \pm 2\hbar$. This is the case for *trigonal* complexes and also, naturally, for complexes of related symmetry such as (*inter al.*) sandwich compounds.[b]

[b] Any *non-degenerate* orbital will have $\ell_z = 0$ (in first order).

(b) Effects of the spin-orbit interaction

(i) first-order spin-orbit coupling

Where a complex can display first order orbital angular momentum, there is the prospect of a directly induced splitting of an orbitally degenerate state under the spin-orbit interaction [§ 2.5]. This is illustrated by Fig. 5.44. Ordinarily, such a complex will be simultaneously susceptible to Jahn-Teller distortion of some kind. These two effects, invariably in conflict, commonly have a marked influence on paramagnetic properties, especially their temperature dependence.

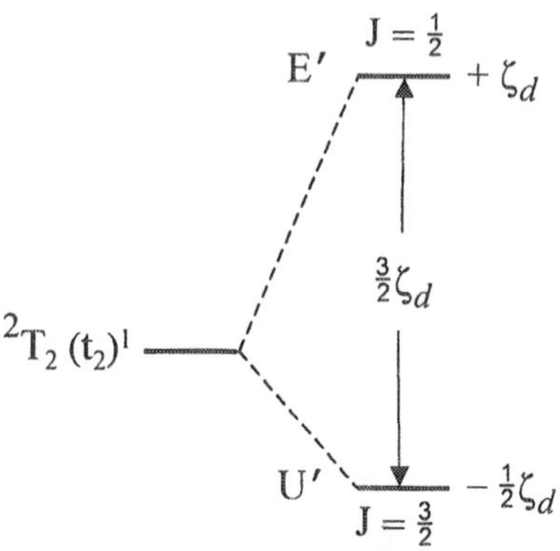

Fig. 5.44 Multiplet splitting in a $(t_{2g})^1$ octahedral complex.

[Recall and contrast Fig. 2.17 for the case $^2P\ (p)^1$. See also Fig. 5.54.]

(ii) second order orbital angular momentum

The orbital angular momentum is never *entirely* quenched, even in spatially non-degenerate states. The spin-orbit coupling invariably regenerates some orbital momentum by mixing into the ground state components of particular, degenerate excited states. If a complex lacks a symmetry axis of order $n \geqslant 3$, so that the degeneracy of the *d* orbitals is completely destroyed, mixing under the spin-orbit interaction is the only source of orbital angular momentum in the ground state.[b]

[b] – to be distinguished carefully from that associated with TIP which derives from mixing by the applied magnetic field [recall § 3.3(2)].

(iii) zero-field splitting

When $S \geq 1$, *while the orbital momentum is quenched in first order*, the spin-orbit interaction will in general induce a small so-called **zero field splitting** (ZFS) in the ground state. This reflects the splitting of orbitally degenerate *excited* states due to low symmetry components of the ligand field. In some cases the splitting can be large enough to affect paramagnetic behaviour even at ordinary temperatures. [See section (4c).]

(c) the implications for EPR spectroscopy

There are a number of practical constraints upon EPR measurements on solids. In particular, complexes with first order orbital angular momentum have usually a strong coupling, *via* the spin-orbit interaction, between the electron spin and the vibrations of the metal-ligand bonds and/or those of the crystal lattice. This provides a facile mechanism for what is called *spin-lattice relaxation*, the thermal quenching of the spin excitations occurring in electron resonance. The excited states then have only short life-times and consequently the resonances are commonly broadened beyond detection, except perhaps at very low temperatures when vibrational amplitudes are minimised.[a] Thus low-spin Mn(II) complexes, *e.g.* $[Mn(CN)_6]^{4-}$, with the first order **L** of its $(t_{2g})^5$ ground configuration (a hole in $(t_{2g})^6$), yield EPR spectra of adequate resolution only at temperatures ≤ 20 K.

a EPR signals from solid samples are otherwise broadened by the magnetic fields due to neighbouring paramagnetic centres. The effect may be reduced by dilution (where practicable) in a diamagnetic, isomorphous lattice. Thus $[IrCl_6]^{2-}$ and $[ReCl_6]^{2-}$ may advantageously be investigated in K_2PtCl_6. Moreover, EPR may prove undetectable when the g-factor is extremely anisotropic and the molecular orientations disordered.

On the other hand, EPR is often observable at ordinary temperatures when the orbital angular momentum is fully quenched by the ligand field – as in the case of most *high*-spin Mn(II) and Cu(II) complexes, for example.

However, when $S \geq 1$, a large ZFS may prevent the observation of EPR by normal means.

2. The spin-orbit coupling constant ζ_{nd}

As foreshadowed by § 2.5(3), the coupling constant ζ_{nd} increases across each transition series and also on proceeding down each of the subgroups.[b] The trend on crossing the first transition series is illustrated by the plots of ζ_{3d}/Z for the M^{2+} species in Figs. 5.44. It is instructive to compare Fig. 5.44(b) with Fig. 2.18; as there, the $\langle r^{-3}\rangle$ data are theoretical.

b The spin-orbit interaction becomes so large in the third transition series that Russell-Saunders coupling breaks down [recall § 2.5(5)].

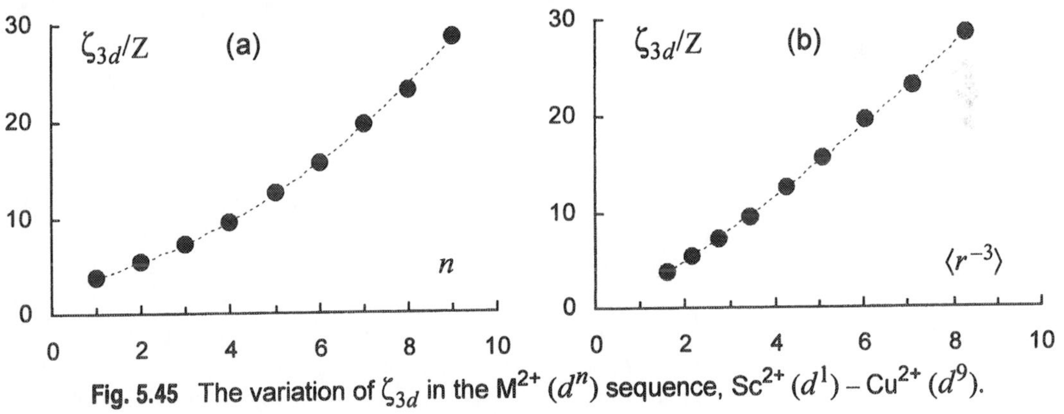

Fig. 5.45 The variation of ζ_{3d} in the M^{2+} (d^n) sequence, Sc^{2+} (d^1) – Cu^{2+} (d^9).

The variation in ζ_{3d} across the transition series is considerable. Thus $\zeta_{3d} = 120$ cm^{-1} for Ti^{2+}; but the parameter increases markedly with nuclear charge (+Ze) to 829 cm^{-1} in the case of Cu^{2+}.

The trend in ζ_{3d} down the subgroups (Table 5.12) is less regular, there being a particularly large increase between the second and third series, following the intervention of the Lanthanide elements.[a]

ζ_{nd} also increases, and in a linear fashion, with external charge q on an atom, the result simply of progressive radial contraction of the d-orbitals. In

Table 5.12

Comparison of selected ζ_{nd} data (/cm^{-1}) in the different transition series.]

ζ_{3d}	Cr^{3+}	270	Fe^{3+}	460
ζ_{4d}	Mo^{3+}	818	Ru^{3+}	1180
ζ_{5d}	W^{3+}	2500*	Os^{3+}	3500

* – or larger.

complexes, ζ_{nd} is invariably reduced by covalency in line with the orbital reduction effect [section (1a)].

3. Characterisation of ground electronic states

The ground electronic terms in the ligand field may be described by an extension (or generalisation) of the Russell-Saunders scheme [§ 2.4].

(a) Octahedral and tetrahedral complexes

Complexes of cubic symmetry have configurations of the general form $(t_2)^a(e)^b$. The twin aspects are considered in turn.

$(t_2)^a$ configurations have the various ground states pictured in Fig. 5.46, distinguished by an extension of the notation ^{2S+1}X for the terms of atomic configurations. The label X = T indicates a three-fold spatial degeneracy analogous to that of an atomic P term.[a] The $^2T_2$ $(t_2)^1$ and $^3T_1$ $(t_2)^2$ states are comparable to ^2P (p^1) and ^3P (p^2) in an atom. That the ground terms for $(t_2)^5$ and $(t_2)^4$ are the same as those of $(t_2)^1$ and $(t_2)^2$, respectively, follows from the *hole theorem* [§ 2.4(3)].

a Three different orbital arrangements are possible for both $(t_2)^1$ and $(t_2)^2$ (when the electrons are in different orbitals).

The distinction between T_2 and T_1 is not of great importance here.

The $(t_2)^3$ configuration is special in that, for maximal S = 3/2, the orbital arrangement is unique. This is conveyed by the symbol X = A (meaning no degeneracy). The ^4A term is analogous to the atomic term ^4S (p^3), pictured in Fig. 2.11 (along with the ^3P (p^2) term). Fig. 5.46 does not include the closed shell configuration $(t_2)^6$, also non-degenerate with the designation $^1A_1$ (comparable with ^1S (p^6) in an atom).

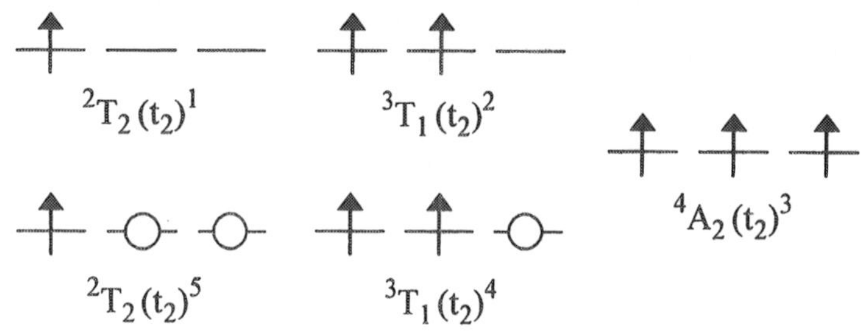

Fig. 5.46 The ground terms ^{2S+1}X of $(t_2)^a$ configurations encountered in cubic complexes. [M_S = +S components.]

The ground terms of $(e)^b$ configurations are portrayed in Fig. 5.47, again excluding the case of the filled subshell – here $(e)^4$ with, like $(t_2)^6$, the unique term $^1A_1$. The symbol X = E indicates an orbital degeneracy of *two.*

The t_2 levels in Fig. 5.46 may be thought of as t_2^+, t_2^- & t_2^0, analogous to p_-, p_+ & p_0, rather than d_{xz}, d_{yz} and d_{xy}.

The ground states of complexes with both the t_2 and e subshells occupied are obtained on combining the attributes of the ground terms of $(t_2)^a$ and $(e)^b$ whilst ensuring a maximum coupled spin S (*i.e.* the sum of their S values, S_a and S_b). For example, $^3T_1$ $(t_2)^4$ and $^3A_2$ $(e)^2$ couple to give $^5T_2$ $(t_2)^4(e)^2$, the three-fold degeneracy originating in the ground state of $(t_2)^4$. This may be expressed symbolically by

$$^3T_1\,(t_2)^4 \otimes {}^3A_2\,(e)^2 = {}^5T_2\,(t_2{}^4e^2)$$

where ⊗ denotes coupling that ensures maximal S. Similarly, $^4A_2$ $(t_2)^3$ and ^2E $(e)^1$ are merged to form a ground term ^5E $(t_2^3e^1)$, the two-fold degeneracy being due to single occupancy of the e subshell. Table 5.13 summarises the outcome overall.

The T ground states, which involve unsymmetrical occupancy of the t_2 subshell (Fig. 5.46), display first order orbital angular momentum resembling that of atomic P states, though with the sign of **L** reversed (*cf.* eqn. 5.10),

$$\mathbf{L}(t_2^{\,n}) = -\kappa\mathbf{L}(p^n) \tag{5.11}$$

(n corresponding to a above).[a] The orbital angular momentum is however quenched (in first order) in both E and A ground terms because they derive from configurations in which the unpaired electrons reside in the e subshell (Fig. 5.47) or, in the special case of $^4A_2\,(t_2)^3$, because spin correlation determines that each of the $t_2^{\pm}$ and t_2^{0} orbitals is singly occupied (Fig. 5.46) when the ℓ_z happen to sum to $L_z = 0$ exactly.

$^2E\,(e)^1$

$^3A_2\,(e)^2$

$^2E\,(e)^3$ $[M_S = S]$

Fig. 5.47 The ground states of the configurations $(e)^b$ in cubic complexes [*cf.* Fig. 5.45)].

[a] Only T states are susceptible to (first order) multiplet splitting. However, both T and E states are subject to Jahn-Teller splitting [section 5(1)], which can be particularly large in the latter case.

Table 5.13 The ground terms[#] of transition metal complexes having various electron configurations $(t_2)^a(e)^b$. X is the orbital attribute of the term symbol ^{2S+1}X. (See text.)

Octahedral fields				Tetrahedral fields			
high-spin complexes							
t_2	e	S	X	e	t_2	S	X
1		1/2	T_2	1		1/2	E
2		1	T_1	2		1	A_2
3		3/2	A_2	2	1	3/2	T_1
3	1	2	E	2	2	2	T_2
3	2	5/2	A_1	2	3	5/2	A_1
4	2	2	T_2	3	3	2	E
5	2	3/2	T_1	4	3	3/2	A_2
6	2	1	A_2	4	4	1	T_1
6	3	1/2	E	4	5	1/2	T_2
6	4	0	A_1	4	6	0	A_1
low-spin complexes							
t_2	e	S	X	e	t_2	S	X
4		1	T_1	3		1/2	E
5		1/2	T_2	4		0	A_1
6		0	A_1	4	1	1/2	T_2
6	1	1/2	E	4	2	1	T_1

[#] As stressed earlier, the subscripts to the term symbols A and T are of little consequence to subsequent developments. They are given merely for completeness.

The really important distinction is that between T states on the one hand, and A and E states on the other.

In the case of *octahedral* complexes, the complete term symbols incorporate also the subscript g – as in $^3T_{1g}\,(t_{2g})^2$, for example.

The *complementary* configurations

$$(t_2)^a(e)^b \equiv (t_2)^{6-a}(e)^{4-b}$$

always have the same pattern of terms.

(b) Uniaxial complexes

The ground terms of *axially symmetric* complexes (C_n or S_n, with $n \geqslant 3$) may be identified in an analogous fashion. Orbital degeneracy is limited to *two*, originating in $(e)^1$ and $(e)^3$ aspects of the ground configuration, each contributing a ^2E component.[b] The E states of uniaxial complexes have residual first order **L** and may therefore display multiplet splittings (unlike the E states of cubic complexes); they are also vulnerable to competitive Jahn-Teller distortions.

[b] Sandwich compounds having ^2E ground states are labelled # on the RHS of Tables 5.10 and 5.11 [section 5(2d)].

The *tbp* species of Fig. 5.41 provide another useful illustration. As the reader may care to verify, the ground states of the Co(II) and Ni(II) complexes are of the type ^{4}A and ^{3}E, respectively.

The ground term of $(e)^2$ is non-degenerate $^3A_2$, exemplified by the *tbp* system of Fig. 5.39.# Single occupancy of a non-degenerate orbital contributes a non-degenerate element to the ground term – *e.g.* $(a_1)^1$ gives $^2A_1$ – with simply the symmetry of the orbital concerned. Co(II) phthalocyanine (Fig. 5.37) is one instance. Closed-shell aspects of the electron configuration contribute totally symmetric $^1A_1$ and may be discounted.

4. Complexes with L quenched in first order

$ Such complexes are characterised by magnetic moments m_{eff} *that are constant* down to (usually very) low temperatures.

Attention is now focussed on complexes having only second order orbital angular momentum and, in particular, on *cubic* species with E or A_2 ground terms (Table 5.13).$

(a) The g-factors from EPR

It is appropriate to concentrate initially on the implications of *second order* orbital angular momentum for the *apparent* electronic spin g-factor.

(i) Tetrahedral d^1 complexes

Consider first the elementary case of one-electron tetrahedral species, exemplified by VCl_4, having the ^2E $(e)^1$ ground state. Here the spin-orbit coupling generates orbital angular momentum in the e subshell through admixture of the spin-orbitals of the excited t_2 subshell,[a] lying at an energy Δ above the ground state. The mixing coefficient κ of Fig. 2.21 proves to be exactly $-\sqrt{6}\zeta/2\Delta$; note the dependency on ζ/Δ and also the *minus* sign.

The (second order) orbital momentum is due exclusively to incidental mixing of the $d_{x^2-y^2}$ and d_{xy} orbitals, whereby a little of the angular momentum characteristic of the $d_{\pm 2}$ set is resuscitated.

The perturbation results in an orbital angular momentum, which is isotropic and *opposes that due to the spin*, given by (so it can be shown)

$$\ell_z = \mp(2\zeta/\Delta)\hbar \qquad (5.12)$$

the axis of quantisation z being freely disposable. This induced angular momentum is manifested in an *effective* spin g-factor,

$$g = g_e - 4\zeta/\Delta \simeq 2(1 - 2\zeta/\Delta) \qquad (5.13)$$

– also isotropic though, in reality, low symmetry effects invariably lead to some anisotropy. That the (average) g-factor should here be less than g_e is confirmed by the contents of Tables 5.14 and 5.15.

Table 5.14

Electron spin g-values g_{av} for some tetrahedral complexes of V(IV), $(e)^1$ [*cf.* Table 5.15].

VCl_4	1.906
$[VO_4]^{4-}$	1.946 #
$V(O^tBu)_4$	1.962
$V(CH_2SiMe_3)_4$	1.968
$V(NEt_2)_4$	1.976

$V^{4+}/Ca_3InGe_3O_{12}$ (a *garnet*)

Table 5.15

Electron spin g-values g_{av} for some tetrahedral d^1 oxyanions.

$[CrO_4]^{3-}$	1.958	$[MnO_4]^{2-}$	1.968 [a]
$[MoO_4]^{3-}$	1.920	$[TcO_4]^{2-}$	1.964 [b]
$[WO_4]^{3-}$	1.675	$[ReO_4]^{2-}$	1.763

Moreover, $[TiO_4]^{5-}$ in α-quartz has $g_{av} = 1.945$.
[a] diluted in $BaSeO_4$; [b] in $SrTcO_4$. The other $[MO_4]^{x-}$ species were investigated as dopants in $CaWO_4$.

Such g-data also demonstrate the expected *inverse* dependency of $g_e - g$ on the orbital splitting Δ. Note, in particular, the decrease of $g_e - g_{av}$ with increasing Δ in the oxyanion sequence $[TiO_4]^{5-} \gtrsim [VO_4]^{4-} > [CrO_4]^{3-} > [MnO_4]^{2-}$, this despite the expected increase in ζ_{3d}.

Analogous complexes of the heavier transition elements usually show larger deviations from g_e because ζ_{nd} tends to increase more than Δ – *e.g.* the complex $Nb(NEt_2)_4$ displays $g_{av} = 1.952$ as compared with $g_{av} = 1.976$ for its V(IV) analogue (Table 5.14). However, the ratio ζ/Δ may not increase *greatly* between the first and second transition series because Δ increases as well as ζ_{nd}. As illustrated by the g_{av} data in Table 5.15, the effect of the escalating ζ/Δ is more noticeable on moving to the third transition series where ζ_{5d} is particularly large (*cf.* Table 5.12) in relation to Δ.

Table 5.16

Electron spin g-values g_{av} for some octahedral d^9 complexes

$[Cu(OH_2)_6]^{2+}$	2.220
$[Cu(bpy)_3]^{2+}$	2.117
$[Ag(OH_2)_6]^{2+}$	2.132
$[Ag(bpy)_3]^{2+}$	2.086
$[Ni(bpy)_3]^{+}$	2.136

[The g-values vary somewhat with chemical environment.]

(ii) The general formulation of g_{av} (many-electron systems)

The expression (5.13) for the isotropic g-factor in tetrahedral d^1 complexes may be generalised, assuming Russell-Saunders coupling, to embrace any cubic species having either an E or A ground term (Table 5.13). The second order orbital momentum induced by the spin-orbit coupling is always essentially of the form $L_z = \mp(c\lambda/\Delta)\hbar$, the parameter c distinguishing individual cases. The g-factor is accordingly expressible as

$$\boxed{g \simeq 2(1 - c\lambda/\Delta)} \qquad (5.14\text{-a})$$

If the electron configuration is again of the general form $t_2)^a(e)^b$ then it transpires that

$$c = \tfrac{2}{3}\,|2a - 3b| \qquad (5.14\text{-b})$$

As in the case of atoms, $\lambda = \pm\,\zeta/2S$ (eqn. 2.45), the minus sign applying if the d subshell is more than half-filled.[a] Thus $g > g_e$ when $a + b > 5$.

The prototype eqn. 5.13 for $(e)^1$ is quickly recovered: there $a = 0$ and $b = 1$, so that $c = 2$ and $g = 2(1 - 2\lambda/\Delta)$; then, with $\lambda = +\zeta$, $g = 2(1 - 2\zeta/\Delta)$. In the related case of octahedral d^9 complexes, ground state ${}^2E_g$ $(t_{2g})^6(e_g)^3$, $c = \tfrac{2}{3} \times (12 - 9) = 2$ also; but, since λ is negative, $g = 2(1 + 2\zeta/\Delta)$, as evidenced by Table 5.16. This could have been deduced immediately from the expression for $(e)^1$ by invoking electron-hole equivalence for the d subshell. High-spin d^5 systems, with ${}^6A_1$ ground states, have $c = 0$ and hence $g = g_e$, the free spin value.[b] Comprehensive application of (5.14) yields the compendium of Table 5.17.

[a] The general formula, as expressed here, does not cover the $(e)^4$ and $(t_2)^6$ configurations. But, obviously, $g = 0$ in these closed-shell cases

[b] The spin-orbit interaction does not perturb the ${}^6A_1$ state in second order. In third order, however, it causes a *very* small zero field splitting (*e.g.* 10^{-2} cm^{-1}).

Table 5.17

octahedral	tetrahedral	term	c
$(t_{2g})^3$	$(e)^4(t_2)^3$	${}^4A_2$	4
$(t_{2g})^3(e_g)^1$	$(e)^3(t_2)^3$	${}^5E$	2
$(t_{2g})^3(e_g)^2$	$(e)^2(t_2)^3$	${}^6A_1$	0
$(t_{2g})^6(e_g)^1$	$(e)^3$	${}^2E$	6
$(t_{2g})^6(e_g)^2$	$(e)^2$	${}^3A_2$	4
$(t_{2g})^6(e_g)^3$	$(e)^1$	${}^2E$	2

Values of the quantity c (eqn. 5.14-b) for the E and A ground states of cubic complexes.

The configurations are listed according to the electron-hole equivalence of the octahedral and tetrahedral cases.

c is dimensionless. Remember that λ in eqn. 5.14-a changes sign when $a + b > 5$.

The scope of the formula (5.13) is further illustrated by the selection of g-data in Tables 5.18 and 5.19[a] where, again, the influence of the λ/Δ term is all too apparent. The d^3 octahedral case (with $g < g_e$) may be instructively contrasted with that of tetrahedral d^7, also involving a $^4A_2$ ground term. Not only is the orbital contribution *positive* in the latter case (*i.e.* $g > g_e$) but, reflecting the generally weaker ligand field, it is typically much larger in absolute terms. Thus $[CoCl_4]^{2-}$, for instance, exhibits a $g_{av} = 2.34$.[b]

a As elsewhere, the g_{av} of discrete complex ions will vary slightly with counter-ion.

b – from the EPR of $(Et_4N)_2[ZnCl_4]$ doped with Co(II). In the compound Cs_3CoCl_5 the g-value of $[CoCl_4]^{2-}$ is instead 2.29.

Table 5.18 Average g-values for a variety of d^3 cations with essentially octahedral coordination – ground term $^4A_{2g}\,(t_{2g})^3$.

$[Cr(OH_2)_6]^{3+}$	1.977	$[V(OH_2)_6]^{2+}$	1.973	Mn^{4+}/SnO_2	1.989
$[Cr(NH_3)_6]^{3+}$	1.984			Tc^{4+}/SnO_2	1.970
$[Cr(CN)_6]^{3-}$	1.991	$[V(CN)_6]^{4-}$	1.988	Re^{4+}/SnO_2	1.696
$[Mo(OH_2)_6]^{3+}$	1.951			$Tc^{4+}/K_2[PtCl_6]$	1.990
				$Re^{4+}/K_2[PtCl_6]$	1.814

Table 5.19 Average g-values for selected 6-coordinate Ni(II) species – d^8 $^3A_{2g}\,(t_{2g})^6(e_g)^2$.

Ni^{2+}/ZnF_2	2.33
$[Ni(OH_2)_6]^{2+}$	2.24
$[Ni(NH_3)_6]^{2+}$	2.16

Comparably large, positive deviations from g_e are observed in octahedral d^8 complexes – in particular, those of Ni(II) (Table 5.19) – even though Δ is generally much greater for 6-coordination. These are attributable to an almost two-fold increase in $\lambda = -\zeta/2S$ on going from Co(II) to Ni(II).[c]

c Thus the complexes $[NiCl_6]^{4-}$ & $[CoCl_4]^{2-}$ have $\Delta \sim 7300$ and 3130 cm^{-1}, respectively. However, the spin-orbit coupling constants are

	ζ_{3d}	λ	$[cm^{-1}]$
Ni^{2+}	649	−324	$(= -\zeta/2)$
Co^{2+}	533	−178	$(= -\zeta/3)$

Well-founded g-data for compounds of the heavier transition metals are too sparse for wide-ranging comparisons to be made. However, the very limited information concerning octahedral d^3 species in Table 5.18 is consistent with the "vertical" trend in g_{av} at constant N (also oxidation state),

$$3d^N \gtrsim 4d^N > 5d^N$$

(the order of increasing $-\Delta g \propto \zeta/\Delta$) as noted earlier for tetrahedral d^1 oxy-anions (Table 5.15) and related species.[d]

d The particularly low values of g_{av} reported for Re(IV) $5d^3$ are dictated by the very large spin-orbit coupling constant ζ_{5d} (~ 3000 cm^{-1} or more in the free ion).

(b) The temperature–invariant m_{eff}

In simple paramagnets, the modified g_{av} should be reflected straightforwardly in magnetic moments $m_{eff} = g_{av}\sqrt{[S(S+1)]}\,\mu_B$. Thus the data for various octahedral d^3 complexes in Table 5.3 are almost invariably *smaller* than the spin-only expectation (3.88 μ_B). Moreover, the m_{eff} reported for tetrahedral d^7 species (with again S = 3/2), almost all complexes of Co(II) (Table 5.20), are consistently *larger* than the spin-only value. The same is true of octahedral d^8 complexes (where $m_{eff} > 2.83\ \mu_B$ always): thus $[Ni(OH_2)_6]^{2+}$ has $m_{eff} \sim 3.14 - 3.24\ \mu_B$, depending on the partner anion.

However, the effect of variations in ζ/Δ is often less evident regarding trends in m_{eff} values than in the case of the g-values. Typically this is because, in the analysis of the susceptibility data, the first order moment is not distinguished from that due to TIP [recall § 3.4(1)]. If the d subshell is less than half-full, the (negative) orbital contribution to m_{eff} is significantly obscured by the (necessarily positive) TIP moment. For instance, VCl_4 is reported to display an m_{eff} of 1.72 μ_B, barely less than the spin-only figure of 1.73 μ_B. But the g_{av} = 1.906 observed by EPR (Table 5.14) implies an

$m_{eff} = 1.65\ \mu_B$. The difference can be explained by a TIP contribution, with magnetisability $\alpha \approx 1.90 \times 10^{19}\ \mu_B^2$, which at 300 K amounts to *ca.* 8 % of the susceptibility and 4 % of the *apparent* magnetic moment.[a]

Meaningful trends in m_{eff} data with ζ/Δ are nonetheless sometimes discernible, even when $N < 5$. The d^3 case history (recall Table 5.3) includes many examples of octahedral complexes from the second and third transition series. In keeping with the available g-data (see above), complexes of the *third* row elements invariably show distinctly smaller m_{eff} than those of the earlier transition metals. Compare, for example, $[Re(NCS)_6]^{2-}$ with $m_{eff} = 3.36 - 3.60\ \mu_B$, depending on partner cation, and the following complexes:

$[V(NCS)_6]^{4-}$ 3.83 μ_B, $[Cr(NCS)_6]^{3-}$ 3.79 μ_B & $[Mo(NCS)_6]^{3-}$ 3.70 μ_B.

An inverse correlation with Δ is usually clear enough when, on the other hand, the d subshell is more than half-occupied and (with $g_{av} > g_e$) the orbital contribution to m_{eff} and the TIP moment are both positive, each varying as $^1/_\Delta$ (though in differing proportion). The data for tetrahedral Co(II) complexes assembled in Table 5.20 provide an elegant illustration.[b]

It should be reiterated, finally, that the constancy of m_{eff} characteristic of such systems is seldom maintained at very low temperatures; this is because of fine structure in the ground state due to minor distortion – whether the direct consequence, as in the case of E terms, or the indirect effect (zero-field splitting) observed in A ground terms. Moreover, fine splitting of the ground term may induce significant magnetic anisotropy at higher temperature even when both m_{eff} and g_{av} are sensibly constant.

Table 5.20 m_{eff} values ($/\mu_B$) at *ca.* 300 K for some tetrahedral complexes of Co(II) d^7 – ground state $^4A_2\ (e)^4(t_2)^3$ – compared with the orbital splitting Δ_t (in cm^{-1}). $

	m_{eff}	Δ_t		m_{eff}	Δ_t
$[Co(NCS)_4]^{2-}$	4.36 – 4.53	4550	$[CoCl_4]^{2-}$	4.66 – 4.80	3130
$[Co(NCO)_4]^{2-}$	4.38	4150	$[CoBr_4]^{2-}$	4.80 – 4.87	2850
$[Co(N_3)_4]^{2-}$	4.47	3920	$[CoI_4]^{2-}$	4.87 – 5.01	2650
$[Co(SPh)_4]^{2-}$	4.50	3870			

The spin-only value of m_{eff} is 3.88 μ_B.
m_{eff} varies considerably with counter-cation.

(c) Zero field splitting in ^{2S+1}A ground states ($S \geq 1$)

A complex having a spatially *non-degenerate* ground state but with $S > \frac{1}{2}$ is susceptible to a splitting (usually denoted D)# under the spin-orbit coupling acting in second (or third) order. This is often so large as to preclude routine EPR measurements.

Certain complexes of essentially *cubic* symmetry have ground terms of the type $^3A_2$ and $^4A_2$ (Table 5.13) with **L** and also, accordingly, the spin-orbit interaction, quenched in first order. However, *perfectly* cubic symmetry is rarely maintained even in such instances, and the zero field splitting is

[a] For a ground configuration $(t_2)^a(e)^b$, the magnetisability can be shown to be

$$\alpha = \frac{4\mu_B^2}{3\langle\Delta E\rangle}|2a - 3b|$$

where $\langle\Delta E\rangle$ is the average energy of the excited states mixed into the ground state by the applied magnetic field.
[Griffith (1961), § 10.4(1).] The unit of α is $J^{-1}\,A^2\,m^4$.

In the $(e)^1$ case, $\alpha = 4\mu_B^2/\Delta$ which for VCl_4, with $\Delta = 9000\ cm^{-1}$, amounts to *ca.* $2.24 \times 10^{19}\ \mu_B^2$. $\chi_{TIP} = \mu_0 N_A \alpha$ is therefore expected to be $0.15 \times 10^{-8}\ m^3\ mol^{-1}$. The experimental figure is $0.125 \times 10^{-8}\ m^3\ mol^{-1}$.

[b] $[CoCl_4]^{2-}$ was discussed briefly in § 3.4(1), where the experimental χ_{TIP} was quoted as $0.85 \times 10^{-8}\ m^3\ mol^{-1}$. Theoretically $\alpha = 8\mu_B^2/\Delta$ so, with $\Delta = 3130\ cm^{-1}$ (Table 5.20), $\chi_{TIP} = \mu_0 N_A \alpha = 0.84 \times 10^{-8}\ m^3\ mol^{-1}$.

$ The same trend with Δ is naturally observed in tetrahedral complexes of Fe(II), ground term $^5E\ (e)^3(t_2)^3$. Thus the tetrahalides $[FeX_4]^{2-}$ have the m_{eff}: Cl 5.39 μ_B, Br 5.43 μ_B and I 5.58 μ_B.
[The data relate to $(N^nPr_4)^+$ salts. The spin-only expectation for m_{eff} is here 4.90 μ_B.]

Not to be confused with D as used to signify χ_{TIP} in the extended Curie-Weiss law (3.22).

attributable to low symmetry splittings transferred from orbitally degenerate *excited* states through admixture by the spin-orbit coupling. It is expressible as a sum of terms of the form $\zeta^2/\Delta E$, where ΔE denotes the energy of the excited state involved.

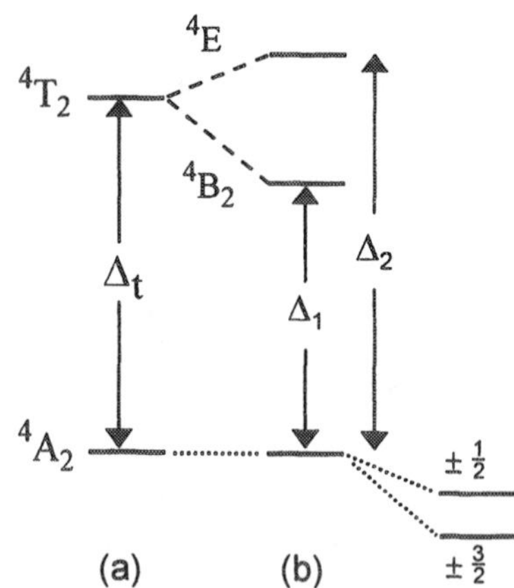

Fig. 5.48 Zero field splitting of the $^4A_2$ ground level of a tetrahedral d^7 complex with a tetragonal distortion. (a) tetrahedral case; (b) tetragonal (dihedral) geometry. (Not to scale.)

The zero field splitting is here actually

$$D = 8\lambda^2\left(\frac{1}{\Delta_1} - \frac{1}{\Delta_2}\right)$$

(though this is often expressed as $2D$ in the literature).

The phenomenon is illustrated in Fig. 5.48, which concerns tetrahedral d^7 complexes – and therefore Co(II) especially – with the spin quartet ground term $^4A_2$, configuration $(e)^4(t_2)^3$. The excited $^4T_2$ state, the lowest quartet of $(e)^3(t_2)^4$, suffers a splitting $\Delta_2 - \Delta_1$ through (commonly) a modest tetragonal distortion of the complex. This is, in effect, communicated by the spin-orbit interaction which (acting in second order) induces a splitting D of the ground state into two Kramers doublets, $M_S = \pm\frac{1}{2}$ and $\pm\frac{3}{2}$. The relative energies of the latter states, *i.e.* the sign of D, depends on the nature of the distortion. The complex anions $[CoCl_4]^{2-}$ and $[CoBr_4]^{2-}$, present in the salts Cs_3CoX_5, have splittings of some 8.6 cm^{-1} and 10.7 cm^{-1}, respectively; in $Cs_2[CoCl_4]$, however, the tetrachloride complex shows $D = -9.7$ cm^{-1}, the ground state being that with $M_S = \pm\frac{1}{2}$ (in contrast to Fig. 5.48). The zero field splitting is comparatively large in these complexes because Δ_t is small, *e.g.* 3130 cm^{-1} in $[CoCl_4]^{2-}$, while ζ_{3d} is substantial, *ca.* 530 cm^{-1} in the free Co^{2+} ion.

Octahedral complexes of d^3 cations – Cr(III), most prominently – with the ground term $^4A_{2g}$ $(t_{2g})^3$, are naturally similar but display considerably smaller ZFS; for example, $D = 0.59$ cm^{-1} in Cr(*acac*)$_3$ and a mere 0.00495 cm^{-1} for Cr^{3+} doped into the complex salt $[Co(en)_3]Cl_3.NaCl.6H_2O$. Zero field splittings are usually larger in complexes of the heavier transition elements due to the stronger spin-orbit interaction: thus Mo(III) complexes commonly display splittings an order of magnitude greater than those found in comparable Cr(III) species.

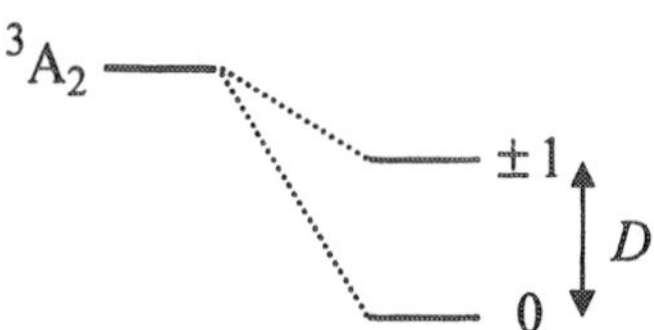

Fig. 5.49 Zero field splitting of the $^3A_2$ ground level of an octahedral d^8 or tetrahedral d^2 complex.

Octahedral d^8 and tetrahedral d^2 complexes, both with $^3A_2$ ground states (S = 1) – *viz.* $^3A_{2g}$ $(t_{2g})^6(e_g)^2$ and $^3A_2$ $(e)^2$, respectively – have zero field splittings, into $M_S = \pm 1$ and 0 components (Fig. 5.49); as with of $^4A_2$ complexes, however, the sign of D will vary. The Ni(II) species in $[Ni(Opy)_6]$-$(ClO_4)_2$ and $[Ni(OH_2)_6](SnCl_6)$, both with the *singlet* ($M_S = 0$) ground state, display splittings of 4.35 cm^{-1} and 0.40 cm^{-1}, respectively.[a] In the salt $[Ni(OH_2)_6](ZrF_6)$ the Ni(II) aquocation has instead the *doublet* ($M_S = \pm 1$) ground state, with $D = -0.61$ cm^{-1}. The tetrahedral d^2 case is illustrated by $[FeO_4]^{2-}$ which, diluted in K_2CrO_4, displays a splitting $D = 0.070$ cm^{-1}.

High-spin d^5 complexes, with the ground term $^6A_1$, are a special case in that ZFS requires spin-orbit coupling acting in *third* order. Accordingly, the splittings observed are always very small, typically ~ 10^{-2} cm^{-1} or less.[b]

Zero field splitting is not usually apparent in the *average* m_{eff} observed, except at very low temperature (though its effects are not always easy to distinguish from the influence of weak magnetic ordering). Any zero field splitting will however introduce *anisotropy* in the system that is frequently evident at ordinary temperatures. Thus the susceptibilities of the $[CoX_4]^{2-}$ complexes (X = Cl & Br) mentioned above display an anisotropy of about 7 – 9 % at *ca.* 300 K, increasing some five-fold as the temperature is lowered to 75 K.# But the average m_{eff} is constant throughout this temperature range. The anisotropy may be exploited, through the theory, to determine the

a When the singlet is the more stable, the system is *non-magnetic* [§ 3.2(2b)] at very low temperatures, $T < D/k$.

b Zero field splittings are also observed in compounds of Gd(III), ^8S $(4f^7)$, where they are typically ~ 0.5 cm^{-1}.

See facing page.

sense of the zero field splitting (*i.e.* the sign of D).

At very low temperatures, $T \lesssim D/k$, the susceptibility of a magnetically dilute species will show a considerable departure from Curie behaviour. Consider, for example, the sandwich compound $(\eta^5\text{–}C_5H_5)_2Ni$ (*nickelocene*), which has a $^3A_2$ ground state (Figure 5.50) but yields no EPR spectrum. This complex has a susceptibility following the Curie-Weiss law down to about 70 K, with $m_{eff} = 2.89 \pm 0.15\ \mu_B$; substantial deviation from this simple trend is however observed in the range 70 – 6.5 K. Detailed analysis of $\chi(T)$ in this lower range demonstrates that the ground state is the singlet ($M_S = 0$) and allows the estimate $D \cong 26\ cm^{-1}$ for the ZFS, and also the (otherwise indeterminate) g-factors, $g_{\parallel} = 2.00$ and $g_{\perp} = 2.06$, for the upper ($M_S = \pm 1$) doublet. Working backwards, the g-values imply a first order m_{eff} of 2.89 μ_B at higher temperatures.

If the ground doublet is $M_S = \pm\frac{3}{2}$ (as in Fig. 5.48), then $m_{\parallel} > m_{\perp}$ with values diverging progressively in the range 300 → 75 K.

$m_{\perp}$ is instead greater than $m_{\parallel}$ when $M_S = \pm\frac{1}{2}$ is the ground doublet.

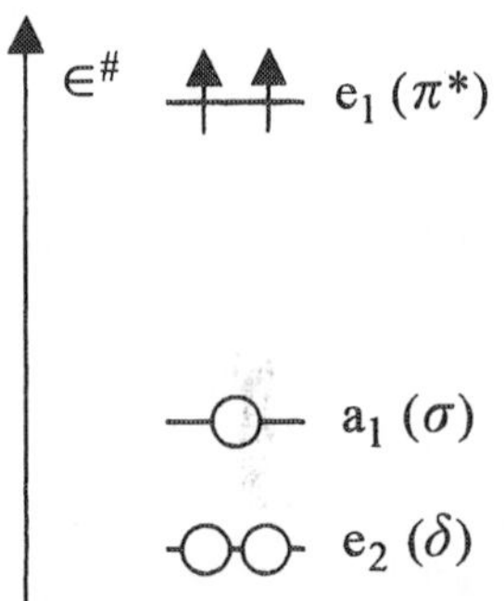

Fig. 5.50 Electronic structure of $(\eta^5\text{–}C_5H_5)_2Ni$, ground term $^3A_2$. (Recall Table 5.10.)

The $^4A_2$ species $[Cr(NH_3)_6]^{3+}$ occurs in a variety of crystals, including the complex salt $[Cr(NH_3)_6](ClO_4)_2Br.CsBr$, where the Cr(III) cations are so widely separated that magnetic interaction between them is negligible even at *extremely* low temperature. Remarkably (to this Chemist, anyway), the susceptibility has been investigated between 4.2 and 40 mK, and its variation indicates a ZFS of 0.37 cm^{-1}. In favourable cases, zero field splittings may otherwise be determined *via* specific heat measurements or *directly*, when comparatively large, by far infra-red (Fourier transform) spectroscopy.

5. Systems with temperature-dependent m_{eff}

Transition metal complexes with residual first order orbital momentum – *viz.* cubic species having T ground terms or uniaxial complexes with E ground terms [section 3] – typically display effective magnetic moments that vary with temperature *even about "room" temperature or above.* This can prove important diagnostically.

When **L** is only partially quenched in the ground term, there is normally a first order multiplet splitting,[a] moderated in varying degree by the effect of Jahn-Teller (and other) distortions. In general, the fine structure exhibited by the ground term involves energy spacings comparable in magnitude to the Boltzmann energy kT, with the result – whatever the detail – that m_{eff} is itself a function of temperature, and the dependence of the susceptibility on temperature correspondingly complex

a – *e.g.* Fig. 5.44 concerning the $^2T_2$ $(t_{2g})^1$ ground state. In octahedral Ti(III) complexes, the multiplet splitting $\frac{3}{2}\zeta_{3d}$ is expected to be $\leq 233\ cm^{-1}$. This may be compared with a Boltzmann energy $kT \approx 209\ cm^{-1}$ at 300 K.
[$\zeta_{3d} = 155\ cm^{-1}$ for the free Ti^{3+} ion, while $kT = 0.6952\ T\ cm^{-1}$.]

(a) Complexes of d^7 and d^8 cations

Both octahedral complexes of Co(II), with ground term $^4T_{1g}\ (t_{2g})^5(e_g)^2$, and tetrahedral complexes of Ni(II), $^3T_1\ (e)^4(t_2)^4$, display magnetic moments declining in value as the temperature is lowered. In the particular case of $[NiCl_4]^{2-}$ (Figure 5.51), the m_{eff} apparently falls from about 3.85 μ_B at 300 K to ~ 3.3 μ_B at 80 K, and approaches ~ 1.0 μ_B at 4.2 K (the precise figures depending slightly on the partner cation).

The simple notion that m_{eff} should be *greater* than the spin-only value (here 2.83 μ_B) for complexes with the d subshell more than half-filled is thus untenable *in general* – though it frequently happens to be true at ordinary temperatures.[b] As for high-spin octahedral Co(II) complexes, the (average)

b Equally, when $\mathbf{L} \neq 0$, m_{eff} is not *necessarily* less than the spin-only value when there are fewer than five d electrons present. Thus, complexes of Ti(III) and V(IV) frequently have m_{eff} in excess of 1.73 μ_B at around 300 K.

magnetic moment is *never* less than the spin-only figure (3.88 μ_B); but it can be as large as 5.5 μ_B at *ca.* 300 K.

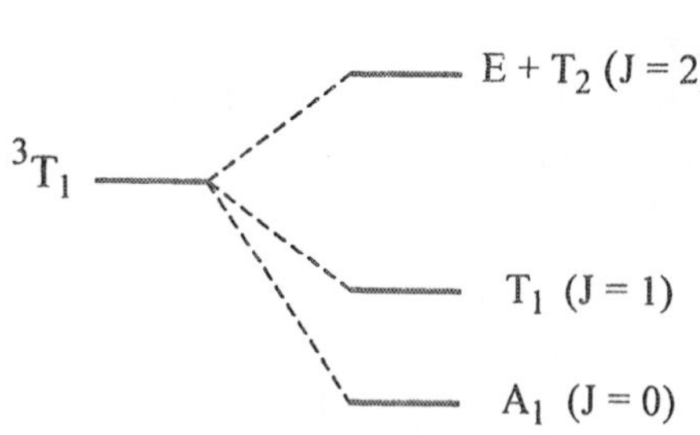

Fig. 5.52 Multiplet splitting of the $^3T_1$ ground state of a cubic complex with the $(t_2)^4$ ground configuration. [*cf.* Fig. 2.20]

The multiplets have an energy width of $3\lambda = \frac{3}{2}\zeta \leq 973$ cm^{-1} in tetrahedral Ni(II) complexes, since $\zeta_{3d} = 649$ cm^{-1} in free Ni^{2+}. [The effective ζ (or ζ_{eff}) is invariably reduced in a complex because of covalency.] The T_1 and A_1 multiplets have a separation of $\lambda = \frac{1}{2}\zeta_{eff} \leq 324$ cm^{-1}.

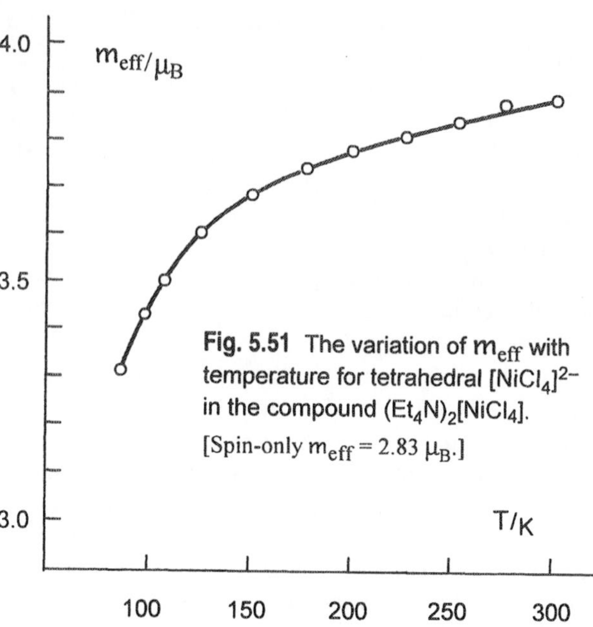

Fig. 5.51 The variation of m_{eff} with temperature for tetrahedral $[NiCl_4]^{2-}$ in the compound $(Et_4N)_2[NiCl_4]$. [Spin-only $m_{eff} = 2.83$ μ_B.]

In contrast, as explained in the previous section, octahedral d^8 and tetrahedral d^7 complexes, with respectively the ground terms $^3A_{2g}$ $(t_{2g})^6(e_g)^2$ and $^4A_2$ $(e)^4(t_2)^3$, show essentially constant m_{eff}, certainly above 80 K. For instance, the complex $[NiCl_6]^{4-}$, as studied in the $CsNiCl_3$ lattice, has $m_{eff} = 3.40$ μ_B invariant above *ca.* 50 K. The bulk susceptibilities of compounds containing the $[CoCl_4]^{2-}$ anion similarly conform to the Curie-Weiss law (at least down to 80 K), with $m_{eff} \approx 4.7$ μ_B [Table 5.20] and small Weiss temperatures ($\theta \sim 10$ K, or less).

Whether or not m_{eff} varies with temperature can be a strong pointer to the geometry of a complex, when this is unknown.

Thus, as in many other cases, the paramagnetic properties of both Co(II) and Ni(II) are highly dependent on the symmetry of the ligand field. Any paramagnetic compound of Co(II) *not* conforming with the Curie-Weiss law (with a modest θ), and with $m_{eff} > 3.88$ μ_B at ordinary temperatures, almost certainly contains an octahedral complex; tetrahedral coordination, or either of the alternative *tbp* and *sqp* modes of 5-coordination, would lead to a ^4A ground term and therefore a T-invariant m_{eff} (> 3.88 μ_B). However, any compound of Ni(II) with $m_{eff} > 2.83$ μ_B at *ca.* 300 K, but also changing significantly with temperature, is most unlikely to involve octahedral coordination. The complex concerned must have a spatially degenerate ground term and, therefore, is probably either tetrahedral ($^3T_1$ ground term) or 5-coordinate (^3E) [recall section (3b)].

(b) Further to complexes involving the $(t_2)^4$ configuration

(i) tetrahedral d^8 species

The temperature dependence of m_{eff} in Fig. 5.51 reflects the varying populations of the multiplet levels (Fig. 5.52), their magnetic moments differing because they involve distinct relative orientations of **L** and **S** (recall § 4.1).

The ensemble will in general have a substantial thermal "spread" among the multiplet states.

The splitting of the $^3T_1$ ground term of a tetrahedral d^8 complex under the spin-orbit coupling is actually such (Fig. 5.52) that its magnetic moment would tend to zero at very low temperatures were magnetic ordering not to intervene beforehand. It follows from the sign reversal in eqn. 5.9 that the splitting pattern for $^3T_1$ $(t_2)^4$ is essentially the same as that in the atomic p^2 case (Fig. 2.20), the three multiplets realised being formally equivalent to the atomic states distinguished by J = 0, 1 & 2.# The non-degenerate ground state A_1 (J = 0) is "non-magnetic" in the special sense of having no *first* order magnetic moment, because the orbital angular momentum (with L ~ 1) and the spin (S = 1) have an anti-parallel configuration, a situation reminiscent of the ground state of f^6 [§ 4.3(4)].

The only qualitative difference is that, in the ligand field, the J = 2 state is split slightly in second order into multiplets E and T_2.

In **j–j** *coupling*, the ground multiplet (J = 0) would be viewed as a filled $j = 3/2$ spin-orbital ($\omega = 2j+1 = 4$).

A ground state may otherwise be non-magnetic because of cancellation of the separate orbital and spin moments, these proportional to **L** and g_e**S**, respectively [section (d)].

(ii) low-spin octahedral d^4 complexes

The energy level scheme of Fig. 5.52 applies equally to octahedral species having the $^3T_{1g}$ $(t_{2g})^4$ ground term. The well documented case histories relate to complexes of the "heavier" d^4 cations, *e.g.* Re(III) and Os(IV). There the strong spin-orbit coupling causes a sufficiently large $T_1 - A_1$ energy separation that only the A_1 ground *singlet* is thermally populated. Since this is non-degenerate such complexes can show only TIP, the result primarily of mixing with the T_1 multiplet under the applied magnetic field. Thus, in a recent study, the complex $[ReH_6]^{3-}$ was found to display an essentially constant, though small paramagnetic susceptibility of ~ 3.1 x 10^{-9} m^3 mol^{-1} throughout the temperature range 300 – 3.5 K (equivalent to an m_{eff} of 0.77 μ_B at 300 K). Larger, also constant susceptibilities are observed in hexahalide complexes of the third transition series, such as $[OsX_6]^{2-}$ (with *apparent* m_{eff} at *ca.* 300 K ranging from 1.3 to 2.3 μ_B).

Needless to say, perhaps, such species fail to display EPR.

This TIP term may be calculated *via* eqn. 5.16 below, the result being

$$\chi_{TIP} = \mu_0 N_A |\alpha_0'| \mu_B^2 = 24\mu_0 N_A \mu_B^2/\zeta$$

There is a further, minor contribution from mixing with the J = 0 multiplets of excited terms.

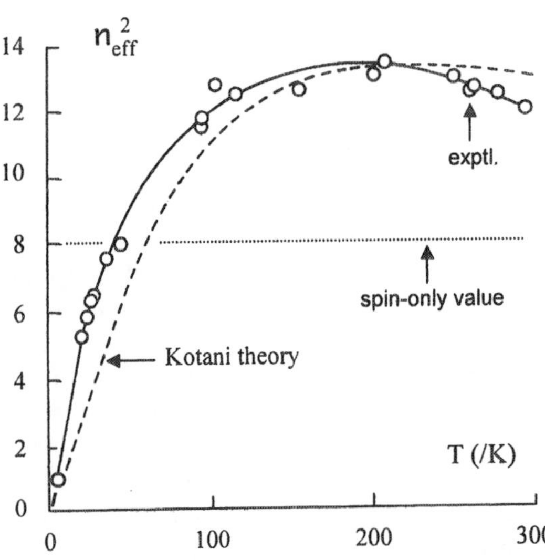

Fig. 5.53

Application of the simple Kotani model to the complex $[Mn(CN)_6]^{3-}$, ground term $^3T_{1g}$ $(t_{2g})^4$.

The data relate to the salt $K_3[Mn(CN)_6]$. The theory assumes that ζ = 360 cm^{-1}, the value for free Mn^{3+}.

[After Cooke & Duffus 1955.]

Analogous complexes in the chemistry of the first and second transition series show quite different behaviour because the multiplet width is much smaller and the ground singlet not therefore thermally isolated, except at low

Table 5.21

m_{eff} data ($/\mu_B$) at about 300 K for some 6-coordinate low-spin d^4 compounds.

$[V(bpy)_3]I$	2.8
$[Cr(bpy)_3]^{2+}$	2.92 – 3.38*
$[Cr(phen)_3]^{2+}$	2.79 – 3.12*
$K_4[Cr(CN)_6].2H_2O$	3.40
$K_3[Mn(CN)_6]$	3.50

* – varying with counter-anion
[S = 1; spin-only m_{eff} = 2.83 μ_B.]

temperatures. Table 5.21 gives the m_{eff} observed at ordinary temperatures for selected complexes from the first series (all necessarily involving *strong* ligands). Complexes of $4d^4$ ions, *e.g.* those of Mo(II), Tc(III) and Ru(IV), *always* low-spin, typically display m_{eff} in the range 2.7 to 3.1 μ_B – as expected, given an increase in ζ_{nd}, generally somewhat smaller than found in complexes from the first row.

The variation in magnetic moment of the Mn(III) complex $[Mn(CN)_6]^{3-}$ between 300 and 4 K is shown in Fig. 5.53, though it is actually the *square* of $n_{eff} = m_{eff}/\mu_B$ that is plotted. The susceptibility becomes independent of temperature, *i.e.* $n_{eff} \propto \sqrt{T}$, over the lower part of the temperature range.

(c) Kotani theory

Early developments in this area were due to Kotani (1949); a theoretical plot of m_{eff} (or its square) *vs.* T or kT/ζ is known as a Kotani diagram.

The dependency of m_{eff} on temperature in any $^{2S+1}T\ (t_2)^a$ ground term may be determined analytically through adaptation of the atomic theory of § 4.1,2 as applied to $^{2S+1}P\ (p)^a$ systems - due allowance being made for the $t_2 - p$ isomorphism (5.9). Retracing the argument of § 4.1(2), with **L** replaced by **-L**, and putting $L(L+1) = 2$, the expression (4.3) for the g-factor becomes

$$g_J = -1 + (g_e + 1)\left[\frac{J(J+1) + S(S+1) - 2}{2J(J+1)}\right] \tag{5.15}$$

– except that, *for the purposes of this equation only*, $g_J = 5$ when J = 0. It can also be shown that, instead of (4.6),

$$\alpha'_J = -\frac{2}{3\lambda}(g_J + 1)(g_J - 2) \tag{5.16}$$

The second order energy shifts of the multiplets are given by

$$\Delta E(J) = -\tfrac{1}{2}\alpha'_J(\mu_B B)^2$$

while $E_{so}(J) = -\tfrac{1}{2}\lambda[J(J+1) - S(S+1) - 2]$ (5.17)

The *minus* sign in the latter equation confirms that the multiplets of $(t_2)^a$ are indeed inverted in comparison with those of $(p)^a$. (Contrast eqn. 2.42.) The results above allow the calculation of the susceptibility by the general formula (4.14). In the high temperature limit (actually, never closely approached in practice), n_{eff} is given by eqn. 4.8 with L = 1, *i.e.*

$$n_{eff} = \sqrt{[4S(S+1) + 2]} \tag{5.18}$$

The viability of the atomic model is tested in Fig. 5.53; it can be seen to predict the shape of things successfully, including the broad maximum at around 200 K.[a] The agreement with experiment can be made near-exact by extending the theory to allow for, in particular, a small axial distortion and also covalency [recall eqn. 5.10]. The limiting value of n_{eff} at high temperature should be $\sqrt{10} = 3.16$.

[a] The maximum is expected at kT/ζ = 0.45, when n_{eff} = 3.62.

The atomic theory is easily extended to complexes with mixed configurations $(t_2)^a(e)^b$ that also have ground terms of the ^{2S+1}T type (Table 5.13).[b] Important cases include $^5T_{2g}\ (t_{2g})^4(e_g)^2$ and $^4T_{1g}\ (t_{2g})^5(e_g)^2$ – *e.g.* high-spin octahedral Fe(II) and Co(II), respectively. But the theory has its limitations, which may be highlighted through consideration of the elementary (if not altogether simple) case octahedral d^1 complexes.

[b] For further detail on the Kotani model and its elaborations see Griffith (1961) and Figgis & Hitchman (2000), and references therein.

(d) $^2T_2\ (t_2)^1$ species

Table 5.22

Parameters describing the effect of a magnetic field on the multiplets of $^2T_2\ (t_2)^1$.

J	g_J	α'_J
$\frac{1}{2}$	−2	$-8/3\zeta$
$\frac{3}{2}$	0	$4/3\zeta$

The $^2T_{2g}\ (t_{2g})^1$ ground state is formally equivalent to the atomic term $^2P\ (p^5)$ with $\lambda = -\zeta$. The spin-orbit interaction induces a splitting into multiplets with $J = \frac{1}{2}$ and $\frac{3}{2}$ (having symmetry labels E′ and U′, respectively), the latter being the ground state; the multiplet splitting is $\frac{3}{2}\zeta_d$ [Fig. 5.51(a)]. The relevant magnetic parameters, calculated *via* eqns. 5.14 and 5.15, are listed in Table 5.22, and the Zeeman effects anticipated are shown in Fig. 5.54(b,c).

The ground multiplet U′($\frac{3}{2}$), a *quartet*, has g = 0 and is non-magnetic – like the $^2\Pi_{\frac{1}{2}}$ ground doublet of NO [§ 3.4(2)], and for a similar reason.

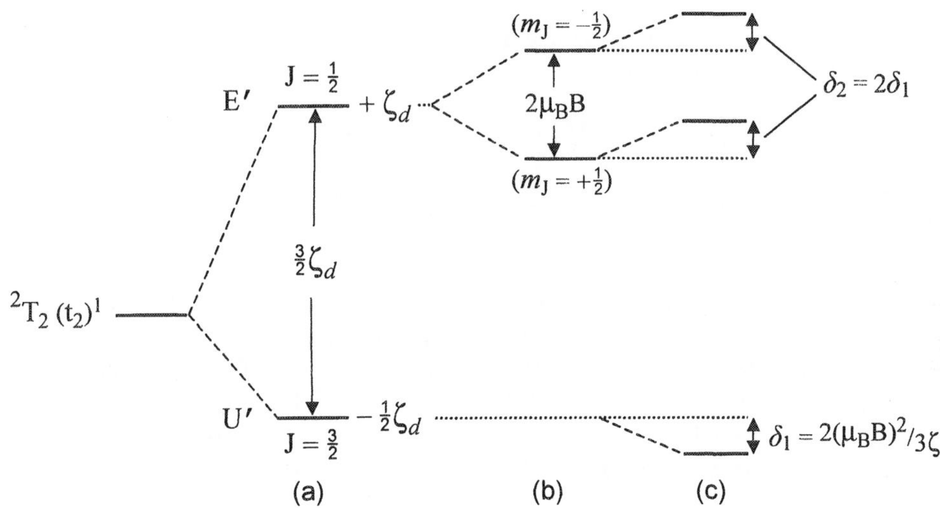

Fig. 5.54 The energetics of a regular $(t_{2g})^1$ octahedral complex in an external magnetic field B: (a) the effect of the spin-orbit interaction; (b) the first order Zeeman splitting; and (c) the second order Zeeman effect. [Not to scale.]

The four components of U′($\frac{3}{2}$) are distinguished by $m_J = \pm\frac{3}{2}$ or $\pm\frac{1}{2}$. That with the maximum J_z, *i.e.* $m_J = +\frac{3}{2}$, may be described simply as an "up-spin" electron ($m_s = +\frac{1}{2}$) occupying the d_{+1} orbital. But, according to (5.9), d_{+1} must here be regarded as carrying an angular momentum of $-\hbar$ units, when

$$\mathsf{m}_z = -(\ell_z + 2s_z)(\mu_B/\hbar) = -[(-1) + 2(\tfrac{1}{2})]\mu_B = 0$$

– which confirms that the (first order) magnetic moment $\mathbf{m} = 0$.[a] The orbital and spin magnetic moments cancel, just as in the ground multiplet of NO and also (though for a different reason) in the ground states of Eu^{3+} [§ 4.3(3)] and octahedral $(t_{2g})^4$ complexes.

Conversely, the magnetic moments are complementary in the excited multiplet E′($\frac{1}{2}$); moreover, the change in sign of ℓ_z in the $d_{\pm1}$ orbitals determines a *negative* g-factor (actually g = −2) and so the lower Zeeman level in Fig. 5.54(b) is that with $m_J = +\frac{1}{2}$.[b]

Implementing eqn. 4.14, while referring energies to that of the ground multiplet, the bulk susceptibility is[c]

$$\chi_{\mathrm{mol}} = \frac{\mu_0 N_A \mu_B^2}{3kT}\left\{\frac{4n_{\mathrm{eff}}^2(\frac{3}{2}) + 2n_{\mathrm{eff}}^2(\frac{1}{2})\exp(-\frac{3\zeta}{2kT})}{4 + 2\exp(-\frac{3\zeta}{2kT})}\right\} \qquad (5.19)$$

The notation U′, E′, *etc.* was introduced in § 4.3(5).

a That $\mathsf{m}_z = 0$ must be equally true of the other three components of U′($\frac{3}{2}$) since m_J merely defines the projection of $\mathbf{m}$ in the *z*-direction.

b In the case of a pure spin state with $S = \frac{1}{2}$, the ground Zeeman level would of course be $\mathsf{m}_S = -\frac{1}{2}$. [§ 3.1(1).]

c – where, recalling eqn. 4.13,

$$n_{\mathrm{eff}}^2(J) = g_J^2 J(J+1) + 3kT\alpha'_J$$

Incorporating the parameters in Table 5.22, while introducing the variable $x = \zeta/kT$, and simplifying,

$$\chi_{mol} = \frac{\mu_0 N_A \mu_B^2}{3kT}\left\{\frac{\frac{8}{x} + (3 - \frac{8}{x})\exp(-\frac{3}{2}x)}{2 + \exp(-\frac{3}{2}x)}\right\} = \frac{\mu_0 N_A \mu_B^2}{3kT}\left\{\frac{8 + (3x - 8)\exp(-\frac{3}{2}x)}{x[2 + \exp(-\frac{3}{2}x)]}\right\} \qquad (5.20)$$

The corresponding formula for the effective Bohr magneton number, $n_{eff} = m_{eff}/\mu_B$, is

A formula due originally to Kotani (1949).

$$n_{eff} = \left[\frac{8 + (3x - 8)\exp(-\frac{3}{2}x)}{x[2 + \exp(-\frac{3}{2}x)]}\right]^{1/2} \qquad (5.21)$$

which is plotted against $x^{-1} = kT/\zeta$ in Fig.5.55.[a] The limiting values of n_{eff} are zero as $x \to \infty$ and $\sqrt{5}$ when $x \to 0$. The former ($kT/\zeta \to 0$) was anticipated (the $U'(\frac{3}{2})$ ground multiplet being identified as non-magnetic);[b] the latter ($kT/\zeta \to \infty$) was also to be expected because, at sufficiently high temperature, all the micro-states of $(t_{2g})^1$ become equally populated and, invoking eqn. 5.18, the magnetic moment *in the ensemble* is simply $m_{eff} = \sqrt{(4.\frac{1}{2}.\frac{3}{2} + 2)}\mu_B = \sqrt{5}\mu_B$. This limiting situation is of no *practical* interest, however; the data available for transition metal compounds relate usually to values of $kT/\zeta \lesssim 1.5$.

a Compare Fig. 3.6 for nitric oxide.

b However, the susceptibility should not vanish. As $x \to \infty$, $n_{eff} \to 2/\sqrt{x}$ and

$$\chi_{mol} \to \mu_0 N_A \mu_B^2 (4/3\zeta).$$

This limiting TIP term could have been deduced immediately from Table 5.22.

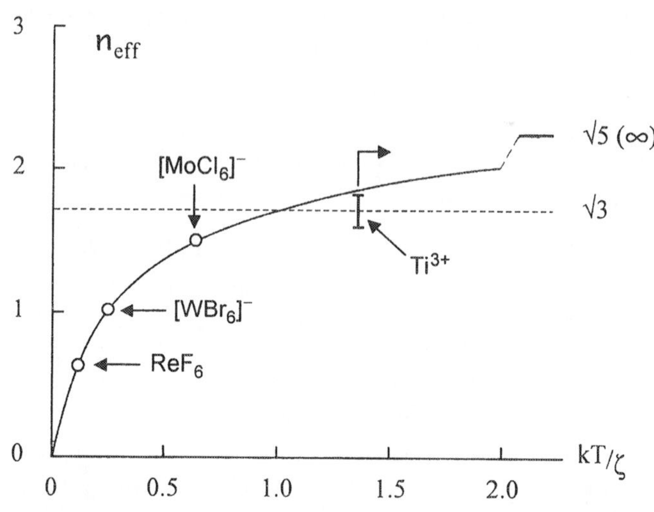

The variation in n_{eff} observed in Ti(III) complexes is represented by a bar. This range bar is drawn at $kT/\zeta = 1.35$, based on $\zeta = 155$ cm^{-1} for the free Ti^{3+} ion. The attached right-directed arrow indicates that, because of covalency (which reduces ζ_{nd} usually), the n_{eff} should really be located at higher kT/ζ.

The variation of n_{eff} among octahedral V(IV) complexes is comparable, but the range bar is positioned at lower kT/ζ.

Fig. 5.55 The effective Bohr magneton number n_{eff} as a function of $kT/\zeta = x^{-1}$ for octahedral $(t_{2g})^1$ complexes (eqn. 5.20), together with selected experimental data for T ~ 300 K. [See text.].

In the first transition series, it so happens that kT/ζ is of the order unity at ambient temperature, when the expected magnetic moment approximates to the spin-only value (Fig. 5.55).[c] The m_{eff} reported for octahedral Ti(III) and

c $n_{eff} = 1.760$, close to the spin-only value of $\sqrt{3}$, when $kT/\zeta = x^{-1} = 1$.

V(IV) range between 1.6 μ_B and 1.85 μ_B, clustering mainly about the value 1.73 μ_B. For example, both $[TiCl_6]^{3-}$ and $[VCl_6]^{2-}$ have $m_{eff} \approx 1.76\ \mu_B$ in their pyridinium salts at around 300 K. But, it must be emphasised, the near spin-only values are essentially fortuitous.

Octahedral V(IV) complexes are however uncommon in comparison with those of Ti(III).

Complexes of the heavy transition elements have generally lower values of kT/ζ (the spin-orbit coupling constant ζ being larger) and, at ordinary temperatures, display magnetic moments that are typically less than the spin-only value (Fig 5.55). Note, in particular, the low moment observed for molecular ReF_6 – *viz.* 0.63 μ_B, apparently, at 296 K – in essence reflecting a huge ζ_{5d} of 3400 cm^{-1} (as determined by optical absorption spectroscopy).[a] Congeneric TcF_6, involving weaker spin-orbit coupling ($\zeta_{4d} \approx 1450\ cm^{-1}$ in free Tc^{6+}), displays an apparent $m_{eff} = 1.12\ \mu_B$ at ordinary temperatures.

[a] This experimental value represents a 23 % reduction as compared with $\zeta_{5d} = 4398\ cm^{-1}$ in the free Re^{6+} ion.

While the Kotani theory provides the larger picture it usually fails, however, in the detail, especially regarding the data for low temperatures. Octahedral $(t_{2g})^1$ complexes do display the anticipated decrease in magnetic moment as the temperature is lowered, but the decline in the experimental m_{eff} invariably proves less than expected (*e.g.* Fig. 5.56), failing to approach zero as $T \to 0$.[b]

[b] Moreover, the energy level scheme on which the theory is based (Fig. 5.54) cannot account for the commonplace observation of EPR at low temperatures (*e.g.* Fig. 5.32). The EPR spectrum of $[Ti(OH_2)_6]^{3+}$ in caesium Ti(III) alum at 4.2 K has $g_{\parallel} = 1.25$ and $g_{\perp} = 1.14$.

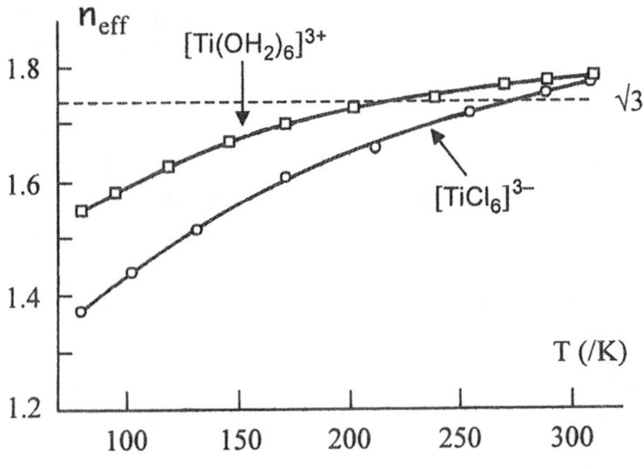

Fig. 5.56 Variation with temperature of the magnetic moments of the Ti(III) complexes in $CsTi(SO_4)_2.12H_2O$ and $(pyH)_3[TiCl_6]$.

By way of comparison, the V(IV) complex in the salt $(pyH)_2[VCl_6]$ has $m_{eff} = 1.76\ \mu_B$ at 300 K, falling to ~ 1.3 μ_B at 80 K.

(e) $^2T_2\ (t_2)^5$ complexes

The multiplet splitting in low-spin octahedral complexes of d^5 cations is the inverse of that in Fig. 5.54, the ground state being the Kramers doublet $E'(\frac{1}{2})$, which should display EPR.[c] The spin-orbit coupling constant λ now being *minus* ζ_d, the formula for n_{eff} in the Kotani approximation may be obtained immediately by substituting $-x$ for x in eqn. 5.21, and so

[c] – at least at low temperatures.

$$n_{eff} = \left[\frac{8 + 3x - 8\exp(-\frac{3}{2}x)}{x[1 + 2\exp(-\frac{3}{2}x)]} \right]^{1/2} \tag{5.22}$$

Its variation with $kT/\zeta\ (= x^{-1})$ is contrasted with that in the $(t_2)^1$ case in Fig. 5.57 The n_{eff} have the same limiting value of $\sqrt{5}$ as kT/ζ becomes indefinitely large (the high T limit), but in the $(t_2)^5$ case n_{eff} tends to $\sqrt{3}$, the spin-only value, rather than zero, when $kT/\zeta \to 0$ (low T limit). The Kotani plot has a maximum at $x = 1.302$ when $n_{eff} = 2.54$.

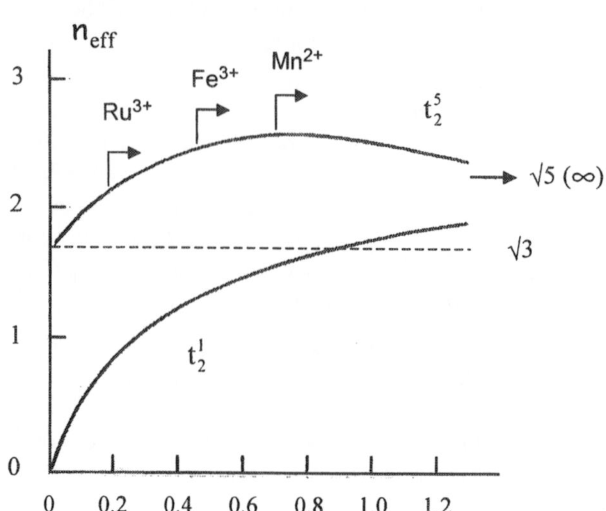

Fig. 5.57 Kotani diagram for octahedral $(t_{2g})^5$ complexes (eqn. 5.12). The plot for the $(t_{2g})^1$ case is included for comparison. The n_{eff} expected for certain cations at 300 K are indicated. As in Fig. 5.55, the theoretical indicators are anchored at values of kT/ζ that are based on free ion values of the ζ_{nd}.

Table 5.23

n_{eff} data (m_{eff}/μ_B) at *ca.* 300 K for a selection of low-spin d^5 compounds that contain octahedral complexes.

Compound	n_{eff}
$V(bpy)_3$	1.93
$[Cr(bpy)_3](ClO_4)$	2.05
$K_4[Mn(CN)_6].3H_2O$	2.18
$K_3[Fe(CN)_6]$	2.25
$[Fe(bpy)_3](ClO_4)_3.3H_2O$	2.40
$Na_3[RuCl_6].12H_2O$	2.24
$Ru(acac)_3$	1.95
$CsRu(SO_4)_2.12H_2O$	1.92
$[Ru(NH_3)_6]Cl_3$	2.13

The effective magnetic moments are predicted to exceed $\sqrt{3}$ μ_B *always*, and this is generally borne out by experiment (Table 5.23). Equally, m_{eff} is never found to be more than 2.54 μ_B.

The generally lower values of m_{eff} observed in complexes of Ru(III), as compared with analogous Fe(III) species, are attributable to the stronger spin-orbit coupling in Ru(III). Such complexes also usually show the expected decrease in m_{eff} as the temperature is lowered: thus, between *ca.* 300 K and 80 K, the moment of $[Fe(en)_3]Cl_3$ falls from 2.45 μ_B to 2.17 μ_B and that of $[Ru(en)_3]Cl_3$ from 2.13 μ_B to 1.88 μ_B. Similarly, $Cs_2[RhF_6]$ has $m_{eff} = 2.01$ μ_B at 295 K, diminishing to 1.76 μ_B at 90 K.

However, as with $(t_{2g})^1$ complexes, the unadulterated Kotani theory fails to provide a thoroughgoing, quantitative account of the wealth and variety of magnetic data available.

(f) The quenching of L, and therefore the spin-orbit coupling, by the Jahn-Teller effect

Compounds expected to possess first order **L** may sometimes nonetheless display constant values of m_{eff} over a wide temperature range. For example, the remarkable d^5 complex Co(1-*nor*)$_4$ has the low-spin ground term $^2T_2$ $(e)^4(t_2)^1$ – assuming a tetrahedral geometry – yet its susceptibility conforms well with the formula $\chi = C/(T-\theta) + D$ [eqn. 3.22], with $m_{eff} = 1.89$ μ_B and $\theta = 0.7$ K, together with a credible value of $D = \chi_{TIP}$. The tetra-alkyl complexes of Cr(III) and Mn(IV), $[CrR_4]^-$ and MnR_4 – both having, if tetrahedral, the ground term $^4T_1$ $(e)^2(t_2)^1$) – are apparently similar.

Most of the time, such behaviour is due to Jahn-Teller distortion [recall § 5.5(1a) *et seq.*]# which lifts the orbital degeneracy of the ground term and suppresses the first order effects of the spin-orbit coupling.

- or, not infrequently, in the case of solids, a distortion engineered by the environment of a complex or required for the efficient long range packing of ions within the lattice.

(i) The case of tetrachlorocuprate(II) $[CuCl_4]^{2-}$

With the ground state $^2T_2$ $(e)^4(t_2)^5$ (were it tetrahedral), $[CuCl_4]^{2-}$ is prone to distort tetragonally to a dihedral shape, intermediate between tetrahedral and square-planar (recall Fig. 5.35). The consequent splitting of the $^2T_2$ term (Fig. 5.58), leaving an orbitally non-degenerate ground state, is sufficiently large that the magnetic moment is invariant in the range 300 – 80 K; in the particular salt $Cs_2[CuCl_4]$ the constant $m_{eff} = 1.93\ \mu_B$.

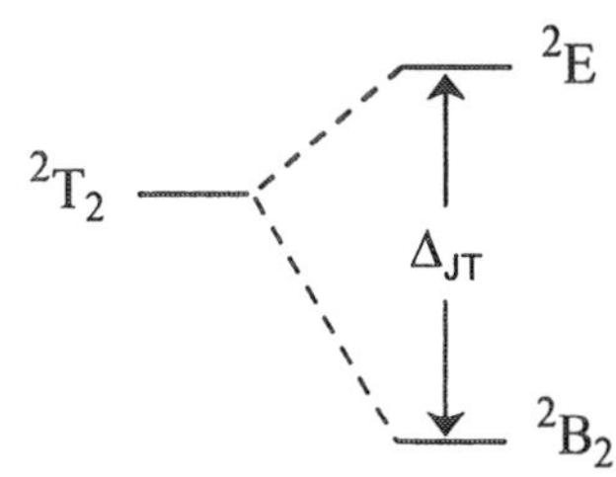

Fig. 5.58 Splitting of the $^2T_2$ ground level of a tetrahedral d^9 complex due to a tetragonal compression.

(The hole is in the d_{xy} orbital, or $d_{x^2-y^2}$, according to the choice of axes.)

Such distortions naturally lead to a substantial anisotropy in the paramagnetism. In the above case the components of m_{eff} parallel and perpendicular to the 4-fold axis are 2.18 μ_B and 1.79 μ_B, respectively, at 300 K.[a]

Were it not for the Jahn-Teller effect, the $^2T_2$ ground term of the tetrahedral complex would be split by the spin-orbit interaction into multiplets J = $\frac{1}{2}$ and $\frac{3}{2}$; through the sign reversal in eqn. (5.8), the J = $\frac{1}{2}$ doublet would be the ground state, as in the atomic p^1 case (Fig. 2.17). The multiplet interval would be $\frac{3}{2}\zeta_{3d} \leq 1243\ \text{cm}^{-1}$ (since $\zeta_{3d} = 829\ \text{cm}^{-1}$ for Cu^{2+}). However, the Jahn-Teller effect, capable here of a splitting (Δ_{JT}) of some 5000 cm^{-1}, dominates the energetics, quenching the *first* order spin-orbit coupling.[b]

The pattern of *d*-orbital energies in $[CuCl_4]^{2-}$ is similar to that shown in Fig. 5.35 (where the occupancies relate to the d^8 configuration), there being an additional, unpaired electron in the $d_{x^2-y^2}$ orbital. In a dihedral complex, the latter orbital belongs to the symmetry species b_2; hence the designation of the ground state as $^2B_2$ in Fig. 5.58.

It is curious that $[NiCl_4]^{2-}$ does not display a comparable tendency to distort, this despite weaker spin-orbit coupling in Ni(II) ($\zeta_{3d} = 650\ \text{cm}^{-1}$ in the free ion). Actually, $[NiCl_4]^{2-}$ shows a minor, tetragonal *elongation* – as do also $[CoCl_4]^{2-}$ and VCl_4. This is apparently large enough in the case of $[NiCl_4]^{2-}$ to prevent m_{eff} tending to zero at very low temperatures.

(ii) Octahedral d^1 complexes

The discrepancies between theory and experiment have again been largely attributed to the Jahn-Teller effect or to low symmetry distortions of different causation, which tend to counteract the spin-orbit coupling. In complexes of the first transition series the spin-orbit coupling may be essentially quenched or, at least, the $U'(\frac{3}{2})$ ground multiplet (Fig. 5.54) experiences a splitting into two doublets with, concomitantly, a partial uncoupling of the spin and orbital magnetic moments.[c] The ground state is then to an extent paramagnetic (in the full, *first* order sense) and may display EPR, if only at low temperature, both the g-factor and the susceptibility naturally being *anisotropic*. The anisotropy may be as pronounced as that observed in inherently axial complexes such as *tris*-chelates (*e.g.* Fig. 5.59).

In the *third* transition series the spin-orbit coupling is so strong that it may instead quench the Jahn-Teller effect. Thus ReF_6, for which ζ_{5d} (or, rather, ζ_{eff}) $\approx 3400\ \text{cm}^{-1}$,[d] apparently retains a regular octahedral shape, this presumably stabilised by the spin-orbit coupling. The molecular ensemble has a paramagnetic susceptibility at 300 K of $0.21 \times 10^{-8}\ \text{m}^3\ \text{mol}^{-1}$ corresponding to a magnetic moment of some 0.63 μ_B (as recorded in Fig. 5.55).

a The EPR spectrum of $[CuCl_4]^{2-}$ yields $g_{\parallel} = 2.384$ and $g_{\perp} = 2.094$. [Detailed analysis indicates an orbital reduction factor $\kappa \approx 0.7$.]

b But the spin-orbit interaction remains important in second order terms, and is responsible for m_{eff} being greater than $\sqrt{3}\ \mu_B$, the spin-only value.

c The upper multiplet $E'(\frac{1}{2})$ is at the same time mixed with components of the ground multiplet.

d This experimental value (from optical spectroscopy) is 23 % less than $\zeta_{5d} = 4398\ \text{cm}^{-1}$ observed in free Re^{6+}.

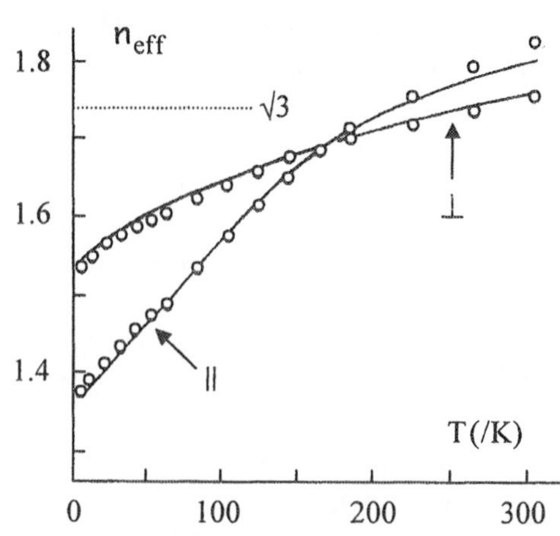

Fig. 5.59 Anisotropy in the bulk magnetism of $[Ti(urea)_6](ClO_4)_3$.

The solid lines are theoretical, based on an elaborated Kotani model allowing for the trigonal symmetry of the Ti(III) complex and covalency. (The theory is not always so successful, however.)

The literature abounds with detailed analyses of paramagnetic data that yield empirical values of the orbital reduction parameter κ and the spin-orbit coupling constant ζ_{eff}, together with energy splittings[a] such as Δ_{JT} in Figs. 5.31 and 5.57. However, such empirical data are sometimes fraught with uncertainty – there often being, for example, doubt as to the uniqueness of the fit, especially as regards susceptibility data. Not infrequently, moreover, the analyses produce estimates of κ and ζ_{eff} that are unrealistically small, *i.e.* they appear to overestimate the extent of covalency.

[a] The low-symmetry splittings can sometimes be characterised directly through spectroscopy.

The most recent theoretical studies demonstrate that the incompetence of an elaborated Kotani theory in such cases is probably due to neglect of the dynamical aspect of the Jahn-Teller effect.[b]

[b] Another complicating factor, in the case of many-electron species, is *configuration interaction*, the mixing of the ground term with terms of excited configurations under the inter-electronic repulsion – in the $(t_{2g})^5$ case, mixing of the $^2T_{2g}$ term of $(t_{2g})^4(e_g)^1$.

(g) The dynamic Jahn-Teller effect

Mention thus far of the Jahn-Teller effect has been of the *static* phenomenon, manifested by distortions from higher symmetry that are evident through structure determination. But the Jahn-Teller effect has also, and more fundamentally, a dynamical aspect: in general, the effect is about vibrational-electronic (*vibronic*) interactions, *i.e.* the breakdown of the famous Born-Oppenheimer approximation.

When the ground state of a complex is Jahn-Teller unstable, the detailed analysis of its magnetic properties by any *static* model of the ligand field may sometimes be little more than a useful analogical device of uncertain physical validity.

The influence of otherwise latent vibronic effects proves important in all complexes suffering Jahn-Teller instability that have been thoroughly investigated, including species as diverse as ReF_6, VCl_4 and cobaltocene $(cp)_2Co$. However, further exploration of this subject is way beyond the scope of the present text. Suffice it to say that, in any complex having an orbitally degenerate ground state, reported estimates of covalency, whether derived from EPR g-values or from susceptibility measurements, should always be viewed with caution.

5.7 Bibliography

A Abragam and B Bleaney, *Electron Paramagnetic Resonance of Transition Ions*, Oxford 1970

C J Ballhausen, *Introduction to Ligand Field Theory*, McGraw-Hill, 1962

C J Ballhausen, *Molecular Electronic Structures of Transition Metal Complexes*, McGraw-Hill, 1979

A Earnshaw, *Introduction to Magnetochemistry*, Academic Press, 1968 #

B N Figgis and M A Hitchman, *Ligand Field Theory and its Applications*, Wiley-VCH, 2000

See also Boudreaux and Mulay (1976), Carlin (1986) and Kahn (1993) – all cited in § 3.11.

J S Griffith, *The Theory of Transition-Metal Ions*, Cambridge 1961

J E Huheey *et al.*, *Inorganic Chemistry*, HarperCollins, 1993

S FA Kettle, *Physical Inorganic Chemistry*, Oxford 1998 #

introductory text

F E Mabbs and D J Machin, *Magnetism and Transition Metal Complexes*, Chapman and Hall, 1973

B R McGarvey, *Electron Spin Resonance Spectra of Transition-Metal Complexes*, vol. 3 of *Transition Metal* Chemistry, Marcel Dekker, 1966

J R Pilbrow, *Transition Ion Electron Paramagnetic Resonance*, Oxford 1990

5.8 Exercises

1. Explain why the following compounds are diamagnetic.\$

   $K_2Cr_2O_7$, $Fe(CO)_5$, $Co(NH_3)_6Cl_3$, K_3RhCl_6, Cs_2CuCl_3, $Mo(O_2CMe)_2$, $KReCl_4$, $Co(C_5H_5)_2BF_4$, Cs_3RhH_4, Cs_4CoH_5, Mg_2FeH_6, $NaV(CO)_6$, Os(*cyclohexyl*)$_4$, $Cs_3W_2Cl_9$, $Fe_2(CO)_9$, $Ti(C_6H_6)_2$, $NOAuF_6$, Na_2O_2.

   \$ – or display only weak, temperature independent paramagnetism.

2. Why does *magnetite* Fe_3O_4 have the inverse spinel structure [§ 3.6(4a)] while Mn_3O_4 and Co_3O_4 are normal spinels?

3. (a) Starting with an octahedral complex ML_6, with the ligands L located on cartesian axes, show that removal of a pair of ligands *trans* to each other on the z-axis produces a d-orbital splitting of the same general form as that in Fig. 5.35(b).

   [See, for example, Kettle (1998) § 7.9.]

   (b) Speculate as to the likely sequence of d-orbital energies in square-planar $[Cu(NH_3)_4]^{2+}$ and $[Ni(CN)_4]^{2-}$.

   (c) Comment on the statement: "Among known complexes $[M(NCS)_4]^{X-}$ all but that of chromium(II) are tetrahedral (or almost so)".

   (d) Evaluate the prospect of a 4-coordinate d^6 complex being more stable with a square-planar, as opposed to tetrahedral geometry.

4. Discuss the following observations.

   (a) The m_{eff} observed for $[Cr(CN)_5(NO)]^{3-}$ and $[Fe(CN)_5(NO)]^{3-}$ show that both complexes have one unpaired electron. Analysis of the (anisotropic) EPR g-values indicates that the odd electron occupies the d_{xy} orbital in the former complex but, instead, the d_{z^2} orbital in the latter.

   $[Cr(CN)_5(NO)]^{3-}$ featured in section 4(5).

   [Take the z-axis to be in the direction of the nitrosyl group.]

   (b) The complex $[MoOCl_5]^{2-}$ also has one unpaired electron (m_{eff} = 1.67 μ_B and g_{av} = 1.947 in its K^+ salt), the components of g showing that this resides in the d_{xy} orbital. The complex $[Mo(NO)Cl_5]^{2-}$ is diamagnetic.

5. Interpret the following information.

   (a) A square coplanar Fe(II) complex shows m_{eff} = 3.9 μ_B at ordinary temperatures.

   [*e.g.* Fe(II) phthalocyanine (Table 5.7) – a chastening test of the spin-only formula.]

   (b) Monomeric $Co(PEt_3)_2(NCS)_2$ has m_{eff} = 2.3 μ_B in the solid state at ordinary temperatures, but its solution in CH_2Cl_2 displays an m_{eff} = 3.50 μ_B.

   $Co(PPh_3)_2(NCS)_2$ shows m_{eff} ~ 4.4 μ_B in both the solid and a variety of solvents.

thf = tetrahydrofuran

(c) The complexes $[MnMe_4]^-$ and $Cr\{(Me_3Si)_2N\}_2(thf)_2$ are paramagnetic, both having $m_{eff} \sim 4.8\ \mu_B$ constant in the range 300 – 85 K.

(d) AgF_3 and AuF_3 are diamagnetic, but the compound Ag_3F_8 has $m_{eff} = 1.92\ \mu_B$ per formula unit. The compounds K_3CuF_6 and KCs_2CuF_6 display m_{eff} of about 2.8 μ_B and 2.6 μ_B, respectively.

mnt = $\{S_2C_2(CN)_2\}^{2-}$

(e) The compound $(Ph_3MeAs)[Fe(mnt)_2]$ shows $m_{eff} = 1.89\ \mu_B$; but in solution it has m_{eff} in the range 3.75 – 4.19 μ_B, depending on the solvent. The neutral complex $Fe(S_2C_2Ph_2)_2$ is diamagnetic.

[*Hint*: Recall Table 5.4. If defeated see Greenwood & Earnshaw (1997), p. 1161.]

6. When cyanide ions are added to the blue-violet aqueous solution of $[Ni(NH_3)_6]^{2+}$, and this is then shaken with benzene, a pale-violet precipitate is obtained. The product shows an effective magnetic moment at ordinary temperatures of *ca.* 2.25 μ_B. Discuss.

7. Comment on the m_{eff} data for selected cobalt compounds given below (all relating to *ca.* 300 K).

* Compare $[Cr(bpy)_3]^-$ with an m_{eff} of 1.83 μ_B.

(a) $Co(bpy)_3$:* 2.23 μ_B *t*-$Co(dppe)_2Cl_2$: 1.97 μ_B

(b) The cobalt-substituted tungstate $K_5CoW_{12}O_{40}$: 5.07 μ_B

(c) $[Co(OH_2)_6]Cl_2$: 4.82 μ_B

dppe = $Ph_2PCH_2CH_2PPh_2$

depe = $Me_2PCH_2CH_2PMe_2$
[See 10(b) below.]

Q = *quinoline*

(d) Some (inter-related) Co(II) complexes:

$Co(PPh_3)_2Cl_2$	4.42 μ_B	$(QH)_2[CoCl_4]$	4.80 μ_B
$Co(PPh_3)_2Br_2$	4.45 μ_B	$(QH)_2[CoBr_4]$	4.87 μ_B
$Co(PPh_3)_2I_2$	4.57 μ_B	$(QH)_2[CoI_4]$	5.01 μ_B

(e) The *tris*-dithiocarbamates $Co(S_2CNR_2)_3$ and $[Co(1\text{-}nor)_4]^+$: diamagnetic (or weakly paramagnetic).

$[Co(S_2CNR_2)_3]^+$: $m_{eff} \sim 2.2 - 2.7\ \mu_B$ $[Co(1\text{-}nor)_4]^-$: 3.18 μ_B

Might any of the above compounds be expected to show a significant variation in m_{eff} on lowering the temperature to (say) 80 K?

diars (structure: benzene ring with two ortho $AsMe_2$ groups)

8. $Mo(diars)_2Cl_2$ is paramagnetic with $m_{eff} = 2.85\ \mu_B$ at 297 K, but $MoCl_2$ and K_2MoCl_4 are diamagnetic or feebly paramagnetic.

9. (a) $Fe(S_2CNEt_2)_2Cl$ and $K_3[Co(CN)_5]$ show m_{eff} of 3.98 μ_B and 1.77 μ_B, respectively, at *ca.* 300 K. Comment. (Both the complexes concerned are *square pyramidal*.).

(b) Rationalise the m_{eff} data for *t*-$M(dmpe)_2X_2$ species in Table 5.24.

Table 5.24
m_{eff} values (/μ_B) for some M(II) complexes *trans*-$M(dmpe)_2X_2$.#

	$M(dmpe)_2Cl_2$	$M(dmpe)_2Me_2$
Ti	2.9	–
V	3.7	3.7
Cr	2.76	2.8
Mn	5.9	2.4

\# *dmpe* = $Me_2PCH_2CH_2PMe_2$
298 K in toluene solution (*via* NMR).

10. Discuss the following observations.

(a) The colour of the silicate mineral *olivine* is due to Fe(II), which occupies octahedral sites and is normally *high*-spin. However, samples of *olivine* from deep beneath the Earth's crust contain *low*-spin Fe(II).

(b) The compound $CrI_2(depe)_2$ has $m_{eff} = 4.87\ \mu_B$ at about 300 K but the moment decreases suddenly to 2.82 μ_B at about 170 K.

(c) The gas-phase UV photoelectron spectrum of $(\eta^5\text{-}C_5H_5)_2Mn$ (*manganocene*) is essentially that of a pure compound. However, the spectrum of the methyl-substituted derivative, $(\eta^5\text{-}C_5H_4Me)_2Mn$, suggests that the vapour is a mixture of two different molecular species.

(d) Compounds of the type *trans*-$[Ni(N\text{–}N)_2X_2]$, where N–N is an alkyl-substituted ethylene diamine and X an anion (*e.g.* halide), are commonly paramagnetic with $m_{eff} \sim 3.0 - 3.2\ \mu_B$.$ In solution many of these show greatly reduced m_{eff} to an extent dependent on both the solvent and the anion X. In some cases m_{eff} is found to vary with temperature.

$ N–N = $R_2NCH_2CH_2NR_2$
One instance is *stilbene diamine* with R = Ph. The compounds are known as *Lifschitz complexes*.

11. The figure below shows plots of the inverse molar susceptibility against temperature for three *tris*-dithiocarbamate complexes of Fe(III). Interpret the essential form of these data. (The solid curves are theoretical.)

[The reader may need to consult section 3(6a).]

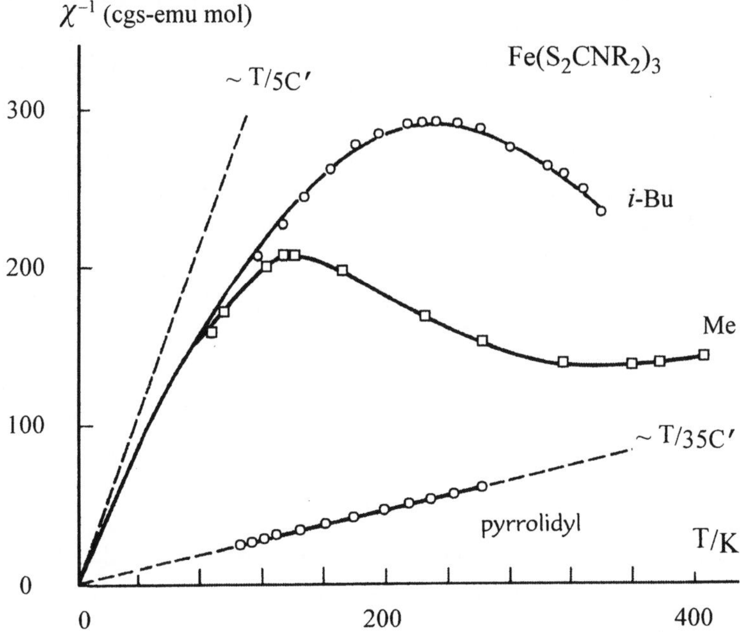

Also, comment on the fact that the related *xanthates*, $Fe(S_2COR)_3$, have invariably $m_{eff} \sim 2.2\ \mu_B$, varying only slightly with temperature.

12. The figure opposite shows the variation of inverse molar susceptibility with temperature for the Co(II) compound $[Co(pmi)_3](BF_4)_2$, where *pmi* is 2-pyridinal-methylimine (a *Schiff base*). The solid curve is theoretical. The data yield $m_{eff} = 4.31\ \mu_B$ at 300 K falling to 2.16 μ_B at 100 K. Discuss.

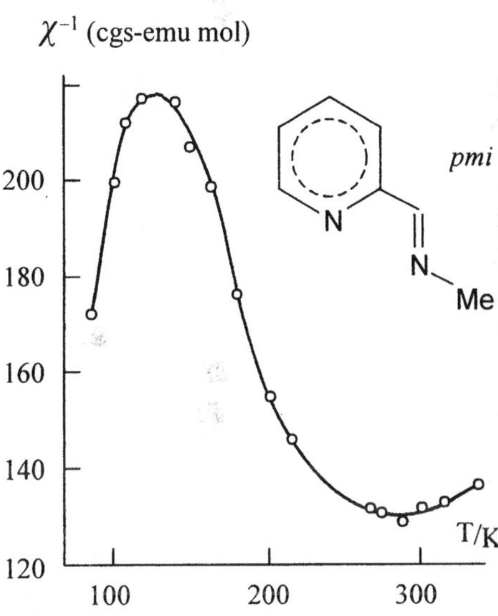

13. The EPR spectrum of the molecule VF_2, isolated in an argon matrix at 4 K, shows eight resonances of equal intensity, each with a symmetrical 1:2:1 splitting. Explain this structure.

[^{51}V (I = 7/2) ~ 100 %; ^{19}F (I = 1/2) 100 %.]

14. Comment on the following g_{av} data from EPR spectra.

(a) V^{2+}/CaO 1.968 Ni^{2+}/CaO 2.327
V^{2+}/MgO 1.980 Ni^{2+}/MgO 2.225 Ni^{2+}/Al_2O_3 2.188
Co^{+}/MgO 2.170 Cu^{3+}/Al_2O_3 2.078

(b) $Cr^{3+}/Al(acac)_3$ 1.982 $Mo^{3+}/Al(acac)_3$ 1.940

(c) $[OCrF_5]^{2-}$ 1.965 $[OMoF_5]^{2-}$ 1.899
$[OCrCl_5]^{2-}$ 1.987 $[OMoCl_5]^{2-}$ 1.948

$LuPO_4$ has the *zircon* ($ZrSiO_4$) structure, with cubic 8-coordination of the metal ion.

$$C' = \frac{\mu_0 N_A \mu_B^2}{3k}$$

$= 1.57 \times 10^{-6}\ m^3\ mol^{-1}\ K$

(d) $Zr^{3+}/LuPO_4$ 1.903 $Hf^{3+}/LuPO_4$ 1.658

15. Interpret the following information.

The magnetic susceptibility of molecular ReF_6 at ~ 300 K is 1.42×10^{-9} $m^3\ mol^{-1}$. That of WF_6 is $-0.67 \times 10^{-9}\ m^3\ mol^{-1}$. Estimate the apparent magnetic moment of ReF_6. [See section 6(5d).]

Doped into MoO_3, Re(VI) displays an axial EPR spectrum with g_{av} = 1.63, similar in appearance to that of $[OReF_5]^-$ (a normal paramagnet with $g_{av} \approx 1.80$). ReO_3 itself shows only a weak TIP.

The compound $CsReF_6$ displays an m_{eff} = 1.53 μ_B at 300 K, falling to 0.63 μ_B at 80 K.

16. In the Kotani theory, the effective Bohr magneton number n_{eff} for an octahedral $(t_{2g})^1$ complex is given by eqn. 5.21, where $x = \zeta/kT$. Show that, at high temperature, n_{eff} approaches the value $[5/(1+x)]^{\frac{1}{2}}$. Hence comment on the observation that many such complexes have susceptibilities conforming to the Curie-Weiss law above *ca.* 100 K.# [Recall Fig. 3.3.]

e.g. the susceptibility of $Cs_2[VCl_6]$ may be simulated by the Curie–Weiss law with n_{eff} = 1.83 and $\theta \approx -138$ K.

17. Verify the contents of Table 5.25, which concerns octahedral complexes with the $(t_{2g})^2$ ground configuration, and show that the effective Bohr magneton number $n_{eff} = m_{eff}/\mu_B$ is given by

$$n_{eff} = \left[\frac{3[5(x+6) + (x+18)\exp(-x) - 48\exp(-\frac{3}{2}x)]}{2x[5 + 3\exp(-x) + \exp(-\frac{3}{2}x)]} \right]^{1/2}$$

What are the limiting values of n_{eff} when x tends to zero and infinity?

Verify, in the case of the corresponding hole configuration $(t_{2g})^4$, that the susceptibility tends to a limiting value of $24\mu_0 N_A \mu_B^2/\zeta$ as $T \to 0$. [*cf.* section 6(5b)(ii).]

Table 5.25

Parameters describing the effect of a magnetic field on the multiplets of $^3T_1\ (t_2)^2$.

J	g_J	α'_J
0	5$	$-24/\zeta$
1	$\frac{1}{2}$	$3/\zeta$
2	$\frac{1}{2}$	$3/\zeta$

[*cf.* Table 5.22] $ special case

Index to tabulated paramagnetic data

Index